Aspects of the Genesis and
Maintenance of Biological Diversity

Aspects of the Genesis and Maintenance of Biological Diversity

Edited by

MICHAEL E. HOCHBERG
JEAN CLOBERT
and
ROBERT BARBAULT

Institut d'Ecologie
Ecole Normale Supérieure
Université Pierre et Marie Curie
Paris, France

Oxford   New York   Tokyo
OXFORD UNIVERSITY PRESS
1996

33014033
DLC

11-27-96

Oxford University Press, Walton Street, Oxford OX2 6DP
Oxford   New York
Athens   Auckland   Bangkok   Bombay
Calcutta   Cape Town   Dar es Salaam   Delhi
Florence   Hong Kong   Istanbul   Karachi
Kuala Lumpur   Madras   Madrid   Melbourne
Mexico City   Nairobi   Paris   Singapore
Taipei   Tokyo   Toronto

and associated companies in
Berlin   Ibadan

Oxford is a trade mark of Oxford University Press

Published in the United States
by Oxford University Press Inc., New York

A catalogue record for this book is available from the British Library

Library of Congress Cataloging in Publication Data

Aspects of the genesis and maintenance of biological diversity/
edited by Michael E. Hochberg, Jean Clobert, and Robert Barbault.
Based on contributions to a series of six workshops held at the
Ecole normale supérieure, Paris, 1993.
Includes index.
1. Biological diversity.   2. Evolution (Biology).   3. Ecology.
4. Biological diversity conservation.   I. Hochberg, Michael E.
II. Clobert, Jean.   III. Barbault, R.
QH541.15.B56A76   1996   333.95'11—dc20   95-33374
ISBN 0 19 854884 2

Typeset by Cotswold Typesetting Ltd, Gloucester
Printed in Great Britain by
Bookcraft Ltd, Midsomer Norton, Avon

Preface

This book is based on contributions to a series of six workshops held at the Ecole Normale Supérieure in Paris in 1993. The main accomplishment of these workshops was the free discussion which took place between British and French participants. In all, there were over 75 presentations, of which 16 comprise the present work.

This book does not attempt to provide a complete synthesis of the subject of biological diversity; rather, it is a transect of some of the most important and interesting topics in biodiversity research, spanning the themes of evolutionary biology, population and community ecology, and conservation. We feel the originality and usefulness of this approach is that the reader has at his or her disposal a continuum of perspectives on biological diversity, flowing in a logical fashion from the genesis of diversity, through to its maintenance, and finally to current diversity patterns and future conservation. We would consider this book to be a success if your research perspectives were to be constructively altered by its contents, be it the questions posed, the methods used, the patterns revealed, or the ideas evoked. Our minimum hope is that you be aware that biodiversity research is moving quickly and usefully on many fronts; it is very likely that important future advances will involve contributors to this volume and the questions they are attempting to resolve.

This compilation stems from the efforts of a number of people.

First, the workshops were made possible by co-funding from the British Council and the French CNRS, and through the particular efforts of Ms. June Rollinson of the British Council and Mme. Françoise Saunier of the CNRS.

Second, each of the workshops from which the overview chapters are based were convened by one British and one French scientist. These were as follows (workshop number, title and conveners): 1, 'Gene flow, polymorphism and the genetic structure of populations', G. M. Hewitt and P. H. Gouyon; 2, 'Population dynamics, extinction and colonisation', J. Clobert and C. Gliddon; 3, 'Co-evolution of host-parasite associations', R. D. Gregory and C. Combes; 4, 'Dynamics and richness of multi-species communities', M. E. Hochberg and B. A. Hawkins; 5, 'Phylogenetic and systematic perspectives in evolutionary biology', J. Jaeger and S. Nee; 6, 'Trophic organisation and food-webs', J. H. Lawton and R. Barbault.

Third, we thank J. H. Lawton for suggesting the precursor of the book title, all of the invited participants for their input during the meetings, and those participants, as well as M. Bruford, H. V. Cornell, D. Couvet, A.

Hallam, R. D. Holt, N. Martinez, Y. Michalakis, D. Pontier, M. Rees, and M. Veuille, for their comments on the manuscripts.

Finally, we are grateful to the staff at Oxford University Press for their efficient help.

Paris, France   M.E.H., J.C. and R.B.
October 1994

Contents

Contributors

André Adoutte Laboratoire de Biologie Cellulaire 4, URA 1134, Université de Paris-Sud, Bât. 444, 91405 Orsay Cedex, France

Robert Barbault Institut d'Ecologie, URA 258, Université Pierre et Marie Curie, Bât. A, 7ème étage, Case 237, 7 quai St. Bernard, 75232 Paris Cedex 05, France

Mark A. Beaumont School of Biological Sciences, Queen Mary and Westfield College, London E1 4NS, England

Michael J. Benton Department of Geology, University of Bristol, Wills Memorial Bld., Queens Road, Bristol BS8 1RJ, England

Jean Clobert Institut d'Ecologie, URA 258, Université Pierre et Marie Curie, Bât. A, 7ème étage, Case 237, 7 quai Saint-Bernard, 75252 Paris Cedex 05, France

Claude Combes Centre de Biologie et d'Ecologie Tropicale et Méditerranéenne, URA 698, Université de Perpignan, Avenue de Villeneuve, 66860 Perpignan Cedex, France

Thierry de Meeûs Laboratoire de Parasitologie Comparée, URA 698, Université de Montpellier II, Place E. Bataillon, 34095 Montpellier Cedex 05, France

Richard A. Ennos Institute of Ecology and Resource Management, University of Edinburgh, Darwin Building, Mayfield Road, Edinburgh EH9 3JU, Scotland

Kevin J. Gaston Department of Entomology, The Natural History Museum, Cromwell Road, London SW7 5BD, England

Chris Gliddon School of Biological Sciences, University of Wales, Bangor LL57 2UW, Wales

Pierre-Henri Gouyon Systématique et Ecologie Végétales, URA 1492, Université de Paris-Sud, Bât. 362, 91405 Orsay Cedex, France

Jeremy J. D. Greenwood British Trust for Ornithology, The Nunnery, Thetford, Norfolk IP24 2PU, England

Richard D. Gregory British Trust for Ornithology, The Nunnery, Thetford, Norfolk IP24 2PU, England

James P. Grover Department of Biology, University of Texas, Box 19498, Arlington, TX 76019-0498, USA

Paul H. Harvey Department of Zoology, University of Oxford, Oxford OX1 3PS, England

Michael P. Hassell Department of Biology, Imperial College at Silwood Park, Ascot, Berks SL5 7PY, England

Bradford A. Hawkins Department of Ecology and Evolutionary Biology, University of California, Irvine, CA 92717, USA

Godfrey Hewitt School of Biological Sciences, University of East Anglia, Norwich NR4 7TJ, England

Michael E. Hochberg Institut d'Ecologie, URA 258, Ecole Normale Supérieure, 46 rue d'Ulm, 75230 Paris Cedex 05, France

Jean-Jacques Jaeger Institut des Sciences de l'Evolution, URA 327, Université de Montpellier II, Place E. Bataillon, Case 061, 34095 Montpellier Cedex 05, France

T. Hefin Jones NERC Centre for Population Biology, Imperial College at Silwood Park, Ascot, Berks SL5 7PY, England

Gérard Lacroix Institut d'Ecologie, URA 258, Ecole Normale Supérieure, 46 rue d'Ulm, 75230 Paris Cedex 05, France

John H. Lawton NERC Centre for Population Biology, Imperial College at Silwood Park, Ascot, Berks SL5 7PY, England

Jack J. Lennon Unit for Ecogenetic Biogeography, Department of Genetics, University of Leeds, Leeds LS2 9JT, England

Françoise Lescher-Moutoué Institut d'Ecologie, URA 258, Ecole Normale Supérieure, 46 rue d'Ulm, 75230 Paris Cedex 05, France

Michel Loreau Institut d'Ecologie, URA 258, Université Pierre et Marie Curie, Bât. A, 7ème étage, Case 237, 7 quai Saint-Bernard, 75252 Paris Cedex 05, France

Robert M. May Department of Zoology, University of Oxford, Oxford OX1 3PS, England

Serge Morand Centre de Biologie et d'Ecologie Tropicale et Méditerranéenne, URA 698, Université de Perpignan, Avenue de Villeneuve, 66860 Perpignan Cedex, France

Sean Nee Department of Zoology, University of Oxford, Oxford OX1 3PS, England

Richard A. Nichols School of Biological Sciences, Queen Mary and Westfield College, London E1 4NS, England

Hervé Philippe Laboratoire de Biologie Cellulaire 4, URA 1134, Université de Paris-Sud, Bât. 444, 91405 Orsay Cedex, France

Roger Pourriot Laboratoire de Géologie Appliquée, Université Pierre et Marie Curie, Tour 26, 4 place Jussieu, 75252 Paris Cedex 05, France

Andrew Rambaut Department of Zoology, University of Oxford, Oxford OX1 3PS, England

François Renaud Laboratoire de Parasitologie Comparée, URA 698, Université de Montpellier II, Place E. Bataillon, 34095 Montpellier Cedex 05, France

Bryan Shorrocks Department of Biology, University of Leeds, Leeds LS2 9JT, England

Glenn W. Storrs Department of Geology, University of Bristol, Wills Memorial Bld., Queens Road, Bristol BS8 1RJ, England

Chris D. Thomas Department of Biology, University of Leeds, Leeds LS2 9JT, England

John R. G. Turner Unit for Ecogenetic Biogeography, Department of Genetics, University of Leeds, Leeds LS2 9JT, England

Philip H. Warren Department of Animal and Plant Sciences, School of Biological Sciences, University of Sheffield, Sheffield S10 2TN, England

Paul H. Williams Biogeography and Conservation Lab, The Natural History Museum, Cromwell Road, London SW7 5BD, England

Introduction

Robert M. May

Underlying all discussions of the diversity of life on earth is a fundamental question, which is rarely asked and which is at present unanswerable. Is ours the only inhabited planet in the universe? Or is the evolution of some form of life a cosmic commonplace, occurring in some appreciable fraction of planetary systems?

One post-Copernican reflex is to deny any such special status for our planet. But the suggestion that we are unique is not necessarily anthropocentric. We could be here, asking these questions, simply because we are the end product of a wildly improbable chain of events, the winners of a lottery against literally astronomical odds. A converse view is that complex self-replicating molecules, followed in time by multicellular organisms, and eventually by self-conscious entities of some kind, are likely to evolve on many planets (or at least many of those with water). Sums of money beyond the dreams of the most scientifically avaricious systematists and ecologists have indeed been spent on radio-astronomical and other quests for signals from inhabited planets more technically advanced than ours. And although few evolutionary biologists are aware of it, there exists a scholarly literature arguing that life might frequently originate, but nevertheless most of the universe may be empty at any one time (thus explaining our failure as yet to detect any interstellar signs of 'others').

All such speculation would, of course, be preempted if we had a firm understanding of how life-forms first began to evolve on our own planet. We could then estimate the fraction of other planets likely to have physical and chemical parameters within the envelope which would permit such evolution. Thence we could calculate whether the total number of such planets in the universe is likely to be much less than one (so that we are the outcome of a unique and improbable accident) or much greater than one, or even perhaps around one or so (itself an *a priori* unlikely outcome of such a calculation). But, although there are an increasing variety of interesting ideas about these early stages in the origins of life (Kauffman 1993; Eigen 1992), we are not yet remotely in sight of possessing the understanding needed for such a calculation.

Diversity in the past

We do, however, have an increasingly good understanding of the patterns of diversity seen in the Earth's fossil record, particularly since the advent of hard-bodied organisms in the Cambrian, some 600 million years ago (mya).

This work is being complemented, at an accelerating rate, by assessments of the phylogenetic histories written in the amino-acid sequences of molecules in living (and in relatively recently extinct) organisms. Yet other methods, such as comparative studies of immune systems among mammals, or even between vertebrates and invertebrates, provide additional and independent ways of reading the records. These empirical advances are being matched by theoretical studies—some analytic, but many dependent upon powerful computers—which illuminate evolutionary processes. For example, a sensibly-structured analysis suggests that it is not all that difficult to evolve an eye reasonably fast by incremental steps, thus laying to rest one of Darwin's ghosts (and a star item in Creationist tracts); see Nilsson and Pelger (1994).

And so to the present book, which has its origins in a series of workshops, devoted to aspects of the genesis and maintenance of biological diversity. The first of the book's three parts deal with inferences that can be drawn from past evolutionary patterns.

Sensibly, the book avoids the still-speculative questions about the literal genesis of diversity which are discussed above, and begins with a chapter by Benton and Storrs (Chapter 1) assessing what we know about the fossil record. These authors build a platform for much of what follows by giving a quantitative assessment of the consequences of the ineluctable incompleteness of this record. They compare the evolutionary order of appearance of organisms as inferred from the geological record with the order of branching in phylogenetic trees inferred from cladistic analysis (based on 'character states') or molecular evidence. The conclusion is that stratigraphic knowledge of the fossil record is improving; the branches of phylogenetic trees are increasingly decorated with fossil ornaments all along their lengths.

Benton and Storrs discuss many possible biases and other problems in the fossil record. They are encouraged by Sepkoski's (1993) comparison between the standard databases for marine animals in 1982 and 1992. Over this 10 year span, there was much revision of the data, and the total number of families increased by 13%. But the shapes of the two plots of diversity (numbers of families) as a function of time (over the past 600 myr) were nearly identical. My worry is, however, of a different kind. Around 95% of the fossil databases consist of shallow-water marine invertebrates (Sepkoski 1992). On the other hand, most of today's known species are terrestrial animals (56% of the total are insects). Thus, I think there remain some questions about the extent to which patterns of diversity in the fossil record are representative of the totality of today's organisms.

Species alive today represent the twig-tips of phylogenetic trees. Nevertheless, molecular phylogenies of extant taxa can tell us a lot about the past mode and tempo of evolution. Philippe and Adoutte (Chapter 2) give a careful discussion of how unresolved nodes—'multifurcations'—in molecular phylogenies may, under appropriate circumstances, be indicative of bursts of speciation. More generally, Harvey, Rambaut and Nee (Chapter 3) show how molecular phylogenies can be combined with mathematical models for species' birth and death rates which may vary over time, in order to 'read' the evolutionary record for a given clade. In the past, such attempts have relied mainly on intuition and verbal arguments, which can result in misleading conclusions. Harvey *et al.* demonstrate, for example, that the record could appear to show increasing rates of speciation (for a phylogeny containing all living members of a clade) or, alternatively, decreasing rates of speciation (for a phylogeny containing a sample from the clade), when in fact the true rates of species' origination and extinction are strictly constant.

Population structure and gene flow are important factors shaping the evolutionary destiny of species. As emphasised here by Nichols and Beaumont (Chapter 4), a knowledge of the distribution of allele frequencies among spatially-structured subgroups of a population or species can be used to make inferences about its evolutionary history. A remarkable book by Cavalli-Sforza *et al.* (1994) gives an excellent illustration of how these techniques may be applied, using various markers within the human gene pool, to make deductions about patterns of movement and intermingling of subgroups, for human history since the neolithic. Hewitt (1989) has similarly illuminated the post-glacial movement of insect species, using such methods. Some of the selective mechanisms which can help to maintain these allelic polymorphisms are surveyed by de Meeûs and Renaud (Chapter 5), in the closing chapter of this first part of the book.

Broad surveys indicate that biological diversity, as measured by the total numbers of species, genera, or families, has on average increased—albeit with many fluctuations and episodes of mass extinction—over the past 600 myr (Sepkoski 1992). Sepkoski suggests, indeed, that the numbers of families of marine invertebrates has, to a very rough approximation, increased linearly over this time. Within the overall totals, the average lifespan of an individual species, from origination to extinction, is around 5–10 myr. This average figure, however, masks much variability within and among groups (see the estimates summarised by May *et al.* 1995). If we compare this 5–10 myr lifespan of an average species with the 600 myr span of the fossil record, we would estimate that today's species represent some 1–2% of those ever to have lived. But if we add to this Sepkoski's (1992) observation that today we are at one end of a roughly linear overall increase in diversity, then today's species are more like 2–4% of those ever alive. And if we further note that all such estimates are based mainly on a fossil record dominated by marine invertebrates, whereas today's biodiversity is dominated by terrestrial insects (which began their radiation only around 450 mya, and which may typically have species' lifespans in excess of 10 myr), then the true number could be 5% or more. For a more detailed review of recent work relating to such questions, see May *et al.* (1995).

Ecological factors influencing biological diversity

In *The Origin of Species*, Darwin showed how natural selection acts on the variation within aggregations of individuals, leading over time to descent with modification, and eventually to new species. Amongst much else, he gave us a vivid image of species being packed into what would now be called 'niche space' like wedges in a barrel. But after we have understood, in broad terms, how species originate, the next large question is: how many species should there be? How many wedges in the barrel? Are they packed tightly, almost filling the barrel, or do they rattle around? What are the distributions in shape and sizes of the wedges? And so on.

Against the background of historical facts about species diversity assembled in Part I of this book, Parts II and III turn toward the underlying mechanisms. Part II focuses on ecological factors which influence the demography of populations, beginning with interactions between pairs of populations as competitors or as prey–predator, and building up to the dynamics of communities, food webs, and ecosystems. Part III deals with broader descriptive patterns, such as the latitudinal gradient in species diversity or species-area relations, and their implications for conservation biology. The pot of gold at the end of this rainbow would be to understand, in fundamental terms, why there are some 700 species of breeding birds in North America, but only 220 or so in the UK, or more globally why there are more than twice as many bird species as mammal species. As will be seen, the rainbow's end is, as yet, far distant.

Ever since the simple but pioneering mathematical models of Lotka and Volterra, going hand-in-hand with the 'bottle experiments' of Park, Gause and others, there have been increasingly sophisticated efforts to understand the biological and physical factors which influence the persistence of single populations, or pairs of interacting populations Here, Shorrocks (Chapter 6) gives an analytic survey of ecological factors which affect the coexistence of competing species, prominent among which are niche separation (resulting from differences in resource use) and spatial heterogeneity. Shorrocks goes further, illustrating how these factors can be teased apart, at least in laboratory experiments with *Drosophila*. Jones and Hassell (Chapter 7) look at a special class of prey–predator systems, namely those with one insect host species attacked by one, two, or more species of parasitoids (dipteran or hymenopteran predators which lay their eggs on or in the host, thereby killing it). Combining theoretical and empirical studies, these authors show that spatial heterogeneity is often critical for such systems to persist. Such heterogeneity may derive from intrinsic environmental factors. Surprisingly, however, spatial structure may be self-organised, in an inherently homogeneous environment, generated by nonlinear local dynamics coupled with local movement; such self-organised spatial heterogeneity can take the form of spatial chaos, or spiral waves, or even apparently static patterns of heterogeneity. These two chapters are representative of a growing body of work on the population biology of ecological systems with a few interacting species. Other recent work avoids dealing explicitly with the population dynamics within individual patches, but rather deals in terms of presence and absence of different species, distributed as 'metapopulations' among many patches. Competitive, mutualistic, or prey–predator interactions can be incorporated as qualitative rules about patch-extinction rates of particular species, dependent on who else arrives in their patch. Such models provide a kind of bridge between the detailed few-population models and laboratory or field experiments, and the much messier world of communities or ecosystems. These metapopulation systems can, moreover, give some disturbing and counter-intuitive insights into the likely consequences of habitat destruction (Nee and May 1992; Tilman *et al.* 1994; Kareiva 1994).

One approach to understanding diversity within communities of interacting species is to focus on the network of links—who eats whom—within a food web. Following May's (1973) studies of stability and complexity in model food webs, there has emerged an increasingly clear picture of the kinds of interactions, and the kinds of food web structures, which make for long-term persistence (Pimm *et al.* 1991). On a more empirical level, systematic compilations of information about real food webs have revealed many approximate but interesting patterns (Cohen *et al.* 1990). Warren (Chapter 8) gives a good account of the state of play, with particular emphasis on attempts to find 'Newtonian explanations' that will make fundamental sense of these 'Keplerian patterns'. As an example of one of the treasures the leprechauns may be guarding at the rainbow's end, a truly fundamental understanding of food web structure would enable us to predict the total number of species present in a particular class of web, initially knowing only the number of species of primary producers. Tentative surveys of the available data suggest the relevant ratio may be very roughly around 10, albeit with considerable variability (Gaston 1992).

Warren's overview of food web structure is fleshed out by Lacroix, Lescher-Moutoué and Pourriot's (Chapter 9) account of trophic interactions in lakes. Grover and Loreau (Chapter 10) add a further important factor, by considering decomposers as inherently part of the web. This introduces constraints (conservation of mass, and energy balance), and can affect the conclusions drawn about food web structure and function. In particular, explicit cycling of nutrients may often tend to make webs more stable against environmental perturbations.

Investigations such as those reported in Chapters 8–10 have begun to elucidate the relations between diversity and ecosystem functioning. How does reduction in the number of species present in a community typically affect its ability to handle environmental change? This is a question of obvious practical concern. Notable are the recent manipulative experiments by Tilman and Downing (1994) and by Naeem *et al.* (1994), which suggest that more diverse associations of plants are better able to cope, respectively, with drought or with CO_2 enrichment.

Broader patterns

Ecologists and biogeographers have identified a variety of patterns in the way diversity is distributed. My short list of such patterns is as follows.

1. For most groups of terrestrial plants and animals, there is a pronounced 'latitudinal species diversity gradient' (LSDG), with markedly more species in equatorial regions, and a systematic decline at higher latitudes. This is particularly notable for tree species, where the enormous diversity of tropical forests gives way to a handful of conifer species at northerly latitudes. The evidence for similar latitudinal gradients among invertebrates in coastal sediments is weak. Recent work, however, finds clear LSDGs for small invertebrates in deep sea sediments (Rex *et al.* 1993; Gage and May 1993). Turner, Lennon and Greenwood (Chapter 11) review possible explanations for such gradients, and indicate how some of these ideas can be tested against data for British birds.

2. Other things being equal, there is a relation between a region's area, A, and the number of species found in it, S. A tenfold reduction in area (as when a reserve is established, and its surroundings modified) roughly halves the number of species. More generally, there is usually a power-law relation of the form

$$S = cA^z \tag{1}$$

where the constant c varies from group to group, and the exponent z is usually in the range 0.2–0.3.

3. There are also broad trends in the relative abundances of species within a community or ecosystem. In old-established communities, these patterns of species relative abundance tend to be more even (often described by a special set of so-called 'canonical' lognormal distributions) than those for early successional or highly disturbed situations. These canonical lognormal patterns can be interpreted as arising from the multiplicative interplay of many ecological and evolutionary factors. In essentials, such an interpretation sees the patterns as expressions of the statistical Central Limit Theorem applied to systems with many species. If one makes the additional assumption that total numbers of individuals, N, scale roughly linearly with area, then the canonical lognormal relation between N and S leads to equation (1) with $z \to 0.25$ for large S. More generally, such discussions of the relative abundances of species lead on to questions about the meaning of commonness and rarity of species (Rabinowitz *et al.* 1986). Too often, conservation literature confuses rarity with endangerment. Although rare species are, indeed, disproportionately represented on lists of endangered species, rarity as such is a natural phenomenon; many rare species persist happily and naturally. Such questions about rarity and persistence underline our need for a better understanding of the ecological and evolutionary forces shaping patterns of species' relative abundance.

4. The numbers of species in different categories of physical size vary systematically: for terrestrial animals, a decrease by a factor 10 in characteristic linear dimensions (or 10^3 in mass) roughly results in 100 times more species. This rough rule holds down to size categories around a few mm. Species numbers fall away below this (May 1978).

5. Patterns in the relations between the body sizes and the geographical ranges of species are only just beginning to receive systematic attention. Fenchel (1993) has suggested that geographical ranges are typically more extensive for relatively large organisms and for microorganisms (protozoa and below) than for mid-size organisms (insects). If true, such patterns— which are entwined with the species-size effects of the preceding paragraph— are relevant, amongst other things, to possible range modifications associated with climate change.

Many other questions cut across the patterns I have just listed. For one thing, any discussion of patterns of diversity in relation to the physical size or geographical range of organisms leads to more general questions about the relations between diversity and life history. Morand (Chapter 13) pursues these ideas, using parasites with direct and indirect life cycles to make his discussion more concrete. For another thing, we must ask to what extent any pattern found for one taxon is representative of that for others (for instance, LSDGs vary in intensity among taxa). Such possible variation from group to group must make us wary of simplistic arguments about identifying diversity 'hotspots' for one group, and then arguing that these will automatically serve larger conservation purposes. Gaston (Chapter 12) develops this important theme, giving examples of similarities and differences in diversity patterns among plants, birds, butterflies, grasshoppers, and other groups of organisms.

From gene pools to phyla

This introductory chapter, like most of the book, tends to focus on diversity at the level of species. It is important to remember, however, that biological diversity exists at many different levels, from the genetic diversity within local populations of a species, or between geographically distinct populations of the same species, all the way up to communities or ecosystems. Depending on the context, any one of this nested hierarchy of levels can be of primary importance.

At the most fundamental evolutionary level, the genetic diversity within species is the raw stuff upon which evolution acts. Ennos (Chapter 14) surveys some of the salient issues connected with genetic diversity, particularly in relation to plant conservation.

At the opposite extreme, diversity can be interpreted at various taxonomic levels, from genus to kingdom. As we ascend this taxonomic hierarchy, we tend to get rather different pictures of diversity. For example, a comparison of land and sea shows only about 15% or fewer of recorded species inhabiting the marine realm. But the sea is increasingly represented as we move to higher taxonomic levels, from genus to phylum. At the level of phylum, or basic body-plan, diversity is much greater in the sea: of the global total of 33 phyla (by one classification), 32 in the sea versus 12 on land; or, at the level of class, 73 animal classes in the sea, 35 in freshwater, and 33 on land (May 1994a).

These different dimensions of diversity being recognised, I believe it nevertheless makes sense to focus mainly on species. Species represent visible and operational units, more easily amenable to study and management than are gene pools or ecosystems. These pragmatic concerns aside, species are arguably the basic coinage of evolution; not for nothing is Darwin's book called *The Origin of Species*. But even as we focus on species, we should maintain peripheral vision of diversity at higher and lower levels of organisation.

How many species today?

The total number of living species which have been named and recorded is around 1.7–1.8 million. There is, however, a problem with synonymy whereby the same species has unwittingly been recorded under different names by different researchers. Documented rates of synonymy in particular groups typically run around 20% (and the true rate could be higher: Solow *et al.* 1994). Thus a better estimate of today's total may be around 1.4 million recorded species (Wilson 1992; Groombridge 1992; May 1994*a,b*).

Setting aside problems of synonymy, the number of named species is a fact. It is, moreover, a very fundamental fact characterising life on earth. I find it amazing that we still have no consolidated, computerised database of all recorded species. There is real need for such a central database, complete with associated information about the physical size, range, life history, conservation status, and other ecological properties of individual species, even if this is very often a guess (and identified as such, with some indication of the precision). The lack of codification of existing information is, *inter alia*, an impediment to unravelling some of the broader patterns in biological diversity, like those listed as numbered points above. My estimate of the cost of creating a database of this kind would be of the order of £10^8, which may seem large by the prevailing expectations of taxonomists. To me it is ironic that we should be spending much larger sums looking for 'intelligent life in the universe', while failing properly to survey the diversity of life on earth.

Some taxonomic groups are much better known than others, reflecting patterns in the taxonomic workforce which derive from intellectual fashions rather than analytic assessments of priorities (May 1994*b*). Bird and mammal species are comparatively well documented; even though 3–5 new bird species and around 10 new mammal species are found each year, such numbers are small fractions of the totals recorded in these classes (approximately 9000 species of birds and 4000 of mammals). The roughly 270 000 recorded species of vascular plants probably represent 90% or so of the true total. But comprehensive explorations of invertebrate groups in previously-unstudied places—tropical canopy insects; deep sea benthic macrofauna; fungi—typically find 20–50%, or even more, of the species thus found are new to science (May 1994*a,b*). Taxonomists are distributed roughly evenly among vertebrates, plants, and invertebrates. But there are roughly 10 plant species for each vertebrate animal species, and conservative estimates suggest around 100 insect species for each vertebrate one. Thus current patterns of knowledge reflect the fact that the average vertebrate species receives 10 times more taxonomic effort than the average plant species, and 100 times more than the average invertebrate (Gaston and May 1992).

One result of this comparative neglect of the most species-rich groups is that the true total number of living species is very uncertain. My guestimate is around 3–8 million (May 1994a,b). Dramatically higher numbers have been proposed: 30 million insects on the basis of studies of beetles in tropical canopies; tens of millions of benthic invertebrates on the basis of a deep-shelf transect off the northeastern USA; 1.5 million fungi on the basis of scaling up the species ratio of fungi to vascular plants in Britain; and others. I am sceptical of all these estimates. Were they correct, comprehensive studies of tropical insects or benthic invertebrates or fungi in previously-unstudied regions should find essentially all species are new; this is not the case. But these extravagant estimates can be defended, and they could be correct. The fact that reasonable estimates vary so widely says a lot about how little we know.

In passing, I deplore the tendency of some conservation literature to cite, uncritically, the most dramatic of these estimates. The reason is presumably rhetorical; if there really are 30 million insects, should we not be devoting correspondingly more scientific effort toward recording them? But I think the uncritical acceptance of such huge numbers does not help the cause of conservation, even in the unlikely event that they are correct. First, if there really are upward of 30 million species, so that recorded numbers represent only 5% or less of the true total, then we have a hopelessly large task. Second, and regardless of philosophical niceties, the practical effect of so many species must be to devalue each individual one. The real task, of course, is to refine our estimates of the true total, uncorrupted by any rhetorical considerations.

How many species tomorrow?

Over the past century, rigorously documented extinctions in well-studied groups—primarily birds and mammals—have run around one species per year. Because tropical species typically receive less attention, true extinction rates of birds and mammals are undoubtedly higher (Diamond 1989). But even one per year among the roughly 13 000 species of birds and mammals translates to expected species' lifetimes, based on documented recent extinction rates, of around 10^4 years. Although seemingly long, this is shorter by a factor of order 10^{-3} than the background average lifespan of 5–10 myr seen in the fossil record. That is, recent extinction rates in well-documented groups have run a thousand times faster than the average background rates.

Looking toward the immediate future, three different approaches to estimating impending rates of extinction suggest species' life expectancies of around 200–400 years. One of these approaches is based on species-area relations, coupled with assessments of current rates of habitat loss. The other two are based in different ways on the International Union for the Conservation of Nature's current catalogue of 'endangered' or 'vulnerable' species, for better-known groups such as birds, mammals, or palm trees (May et al. 1995). Such figures correspond to likely extinction rates of a factor of ten thousand or more above background, over the next century or so. This represents a sixth great wave of extinction, fully comparable with the Big Five mass extinctions of the geological past, but different in that it results from the activities of a single other species rather than from external environmental changes.

As we face this future, we must ask: does it matter more if we lose 25% of all mammal species than if we lose 25% of the vastly more numerous insect species? Or does it matter equally? Or less? There is need not only for more taxonomic information, but also for a 'calculus of biodiversity' based on this information. Such a calculus should, ideally, quantify the taxonomic uniqueness, or amount of 'independent evolutionary history' (IEH), inherent in individual species. Williams (Chapter 15) gives an overview, complete with examples, of first steps towards this calculus. He notes that the conceptual issues are fairly straightforward if we have a phylogenetic tree with quantitative measurements of the branch-lengths (from molecular-clocks or otherwise). In this case, if we can only save, say, half of a group of species, then we would ideally choose the set of species which maximises the total branch-lengths preserved. Things are trickier if we have only qualitative information about the topological structure of the tree. Williams summarises various suggestions for how best to proceed in this event. One possibly over-simplified view is to assign the branches the lengths that are, on average, most likely for this particular topology, and then go forward on this basis. Such a procedure will, of course, often in fact be suboptimal, because the underlying evolutionary tree differs from the statistically 'expected' one. In general, however, extensive theoretical simulations of choices made on a topological basis, from artificially-generated trees whose underlying branch lengths are known, suggest that values assigned in this way are close to the 'true' ones (May and Nee, 1995). Ultimately, our question is how much of the IEH within a group will be preserved if we can only save, say, 10 of 20 species? The simulations referred to above suggest that, for the 10 of 20 case, we can on average preserve 82% of the group's IEH if we have quantitative information about branch lengths, 77% if we have only topological information about the branching structure of the phylogenetic tree, and 63% if we must choose at random (May and Nee 1995; May 1994b). Real situations will obviously involve many other important considerations, including other measures of the relative value of species: in preserving 'ecosystem services', or in possessing unusual behavioural or ecological properties which are not captured by crude measures of genetic distances, or simply in having emotional resonance for large numbers of people. Political and economic realities will often constrain which areas may be preserved. In short, any programme of assessment and quantification of biological diversity needs to go beyond mere species counting, and move towards developing a 'calculus of diversity' along the lines just sketched.

It is very appropriate that the book's final chapter, by Thomas (Chapter 16), deals with a case study of conservation of a butterfly species. Deciding where our priorities lie, and what to save, is one thing. Having the ecological knowledge to implement these wishes can be, as Thomas makes clear, a very different problem.

Coda

Most people have believed that the times in which they lived were special. But our own time is truly singular by any objective measure, in that the scale and scope of human activities have for the first time come to rival those of the natural processes that built and maintain the biosphere as a place where life can flourish. Such measures include the amount of elements such as nitrogen, phosphorus or sulphur that are biologically mobilised, or the inputs of CO_2 or CFCs into the atmosphere, as a result of human activities rather than natural cycles.

From the viewpoint of maintaining biological diversity, I think two global measures of our activities stand above all others. One is the fraction of all primary productivity that is sequestered, directly or indirectly, to human uses: on land, the fraction is around 25–50% (Vitousek et al. 1986); in the sea, a careful recent estimate (based on region-by-region assessments of average food chain lengths) suggests 8% overall, but more like 30–40% in productive coastal regions (Pauly and Christensen 1995). Another measure of our impact lies in the estimated extinction rates discussed above: documented extinctions in well-studied groups over the past century have been above the average background extinction rate seen in the geological record by a factor of 10^3 or more, and estimates suggest an impending acceleration by at least another factor of 10 over the next century or so.

These facts presage a sixth great wave of extinction, fully comparable with the Big Five mass extinctions which punctuate the fossil record, ushering in new eras. The wave on whose breaking tip we stand is importantly different from all previous ones, in that it results not from external environmental events, but rather from the activities of a single other species. Circling back to the beginning of this chapter, I wonder whether such sad events are characteristic of the evolutionary trajectories of inhabited planets. And if the biological diversity of our planet represents a unique event in the universe, then the question is all the more poignant.

References

Cavalli-Sforza, L. L., Menozzi, P. and Piazza, A. (1994). *The History and Geography of Human Genes*. Princeton University Press.

Cohen, J. E., Briand, F. and Newman, C. M. (1990). *Community Food Webs: Data and Theory* (Biomathematics, Vol. 20). Springer-Verlag, New York.

Diamond, J. M. (1989). The present, past and future of human-caused extinctions. *Philosophical Transactions of the Royal Society, B,* **325,** 469–77.

Eigen, M. (1992). *Steps Towards Life: A Perspective on Evolution.* Oxford University Press.

Fenchel, T. (1993). There are more small than large species? *Oikos,* **68,** 375–378.

Gage, J. D. and May, R. M. (1993). A dip in the deep seas. *Nature,* **365,** 609–610.

Gaston, K. J. (1992). Regional numbers of insect and plant species. *Functional Ecology,* **6,** 243–247.

Gaston, K. J. and May, R. M. (1992). The taxonomy of taxonomists. *Nature,* **356,** 281–282.

Groombridge, B. (ed.) (1992) *Global Biodiversity: Status of the Earth's Living Resources.* Chapman and Hall.

Hewitt, G. M. (1989). The subdivision of species by hybrid zones. In *Speciation And Its Consequences* (eds. Otte, D. and Endler, D. A.), pp. 85–110. Sinauer, Sunderland, MA.

Kareiva, P. (1994). Connecting landscape patterns with population and ecosystems processes. *Nature,* **373,** 299–302.

Kauffman, S. A. (1993). *The Origins of Order.* Oxford University Press.

May, R. M. (1973). *Stability and Complexity in Model Ecosystems.* Princeton University Press.

May, R. M. (1978). The dynamics and diversity of insect faunas. In *Diversity of Insect Faunas* (eds. Mound, L. A. and Waloff, N.), pp. 188–204. Blackwell Scientific, Oxford.

May, R. M. (1994*a*). Biological diversity: differences between land and sea. *Philosophical Transactions of the Royal Society, B,* **343,** 105–111.

May, R. M. (1994*b*). Conceptual aspects of the quantification of the extent of biological diversity. *Philosophical Transactions of the Royal Society, B,* **345,** 13–20.

May, R. M. and Nee, S. (1995). Making conservation choices; towards a calculus of biodiversity. (In prep.)

May, R. M., Lawton, J. H. and Stork, N. E. (1995). Assessing extinction rates. In *Extinction Rates* (eds. Lawton, J. H. and May, R. M.), pp. 1–24. Oxford University Press.

Naeem, S., Thompson, L. J., Lawlor, S. P., Lawton, J. H. and Woodfin, R. M. (1994). Declining biodiversity can alter the performance of ecosystems. *Nature,* **368,** 734–737.

Nee, S. and May, R. M. (1992). Patch removal favours inferior competitors. *Journal of Animal Ecology,* **61,** 37–40.

Nilsson, D. E. and Pelger, S. (1994). A pessimistic estimate of the time required for an eye to evolve. *Proceedings of the Royal Society, B,* **256,** 53–58.

Pauly, D. and Christensen, V. (1995). Primary production required to sustain global fisheries. *Nature,* **374,** 255–257.

Pimm, S. L., Lawton, J. H. and Cohen, J. E. (1991). Food web patterns and their consequences. *Nature,* **350,** 669–674.

Rabinowitz, D., Cairns, S. and Dillon, T. (1986). Seven forms of rarity. In *Conservation Biology* (ed. Soulé, M. E.), pp. 182–204. Sinauer, Sunderland, MA.

Rex, M. A., Stuart, C. T., Hessler, R. R., Allen, J. A., Sanders, H. L. and Wilson, G. D. F. (1993). Global-scale latitudinal patterns of species diversity in the deep-sea benthos. *Nature,* **365,** 636–639.

Sepkoski, J. J. (1992). Phylogenetic and ecologic patterns in the Phanerozoic history of marine biodiversity. In *Systematics, Ecology, and the Biodiversity Crisis*, (ed. Eldredge, N.), pp. 77–100. Columbia University Press, New York.

Sepkoski, J. J. (1993). Ten years in the library: how changes in taxonomic data bases affect perception of macroevolutionary pattern. *Paleobiology*, **19**, 43–51.

Solow, A. R., Mound, L. A. and Gaston, K. L. (1994). Estimating the rate of synonymy. (Under review).

Tilman, D. and Downing, J. A. (1994). Biodiversity and stability in grasslands. *Nature*, **367**, 363–365.

Tilman, D., May, R. M., Lehman, C. L. and Nowak, M. A. (1994). Habitat destruction and the extinction debt. *Nature*, **371**, 65–66.

Vitousek, P., Ehrlich, P., Ehrlich, A. and Matson, P. (1986). Human appropriation of the products of photosynthesis. *BioScience*, **36**, 368–373.

Wilson, E. O. (1992). *The Diversity of Life*. Harvard University Press.

Part 1   Evolution: patterns and processes

Overview

Sean Nee, Pierre-Henri Gouyon, Godfrey Hewitt and Jean-Jacques Jaeger

The origins of biodiversity lie, of course, in evolutionary history. In the past, the fossil record has been our only source of information into evolutionary patterns, but fossil data are now being supplemented in an unexpected way: it is coming to be understood that phylogenies constructed from molecular data, of which we have an ever increasing, high-quality supply, also provide information about the evolutionary past. The theory which underpins such inferences, the theory of family trees, is also the basis of contemporary approaches in population genetics to using molecular data in making inferences about population structure, a subject of great importance to biodiversity studies.

What sets the chapters in this section apart from classical studies of the tempo and mode of evolution, such as the work of Simpson (see, for example, Simpson (1945) in references to Chapter 1), is the use of quantitative statistical and theoretical models to both assess the quality of the data, whether molecules or fossils, and to make inferences from such data about evolutionary patterns.

It is, of course, well known that the fossil record is far from perfect. Benton and Storrs (Chapter 1) go beyond this simple observation to describe ways of assessing the quality of the fossil record and ways of filling in gaps. For example, it is reasonable to suppose that a taxon which occurs at levels A and C in a rock column, but not in B, did not cease to exist at level B but was simply not preserved. The ratio of known occurrences and inferred occurrences (2/3 in the above example) provides a measure of the completeness of the fossil record. We would also like to know where we are on the macroevolution learning curve as new knowledge is acquired (i.e., how robust is our current understanding of the main patterns in the history of life?). Fortunately, the general picture of macroevolutionary patterns seems to be stable over time, even though there has been much adjustment in the details; rather like an army, the names change over time but the general appearance remains the same. Finally, if one imagines a cladogram plotted as a phylogeny against a geological time scale, one can ask what proportion of the lengths of the branches actually have fossil representatives. Benton and Storrs show that this proportion has increased through time. Hence, stratigraphic knowledge of the fossil record is improving. The authors point out that this is possibly the first time that improvement in knowledge in science has been quantitatively measured.

As noted above, the fossil record is no longer our only source of information into macroevolution. Traditionally, phylogenies of extant taxa have only been used to make inferences about the genealogical relationships amongst clades. The two chapters by Philippe and Adoutte and by Harvey, Rambault and Nee, discuss how phylogenies can also be used to make inferences about the evolutionary processes which have given rise to contemporary diversity. To fully exploit this possibility, it is necessary that the phylogeny come equipped with an approximate relative time axis. This can be provided by fossils, but, more commonly, the existence of approximate molecular clocks can be exploited to infer the relative ordering of the nodes.

Philippe and Adoutte (Chapter 2) argue that unresolved nodes in molecular phylogenies, rather than being merely an irritant, in fact may imply a burst of cladogenesis and be the signature of a historical radiation. Two uninteresting reasons for an unresolved node must first be excluded, though, before one can be confident of having identified a radiation. First, is a long enough molecular sequence being studied? Depending on the length of the sequence being used, and the mutation rates, if cladogenesis occurs too rapidly then the pattern of branching events will not be recorded. Philippe and Adoutte quantify the relationship between the length of sequence being studied and the temporal resolution of which it is capable. Second, if sites are highly variable, then they may have accumulated so many multiple substitutions that they are rendered uninformative—the sequence consists solely of noise. This theme, 'is the molecule suited to the question?', arises again in the chapter by Nichols and Beaumont. Having eliminated these potential sources of artefact, Philippe and Adoutte provide examples in which the molecular phylogeny provides good evidence for a burst of cladogenesis.

It is remarkable that information drawn solely from extant species can illuminate such macroevolutionary events, and this possibility is exploited further by Harvey, Rambault and Nee (Chapter 3). These authors start from the assumption that one has a good, resolved, molecular phylogeny with a time axis and address the question of what inferences can be made about the tempo and mode of the evolution of the clade from the rate of appearance of lineages in the phylogeny. The major difficulty involved in making such inferences is that appearances can be highly deceptive. For example, if the lineages of a growing clade actually appeared at a constant rate over evolutionary history, then the molecular phylogeny would suggest either that they had been appearing at an accelerating rate, if the phylogeny includes all the extant representatives of the clade, or at a decelerating rate, if the phylogeny consists only of a sample of the clade. To assist researchers in getting a feel for the signatures different processes leave in molecular phylogenies, the authors describe a set of user-friendly computer packages which will allow the simulation of phylogenies under a variety of hypotheses about the processes that may have led to their creation. The packages also will enable researchers to analyse their own data with recently available theoretical tools.

Patterns of gene flow amongst populations of a single species are important for genetic diversity, population structure, and ultimately population persistence. Among numerous other topics, Nichols and Beaumont (Chapter 4) discuss the utility of Wright's popular F_{st} statistic for making inferences about population structure. Although superficially a departure from the specific concerns of the previous chapters, there is indeed a very deep connection. The dominant contemporary theoretical approach to understanding the behaviour of this and other statistics in population genetics is based on explicit consideration of the genealogies, the family trees, of the genes in a sample. The chapter of Harvey, Rambault and Nee supposed that these genealogies are known, and the chapter of Philippe and Adoutte addressed the question of how much sequence information, and what sort, is required to reveal the genealogy. Nichols and Beaumont's contribution makes it clear that many weaknesses of the F_{st} statistic as estimated from allele frequency data, for example, arise from the fact that such data retain little information about their family tree. Another problem with such summary statistics is that, because they are single numbers which are ultimately a function of numerous variables, a variety of extremely different scenarios are compatible with the same estimate of F_{st}. Nevertheless, such summary statistics can still be useful when used in conjunction with other information about the species in question. This chapter provides an illuminating overview of many of the important themes in contemporary population genetics, such as the development of theory in the context of non-equilibrium populations.

While the previous chapters take the existence of genetic variation as a given, the final chapter by deMeeûs and Renaud (Chapter 5) looks at what maintains genetic variation. In particular, they review models of selectively maintained polymorphisms in heterogeneous environments. The genetic diversity maintained by habitat heterogeneity is probably the most important from the point of view of the present volume; after all, neutral polymorphism, for example, is irrelevant for the relationships of species to us and the rest of the environment, although useful for inference about other features of interest. The theoretical framework they describe is important for understanding the genesis of diversity though speciation and evolution into diverse ecological roles, as well as the consequences of habitat simplification and fragmentation.

These chapters describe new approaches to acquiring insight into the origins and structure of contemporary biodiversity. There is, nevertheless, a long road ahead before reaching the asymptote on our learning curve of the evolutionary mechanisms responsible for the genesis of biological diversity. One of the major obstacles, highlighted by this volume, will be the integration of evolutionary perspectives with those of ecologists and conservation biologists. We need to know to what extent evolutionary processes can or cannot be understood divorced from an ecological context (that is, to what extent are changes in allele frequencies important to species distributions and abundances, and *vice versa*), and what evolutionary biology teaches us about the future fate of biological diversity in the face of its human-mediated destruction.

1

Diversity in the past: comparing cladistic phylogenies and stratigraphy

Michael J. Benton and Glenn W. Storrs

Introduction

The study of biodiversity involves the dimension of time, whether short- or long-term. Most research in biodiversity is based on observations on human time scales, either experimental protocols lasting for one or two years, or data collation from historical records spanning back a few hundred years. The scale of regional or global diversity change at present is great enough, however, to require comparison with information based on geological time scales of thousands or millions of years. There are many problems in bridging the gap from experimental and field-based studies to the palaeontological work which extends into the depths of the geological past. Many of these problems, often gathered together under the heading 'the incompleteness of the fossil record', are geological, and they will be outlined briefly. Other issues are, however, more biological, and are concerned with phylogenetic reconstruction; these may offer insights into understanding the past history of life.

The value of the fossil record in giving a clear account of evolutionary history has been questioned. Charles Darwin hoped that, over time, more and more fossils would be found which would fill in all the 'missing links' and give a full picture of the history of life. By 1866, some of the first phylogenetic trees based explicitly in evolution were published (Haeckel 1866). Until recently, phylogenetic trees were composed by taking account not only of the morphology of organisms, but also their place in geological time, and hence it was not possible to test the nature of phylogenies and the fossil record directly. However, the development of cladistic techniques has opened up the possibility of testing the pattern of evolution. Cladograms based on morphological and/or molecular data involve no direct measure of the age of fossils: fossils are included as terminal taxa, side by side with living taxa. The divorce of phylogeny reconstruction from stratigraphic evidence opens up exciting possibilities for testing the two sets of evidence against each other, and for moving to fill some of the inevitable gaps in the fossil record.

The quality of the fossil record

It has been asserted that the fossil record is too incomplete for it to yield any useful macroevolutionary results. A lesser claim is that cladistic analyses should be based solely on living forms, since the fossils are both incomplete morphologically and they represent an incomplete sample of all the fossils that ever lived (e.g. Goodman 1989; Hennig 1981; Løvtrup 1977; Nelson 1969; Patterson 1981). This attitude has been opposed by palaeontologists and by many biologists (Gauthier *et al.* 1988*a*; Hecht 1976; Norell and Novacek 1992*a*; Schaeffer *et al.* 1972; Schoch 1986; Smith 1994) on the grounds that

(1) *some* evidence is better than none;

(2) fossils include a sample of the majority of species that have ever lived;

(3) most fossils represent morphologies that are quite unknown today, and they greatly enrich the content of a phylogeny;

(4) fossils may be placed more or less precisely in time, which provides good cross-evidence for the order and age of branching points; and

(5) for many groups, fossils in practice offer as much morphological data as do museum specimens of modern representatives.

The incompleteness of the fossil record may be ascribed to many factors of the organisms themselves, of their habitats, of later changes within the lithifying rocks, and of the ways in which palaeontologists work (Paul 1990; Raup 1972; Sheehan 1977; Signor 1990). Soft-bodied organisms are less likely to be preserved than those with hard parts. Long-living, rare, organisms are less likely to die and be preserved than short-lived common organisms. However, large organisms have a greater preservation potential than small ones, because they can survive incarceration in fine- and coarse-grained sediments and because they are easier to find. As for habitats, organisms that fly, or live in trees, are less often preserved than those that lurk around ponds and rivers, or live on the sea bed. Subsequent geological history is also important: organisms preserved in ancient rocks are more liable to have been subducted, metamorphosed, or eroded out of existence, than are those in more recent sediments. Human factors are also very important: our knowledge of the fossil record depends critically on the interest people have in particular groups, their geographic location, and their ease of study; these variables have been quantified as 'paleontologic interest units' (Sheehan 1977).

The quality of the fossil record, or parts of it, may be tested by relative and absolute measures. Measures of the relative quality of the fossil record are frequently obtained by palaeontologists (Paul 1982; Benton 1987, 1994), and some approaches to measuring absolute quality have been proposed (Meehl 1983; Maxwell and Benton 1990; Benton and Storrs 1994). Both sets of techniques offer the possibility of enhancing existing palaeontological data and of achieving better estimates of phylogenetic patterns and of past diversities. There is a three-step process: (1) filling the Lazarus gaps; (2) estimating range extensions, based on calculations of confidence intervals; and (3) searching for ghost ranges (cladistically defined minimum-implied gaps).

Filling the gaps

Lazarus gaps

A standard graphical approach in palaeontology is the use of range charts. These show by means of solid vertical lines the known distributions in time of particular fossil species or other taxa. A solid vertical line spanning a time interval of 5 myr might represent a succession of rocks containing densely packed fossils all the way through the sequence, or it might represent simply two point occurrences, each of a single fossil, spaced 5 myr apart in the rock column. Knowledge of the density of packing of the fossils along the range bar can provide useful statistical and predictive information (Paul 1982, 1990).

A first approach is to construct a grid of taxa versus time units, and to record presences and absences (Fig. 1.1(a)). Certain absences can then be determined as apparent rather than real; that is, gaps in the range which have resulted from non-preservation rather than non-existence of the taxon. These are the 'Lazarus taxa' (Jablonski 1986), those which apparently go extinct and then reappear higher up in the rock record. The ratio of known taxa to total known plus assumed taxa gives a minimal measure of relative completeness, the Simple Completeness Metric (SCM) (Benton 1987, 1994). This measure allows one to assess the relative quality of the fossil record either of groups or of time intervals (Fig. 1.1(b)).

Confidence intervals and range extensions

It is impossible to say whether a taxon arose before its first known fossil record, or whether it survived its last. Strauss and Sadler (1989) presented a technique of calculating confidence intervals for the ends of stratigraphic ranges. This was based on the intuitive assumption that recorded total ranges will tend to be more accurate the more closely packed fossils are within the known range. In other words, confidence intervals on both end-points of a range should be very small when fossils are closely packed, but huge when fossils are sparse. The confidence intervals on end-points of a range, expressed as a fraction of that range a, are calculated according to (Strauss and Sadler 1989; Marshall 1990):

$$a = (1 - C_1)^{-1/(H-1)} - 1,$$

where C_1 is the confidence level and H is the number of known fossiliferous horizons (smallest identifiable unit levels in the rocks at which fossils occur).

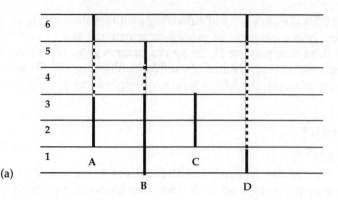

Fig. 1.1   Calculating the effect of Lazarus taxa. (a) A hypothetical range chart of four fossil taxa (A–D) plotted across six time units (1–6); solid lines indicate known stratigraphic ranges, checked lines represent Lazarus taxa, or known intrarange gaps. (b) Calculation of the Simple Completeness Metric (SCM) for the range chart in (a), where the SCM is the ratio of Lazarus gaps to known fossil ranges, calculated by time interval (rows) or by taxon (columns).

At the extreme, where only the terminal points of a range are represented (i.e. $H = 2$), the 95% confidence intervals are more than ten times the observed range ($a = 19$). Even with six point occurrences of fossils through the range, the predicted range extensions must equal the observed range at the 95% confidence level (Fig. 1.2). With more than six records, the error bars diminish, but never quite disappear, reaching negligible values for intensely sampled ranges (e.g., $a = 0.03$ when $H = 100$).

Adding ghost ranges

The ghost range (Norell and Novacek 1992*b*; Norell 1993), or Minimum Implied Gap (MIG) (Benton 1994; Benton and Storrs 1994; Storrs 1994), or Minimal Divergence Time (MDT) (Weishampel and Heinrich 1992), is the difference between the age of the first representative of a lineage and that of its phylogenetic sister. Postulated sister groups in a cladogram arose, by definition, from a single node representing a point in time and hence both

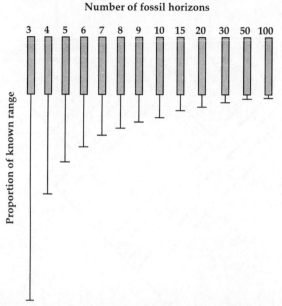

Fig. 1.2 Estimates of true range lengths based on known stratigraphic ranges; 95% confidence intervals are shown. The more densely packed the fossil horizons are within the known range, the shorter is the confidence interval. (Based on information in Marshall 1990.)

taxa should have oldest fossil representatives of the same age. In practice it is rarely the case that known fossil sisters are of the same geological age and the ghost range is a predictable range extension for one of the sister taxa. This depends on the assumptions that (1) the cladogram is a correct representation of relationships, and (2) both sister taxa are correctly assigned stratigraphically.

The fossil record of Triassic sauropterygians (Storrs 1991, 1994), long-necked marine reptiles, is chosen as an example (Fig. 1.3). Individual skeletons are well preserved and rich in osteological characters, which has permitted the production of a cladogram at genus level (Fig. 1.3(a)). When the cladogram is converted into a phylogram, by the addition of stratigraphic information (Fig. 1.3(b)), it can be seen that most genera are point occurrences, known from single geological horizons. Additionally, there are many ghost-range additions: MIGs represent a total of 65 myr, compared to 41.5 myr of known range, giving a value of only 39% completeness (41.5/(41.5 + 65)).

Estimating the absolute quality of the fossil record

How do revisions of data bases affect macroevolutionary conclusions?

The absolute quality of the fossil record has been tested by comparing changes in palaeontological knowledge over research time. The basic

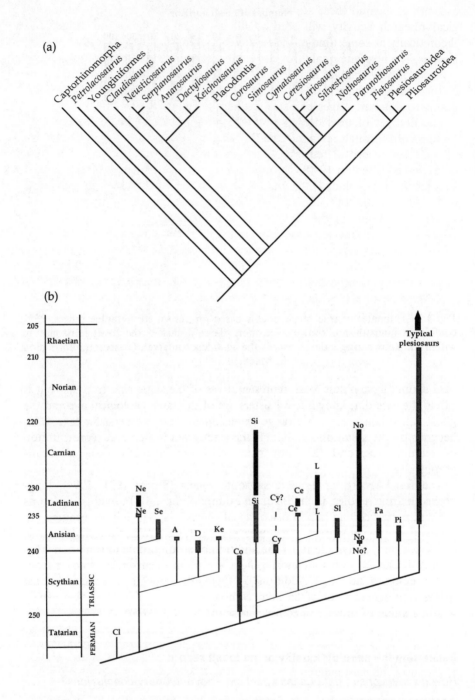

assumption behind these studies is that the cumulative sum of palaeontological research activity will tend to fill gaps and increase knowledge, and theoretically at some time in the future everything that can be known will be known. The absolute quality of the fossil record has been tested in two ways: (1) tests of how changes in palaeontological knowledge affect our perceptions of macroevolutionary patterns and (2) tests of the goodness of fit of cladograms to different stages of palaeontological knowledge.

The first set of tests demonstrate substantial changes in the documented fossil record through research time. For example, palaeontological knowledge of the fossil record of tetrapods has increased most over the past 100 years (Maxwell and Benton 1990), but although the gross number of taxa increased as a result of new finds (Fig. 1.4(a)), other aspects of the taxon lists changed in non-systematic and unpredictable ways. Revisions of stratigraphy, revisions of taxonomy at the alpha level, and broad-scale cladistic revisions of larger groups all affected the taxon range lists, but some ranges were shortened, some increased, and some remained unaltered. Some families and genera disappeared as a result of taxonomic revision, while others appeared, so that there was no overall shift in the results. The main change detected in a comparison of a 1967 data base (Harland *et al.* 1967) with one compiled 20 years later (Benton 1987) was that familial durations of tetrapods had increased marginally (29.1% of families with unchanged range lengths, 44.8% with increased range lengths, 26.1% with decreased range lengths). Increased range lengths would be predicted by the simple rallying cry that 'more palaeontological research means collecting more fossils', but the fact that more than one-quarter of the 515 families analysed showed decreased range lengths was more of a surprise. This was the result of cladistic redefinitions of families and the removal of suspect ancestral taxa from the bottoms of ranges, the latter of which lacked autapomorphies of the family.

A surprising conclusion of this study (Maxwell and Benton 1990) was, however, that although stratigraphic ranges of 70.9% of the tetrapod families had changed in a span of 20 years of research, the macroevolutionary conclusions derived from the data bases altered little. In fact, the phases of diversification and of extinction remained the same (Fig. 1.4(a)). Rates of origination and extinction at particular times also remained in proportion.

Legend to facing page

Fig. 1.3 Calculation of relative completeness of the fossil record of Triassic sauropterygians, marine long-necked reptiles. (a) Cladogram of the genera of Triassic sauropterygians, as well as outgroups, placodonts, and the later clades Plesiosauroidea and Pliosauroidea. (b) Phylogenetic tree of the same sauropterygian genera plotted against a Triassic time scale, with ages in millions of years indicated. Higher taxa and most outgroups excluded. Known ranges are indicated in black, and assumed additional ranges are shown cross-hatched (these are based on cladistic patterns of pairing; see Fig. 1.6). Abbreviations of genera correspond to names given in full shown in (a). (Based on information in Storrs 1991, 1994.)

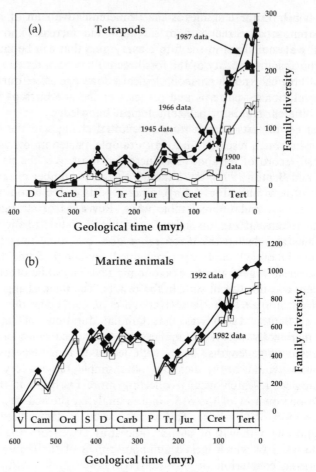

Fig. 1.4 Revisions of data bases have not affected macroevolutionary patterns. Diversity change through geological time for families of tetrapods (a) and marine animals (b), based on data bases published at different times. (Based on data in Maxwell and Benton 1990 and Sepkoski 1993.)

The main effects to be seen were an increase in overall diversity through much of the past 400 million years of tetrapod evolution and a slight sharpening of extinction events.

A similar study (Sepkoski 1993) of the past ten years of change in the standard marine animals data base (Sepkoski 1982, 1992) found turnover in 50% of the data—families had been added and deleted, low-resolution stratigraphic data had been improved, and dates of apparent origination and extinction of families had been altered. However, despite these changes, plots of diversification remained nearly identical but for the fact that the 1992 curve lies 13% higher than the older one (Fig. 1.4(b)). Further, the identity and magnitudes of extinction events remained the same. The main changes were that family range lengths tended to increase (with an approximately equal distribution of earlier originations and later extinctions) and mass extinctions became slightly sharper, with extinctions shifting closer to chronostratigraphic boundaries. Hence, these tests have indicated two facts: (1) palaeontological knowledge is changing (advancing, one hopes) rapidly; (2) despite this, broad-scale macroevolutionary patterns have remained stable. If the fossil record were hopelessly incomplete, the patterns derived therefrom might change wildly as a result of new discoveries.

Has palaeontological knowledge improved?

The second test of the absolute quality of the fossil record compares the quality of different stages in palaeontological research in a quantitative way. Benton and Storrs (1994) compared a sample of 74 tetrapod cladograms (Table 1.1) against two recognised landmark data bases, the *Fossil Record* (Harland *et al.* 1967) and the *Fossil Record* 2 (Benton 1993), using two measures of fossil record quality: (1) the correlation of clade rank and age rank, measured using the Spearman Rank Correlation statistic (SRC, with quality of correlation assessed at confidence levels of $P < 0.05$ and $P < 0.01$); and (2) the Relative Completeness Index (RCI) of all test cladograms when plotted as phylogenies against a geological time scale (measured as the proportion of Minimum Implied Gaps (MIGs), indicated by branching points of sister group pairs, to Simple Range Lengths (SRLs), based on total range lengths represented by fossils).

The results (Fig. 1.5) were mixed: the comparisons of age and clade rank showed no change from 1967 to 1993, but the more precise test of completeness showed a clear improvement. The first test showed no change in the degree of correlation of clade rank and age rank data when the 1967 and 1993 data sets were compared for all 71 comparable cladograms (Fig. 1.5(a),(b)): 43 of the 71 comparisons (61%) showed no change of state, while 28 (39%) changed, but the changes were equally balanced, with 14 'improvements' (i.e. negative to positive correlation; insignificant to significant correlation, at values of $P < 0.05$ or 0.01; correlation at $P < 0.05$ to correlation at $P < 0.01$) and 14 deteriorations. When the RCI values were compared as a whole (Fig. 1.5(c)), however, the means for 1967 and 1993 (67.856, 72.289) show a significant ($P < 0.05$) improvement whether tested parametrically (*t*-test) or non-parametrically (sign test; Wilcoxon signed ranks test).

Hence, the two tests just outlined have shown that knowledge of the fossil record is changing substantially, but that most of the macroevolutionary conclusions based on that record are relatively stable to these statistically non-systematic changes, and the relative completeness of the fossil record is increasing through research time. This may be the first time a quantitative demonstration has been made of improvement in knowledge in palaeontology.

Table 1.1 Seventy-four test-case cladograms used by Benton and Storrs (1994) to test the quality of knowledge of the fossil record in 1967 (data from Harland *et al.* 1967) and 1993 (data from Benton 1993). The match of cladistic node order and stratigraphic position was measured using the Spearman Rank Correlation (SRC) statistic. The quality of the fossil records implied by each phylogeny, for 1967 and 1993 data, was calculated as the Relative Completeness Index (RCI), which is the proportion of the cladistic Minimum Implied Gaps (MIGs) to the known Simple Range Lengths (SRLs) for all taxa in the phylogeny.

TAXON	1967			1993			
	RCI	SRC	n	RCI	SRC	n	
Actinopterygii	−194.7	0.346	9	71.9	0.139	9	(Lauder and Liem 1983)
Amniota 1	84.5	0.928*	6	90.1	0.642	6	(Gardiner 1982)
Amniota 2	75.3	0.957**	10	82.6	0.982**	10	(Gardiner 1982)
Amniota	83.1	0.641*	11	75.3	0.493	11	(Gauthier *et al.* 1988b)
Amphibia	69.3	0.899**	12	80.5	0.860**	12	(Gardiner 1983)
Amphibia	41.1	0.655*	8	66.7	0.673*	10	(Milner 1988)
Amphibia 1	89.1	0.947*	5	95.0	0.772	6	(Panchen and Smithson 1988)
Amphibia 2	84.9	0.418	6	89.8	0.508	6	(Panchen and Smithson 1988)
Amphibia	48.0	0.682	11‡	74.3	0.276	11	(Trueb and Cloutier 1991)
Archosauria	94.7	0.791*	8	95.2	0.657*	9	(Benton and Clark 1988)
Archosauromorpha	84.2	0.151	8	85.4	0.108	8	(Benton 1985)
Archosauromorpha	73.4	−0.620	9	68.1	−0.253	15	(Evans 1988)
Artiodactyla	86.3	0.677*	10§	89.3	0.428	10	(Gentry and Hooker 1988)
Aves	48.5	0.609*	11	77.2	0.720*	11	(Cracraft 1988)
Brontotheriidae	56.1	0.969**	8§	67.7	0.975**	10	(Mader 1989)
Chalicotheriinae	—	—	—	59.3	0.903**	8	(Coombs 1989)
Chalicotherioidea	54.7	0.938*	6§	54.3	0.962**	7	(Coombs 1989)
Crocodylomorpha 1	17.7	0.556**	18§	5.3	0.801**	22	(Benton and Clark 1988)
Crocodylomorpha 2	4.6	0.569*	17§	0.1	0.815**	19	(Benton and Clark 1988)
Diapsida	82.3	0.927**	7	71.0	0.849*	8	(Laurin 1991)
Equidae	58.3	0.975**	12§	59.8	0.942**	12	(Evander 1989)

Taxon							Reference
Eutheria	55.2	0.904**	9	75.6	0.781*	9	(Gregory 1910)
Eutheria	64.1	0.809***	10	81.1	0.590*	10	(Simpson 1945)
Eutheria	44.6	0.646*	12	81.6	0.795**	12	(McKenna 1975)
Eutheria	46.3	0.723**	14	73.1	0.650**	14	(Novacek 1982)
Eutheria	27.4	0.078	9	74.7	−0.211	9	(Miyamoto and Goodman 1986)
Eutheria	53.7	0.857**	12	72.5	0.590*	12	(Novacek and Wyss 1986)
Eutheria	90.6	−0.159	10	74.0	−0.093	10	(Shoshani 1986)
Eutheria	39.7	0.063	13	60.5	−0.031	13	(Novacek et al. 1988)
Eutheria	71.3	0.915**	10	82.0	0.883**	10	(Novacek 1989)
Gnathostomata	96.4	−0.185	7	93.8	−0.286	7	(Rosen et al. 1981)
Gnathostomata	92.6	0.273	8	92.6	0.232	8	(Lauder and Liem 1983)
Hadrosauridae	100.0	—	7§	63.8	0.549	8¶	(Weishampel and Horner 1990)
Hadrosaurinae	100.0	—	5§	73.1	0.803	5¶	(Weishampel and Horner 1990)
Hystricomorpha	67.9	0.288	6	69.1	0.626*	9	(Jaeger 1988)
Lepidosauromorpha	74.1	0.973**	7‡	77.6	0.988***	8	(Benton 1985)
Lepidosauromorpha	63.3	0.876**	12‡	72.5	0.895***	16	(Evans 1988)
Lepidosauromorpha	92.4	0.734*	7	76.6	0.821*	7	(Gauthier et al. 1988a)
Lissamphibia	38.1	0.718	5	71.4	0.754	6	(Bolt 1991)
Mammalia	31.1	−0.012	15	66.7	0.368	15	(Novacek et al. 1988)
Mammalia	56.1	0.460	12	83.9	0.935***	12	(Novacek 1989)
Ornithischia	68.7	0.809	6	68.9	0.927***	7	(Sereno 1984)
Ornithischia	60.1	0.455	8	64.9	0.725*	9	(Sereno 1986)
Ornithischia	60.1	0.455	8	64.9	0.725*	9	(Benton 1990)
Ornithopoda	79.2	−0.487	5	56.6	0.603	6	(Norman 1984)
Perissodactyla	76.9	0.214	12§	97.6	0.428	130	(Hooker 1989)
Primates	49.0	0.376	11§	50.0	0.442	11	(Andrews 1988)
Proboscidea	53.8	0.882**	15§	51.1	0.935**	18	(Tassy and Shoshani 1988)
Ruminantia	59.9	0.582*	16§	67.4	0.606**	17	(Janis and Scott 1988)
Sarcopterygii	94.1	0.588	6	95.0	0.662	7	(Schultze 1987)
Sarcopterygii	95.9	−1.000	5	93.2	−0.973	5	(Chang 1991)

Table 1.1 *Continued*

TAXON	1967			1993			
	RCI	SRC	n	RCI	SRC	n	
Sarcopterygii	94.1	-0.221	6	92.2	-0.387	7	(Forey et al. 1991)
Sauropodomorpha	92.0	0.924**	6	61.3	0.376	9	(Benton 1990)
Sauropterygia	73.1	0.031	6§	94.4	0.651	7	(Storrs 1991)
Squamata 1	41.3	0.291	7	73.4	0.164	7	(Estes et al. 1988)
Squamata 2	82.0	0.782*	8	81.7	0.264	8	(Estes et al. 1988)
Squamata	77.3	0.718	6	64.6	0.638	6	(Rieppel 1988)
Squamata	79.5	0.895**	7‡	64.6	0.749*	8	(Schwenk 1988)
Synapsida	75.2	0.983**	17	73.1	0.986**	17	(Kemp 1982)
Synapsida	69.4	0.958**	16	76.2	0.950**	16	(Gauthier et al. 1988b)
Synapsida	35.4	0.965**	19	64.5	0.967**	19	(Rowe 1988)
Synapsida	71.0	0.957**	20	77.4	0.942**	21	(Hopson 1991)
'tapiroids'	76.0	0.817**	9§	81.1	0.673*	10	(Schoch 1989)
Teleostei	70.6	0.883**	15	75.7	0.827**	16	(Lauder and Liem 1983)
Testudines	79.9	0.250	6	61.4	0.563	6	(Gaffney 1975)
Testudines	53.4	0.236	9‡	58.0	0.748**	11	(Gaffney 1984)
Testudines	83.9	0.388	6	62.0	0.716	6	(Moody 1984)
Testudines	54.7	0.594*	11‡	59.6	0.910**	14	(Gaffney and Meylan 1988)
Tetrapoda	89.7	0.873*	7	89.0	0.883*	7	(Gaffney 1979)
Therapsida	80.2	0.950**	14	91.6	0.905**	14	(Hopson and Barghusen 1986)
Theropoda	84.6	0.730	5	61.1	0.782*	7	(Gauthier 1986)
Theropoda	84.6	0.730	5	40.4	0.476	9	(Benton 1990)
Ungulata	87.4	0.548*	12	85.8	0.553*	14	(Prothero et al. 1988)

n = sample size. ‡ = data from Harland (1967) and Romer (1966), § = data from Romer (1966), ¶ = data from Weishampel et al. (1990), † = data from Prothero and Schoch (1989). * = significant correlation at $P < 0.05$. ** = significant correlation at $P < 0.01$.

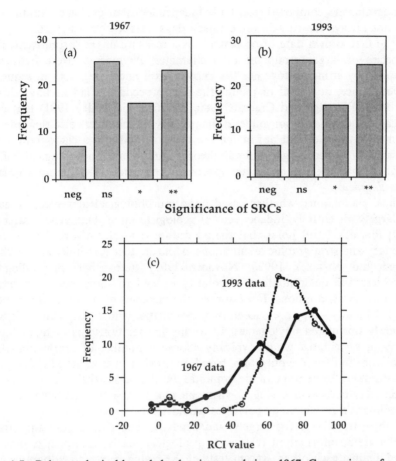

Fig. 1.5 Palaeontological knowledge has improved since 1967. Comparison of a 1967 and a 1993 data base on the fossil record of vertebrates shows no change in the patterns of matching between ordering of branching points based on cladistic and stratigraphic data (a,b), but the relative completeness of phylogenetic trees has improved significantly (c). The first test is Spearman Rank Correlation (SRC) of matching of rank order of origin of taxa, based on cladistic and stratigraphic data, in 71 cladograms of vertebrates. The second test is a comparison of measures of relative completeness of 73 phylogenies, based on comparisons of known Simple Range Lengths (SLRs) and interpreted Minimum Implied Gaps (MIGs) from sister-group comparisons. The first test assesses rank order only, recording simply correlation (* at 95% and ** at 99% significance levels), non-correlation (ns), or negative correlation (neg), and takes no account of the *amount* of difference that exists. The second test is based on a continuous numerical summation of SRLs and MIGs, and offers clear evidence for a significant reduction in the relative *amount* of mismatch between the known fossil dates of origin of sister groups.

Comparing cladograms and stratigraphy

Cladograms are composed essentially independently of geological input. This was not always the case in pre-cladistic days: classically, systematists would use the first known date of occurrence as a measure of the 'primitiveness' of a species or larger group, or of a character. Phylogenies were frequently compiled by stringing together the known fossil and living taxa in sequence of occurrence, in a kind of joining-the-dots procedure. Cladists have shown (Ax 1987; Eldredge and Cracraft 1980; Hennig 1966; Wiley 1981) that geological sequence is a poor guide to the polarity of characters and the order of occurrence in time need not match the order of nodes in a cladogram. Only in cases of superbly good fossil records and poor morphological differentiation may the stratophenetic approach (Gingerich 1985) be appropriate, *faute de mieux*.

Thus, cladograms, whether based upon morphological or molecular data, are largely, or entirely, independent of geological input. One view (Patterson 1981; Platnick 1979) holds that the procedure is wholly divorced from evolutionary and stratigraphic assumptions, while another (Gauthier *et al.* 1988*a*; Norell and Novacek 1992*b*; Novacek 1992) states that the coding of characters and determination of polarity depend to some extent on broad evolutionary assumptions. For example, the members of the outgroup are not selected blindly—they are chosen by a systematist who clearly cannot ignore currently postulated phylogenies. In testing the relationships of mammalian orders, a systematist would typically choose a number of non-mammalian vertebrates to form the outgroup, based on previously established clades such as Chordata, Vertebrata, and Amniota. In this case, the outgroup could legitimately consist of a slug, a virus, an oak tree, and a cabbage, but the determination of character polarities would then be nonsensical.

If they are wholly (or largely) independent, then cladograms and stratigraphic data, information relating to geological time and sequence, may be tested against each other. Recent tests of the match between cladograms and stratigraphic data have suggested that there is strong correspondence. The stratigraphic sequence of first occurrences of fossil vertebrate groups frequently matches the order of branching of cladograms based upon character analysis alone (Gauthier *et al.* 1988*a*; Norell and Novacek 1992*a,b*; Benton and Storrs 1994).

The technique for comparing clade-rank data with age-rank data (Fig. 1.6) involves certain simplifications. Cladistic rank is determined by counting the sequence of primary nodes in a cladogram; nodes are numbered from 1 (basal node) upwards to the ultimate node. As this method cannot cope objectively with complex cladograms comprising several subclades, such cladograms are converted to a hierarchy of nodes along a single branch (Fig. 1.6(a)). This is accomplished (Fig. 1.6(b)) by collapsing each subordinate clade to a single polytomous node originating at the main stem; each polytomous lineage is assessed equally and given equal cladistic rank. Yet, in practice, the oldest

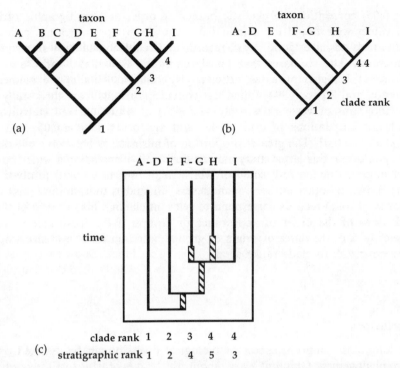

Fig. 1.6 Method for comparison of cladistic and stratigraphic data. A cladogram (a) is reduced to a single stem lineage, or 'Hennigian comb' type of cladogram (b), in order to allow the clade rank to be assessed unequivocally. The polytomes representing collapsed parts of the original cladogram (here numbered 1, 3) are counted as single lineages: to include each separate branch collectively could introduce large amounts of implied gap where information actually exists. The stratigraphic rank is assessed (c) independently by recording the known order of appearance of the taxa. Because of missing early ranges (shown cross-hatched), clade rank does not always match stratigraphic rank.

representative branch is the one chosen for the analysis. In cases where subordinate clades are large, we adopted the convention that the source cladogram is collapsed so as to maximize node number (using the principle of free rotation). The stratigraphic sequence of clade appearance is assessed from the earliest known fossil representative of sister groups. The Minimum Implied Gap (MIG, indicated by cross-hatching in Fig. 1.6(c)) is the difference between the age of the first representative of a lineage and that of its sister, since the oldest known fossils of sister groups are infrequently of the same age. The MIG is a minimum estimate of stratigraphic gap, as the true age of lineage divergence may lie well before the oldest known fossil.

In their small-scale study, Gauthier *et al.* (1988*a*) found that clade rank and age rank are correlated. Using a larger sample, Norell and Novacek (1992*a*) found that 18 of their 24 test cases (75%) gave statistically significant

($P < 0.05$) correlations of cladistic branching order and stratigraphic order. Best values were found for mammalian ungulate groups, which are believed qualitatively to have 'good' fossil records and relatively stable, well-resolved cladograms. The six cases that failed (amniotes, Squamata, hadrosaurs 1, hadrosaurs 2, higher primates, artiodactyls) could not be simply explained.

Benton and Storrs (1994) found less convincing results from their study of 74 cladograms of vertebrates, only 41 (55%) of which showed statistically significant correlations of clade order and age order at $P < 0.05$, and 25 (35%) at $P < 0.01$. The greater proportion of mismatches between clade rank and age rank in this latter study may be the result of including a wider range of cladograms in the test, some of which might not be so well resolved as those chosen by other authors. Nonetheless, all studies to date show that the majority of fossil records correspond to relevant cladograms in terms of their predictions of the order of appearance of groups: if the fossil record were hopelessly bad, the dates of origin of groups based on fossil evidence would show no match to clade ranks at all.

Conclusions

The long-term temporal aspect of biological diversity can be assessed only from phylogenies. Opinions vary about the value of the fossil record in establishing and testing phylogenies and in presenting information about species richness in the past; at one extreme, the historical record is accorded little value because of its supposed incompleteness, and at the other, the palaeontological data are read literally, and with no consideration of how information may have been lost.

There is no doubt that the information content of the fossil record diminishes backwards in time and that no instant in geological time has the potential to be as well understood as the present. However, evidence about phylogeny is available from various independent sources: cladistic reconstruction of phylogeny from morphological characters, stratigraphic information, and molecular phylogeny reconstruction. Statistical methods are available for applying correction factors to particular parts of the fossil record by filling interpolated gaps, estimating terminal confidence intervals, and adding ghost ranges predicted from cladistic phylogenies.

Broad-scale studies of the fossil record have shown that the documented data base has changed substantially during the past 25 years, but that major events appear to be robust enough to be unaffected by the statistical noise produced by the unpredictable vagaries of research. Palaeontological knowledge has improved in the past 25 years, when tested against a constant of cladistic information. The availability of several sets of independent information about the history of life suggests that it will be possible to give quantitative measures of confidence in estimates of diversities in the past.

Acknowledgement

This project was partly funded by a grant from the Leverhulme Trust.

Note added in proof

Since this manuscript was completed, a number of relevant papers have been published. Marshall (1994) has developed a modified approach to the calculation of range extensions (Fig. 1.2). The simple technique shown here is based on an assumption of randomly distributed fossil finds, and hence classical confidence intervals are applicable. His new technique relaxes the assumption of randomness of distribution of fossil horizons, but it is less universally applicable, and there are uncertainties associated with the sizes of the confidence intervals. Huelsenbeck (1994) has presented a test of the fit of cladograms to the stratigraphic record, a stratigraphic consistency index. This technique uses the same approach as has been applied in the present paper, but turns the focus on testing the quality of cladograms instead of the quality of stratigraphic records of fossil occurrences. Finally, Benton and Simms (1995) have shown that the fossil records of continental vertebrates and echinoderms are equivalent in quality when compared to a large sample of available cladograms and molecular trees. This finding validates the use of different kinds of fossil data (continental vs. marine; vertebrate vs. invertebrate) in broad-scale phylogenetic studies.

References

Andrews, P. (1988). A phylogenetic analysis of the Primates. In *The phylogeny and classification of the tetrapods. Volume 2. Mammals* (ed. Benton, M. J.), pp. 143–175. Systematics Association Special Volume 35B. Clarendon Press, Oxford.

Ax, P. (1987). *The phylogenetic system*. Wiley, New York.

Benton, M. J. (1985). Classification and phylogeny of the diapsid reptiles. *Zoological Journal of the Linnean Society*, **84**, 97–164.

Benton, M. J. (1987). Mass extinctions among families of non-marine tetrapods: the data. *Mémoires de la Société Géologique de France*, **150**, 21–32.

Benton, M. J. (1990). Origin and interrelationships of dinosaurs. In *The Dinosauria* (ed. Weishampel, D. B., Dodson, P., and Osmólska, H.), pp. 11–30. University of California Press, Berkeley.

Benton, M. J. (1993). *The Fossil Record 2*. Chapman & Hall, London, 839 pp.

Benton, M. J. (1994). Palaeontological data, and identifying mass extinctions. *Trends in Ecology and Evolution*, **9**, 181–185.

Benton, M. J. and Clark, J. (1988). Archosaur phylogeny and the relationships of the Crocodylia. In *The phylogeny and classification of the tetrapods. Volume 1. Amphibians, reptiles, birds* (ed. Benton, M. J.), pp. 289–332. Systematics Association Special Volume 35A. Clarendon Press, Oxford.

Benton, M. J. and Simms, M. J. (1995). Testing the marine and continental fossil records. *Geology*, **23**, in press.

Benton, M. J. and Storrs, G. W. (1994). Testing the quality of the fossil record: paleontological knowledge is improving. *Geology*, **22**, 111–114.

Bolt, J. R. (1991). Lissamphibian origins. In *Origins of the higher groups of tetrapods* (ed. Schultze, H.-P. and Trueb, L.), pp. 194–222. Cornell University Press, Ithaca, N.Y.

Chang, M.-M. (1991). 'Rhipidistians', dipnoans, and tetrapods. In *Origins of the higher groups of tetrapods* (ed. Schultze, H.-P. and Trueb, L.), pp. 1–28. Cornell University Press, Ithaca, N.Y.

Coombs, M. C. (1989). Interrelationships and diversity in the Chalicotheriidae. In *The evolution of perissodactyls* (ed. Prothero, D. R. and Schoch, R. M.), pp. 438–457. Clarendon Press, New York.

Cracraft, J. (1988). The major clades of birds. In *The phylogeny and classification of the tetrapods. Volume 1. Amphibians, reptiles, birds* (ed. Benton, M. J.), pp. 339–361. Systematics Association Special Volume, 35A. Clarendon Press, Oxford.

Eldredge, N. and Cracraft, J. (1980). *Phylogenetic patterns and the evolutionary process.* Columbia University Press, New York, 349 pp.

Estes, R., Queiroz, K. de, and Gauthier, J. (1988). Phylogenetic relationships within Squamata. In *Phylogenetic relationships of the lizard families. Essays commemorating Charles L. Camp* (ed. Estes, R. and Pregill, G.), pp. 119–281. Stanford University Press, Stanford.

Evander, R. L. (1989). Phylogeny of the Family Equidae. In *The evolution of perissodactyls* (ed. Prothero, D. R. and Schoch, R. M.), pp. 109–126. Clarendon Press, New York.

Evans, S. E. (1988). The early history and relationships of the Diapsida. In *The phylogeny and classification of the tetrapods. Volume 1. Amphibians, reptiles, birds* (ed. Benton, M. J.), pp. 221–260. Systematics Association Special Volume, 35A. Clarendon Press, Oxford.

Forey, P. L., Gardiner, B. G., and Patterson, C. (1991). The lungfish, the coelacanth, and the cow revisited. In *Origins of the higher groups of tetrapods* (ed. Schultze, H.-P. and Treub, L.), pp. 145–172. Comstock Publishing Associates, Ithaca and London.

Gaffney, E. S. (1975). A phylogeny and classification of the higher categories of turtles. *Bulletin of the American Museum of Natural History*, **155**, 387–436.

Gaffney, E. S. (1979). Tetrapod monophyly: a phylogenetic analysis. *Bulletin of the Carnegie Museum of Natural History*, **13**, 92–105.

Gaffney, E. S. (1984). Progress towards a natural hierarchy of turtles. In *Studia Palaeocheloniologica I* (ed. Broin, F. de and Jiménez-Fuentes, E.), pp. 125–131. Studia Geologica Salamanticensia, Volumen Especial 1. Universidad de Salamanca.

Gaffney, E. S. and Meylan, P. A. (1988). A phylogeny of turtles. In *The phylogeny and classification of the tetrapods. Volume 1. Amphibians, reptiles, birds* (ed. Benton, M. J.), pp. 157–219. Systematics Association Special Volume, 35A. Clarendon Press, Oxford

Gardiner, B. (1982). Tetrapod classification. *Zoological Journal of the Linnean Society*, **74**, 207–232.

Gardiner, B. (1983). Gnathostome vertebrae and the classification of the Amphibia. *Zoological Journal of the Linnean Society*, **79**, 1–59.

Gauthier, J. (1986) Saurischian monophyly and the origin of birds. *Memoirs of the California Academy of Sciences*, **8**, 1–55.

Gauthier, J., Estes, R., and Queiroz, K. de (1988a). A phylogenetic analysis of Lepidosauromorpha. In *Phylogenetic relationships of the lizard families. Essays commemorating Charles L. Camp* (ed. Estes, R. and Pregill, G.), pp. 119–281. Stanford University Press, Stanford.

Gauthier, J., Kluge, A. G., and Rowe, T. (1988*b*). Amniote phylogeny and the importance of fossils. *Cladistics*, **4**, 105–209.

Gentry, A. W. and Hooker, J. J. (1988). The phylogeny of the Artiodactyla. In *The phylogeny and classification of the tetrapods. Volume 2. Mammals* (ed. Benton, M. J.), pp. 235–272. Systematics Association Special Volume 35B. Clarendon Press, Oxford.

Gingerich, P. D. (1985). Species in the fossil record: concepts, trends, and transitions. *Paleobiology*, **11**, 27–42.

Goodman, M. (1989). Emerging alliance of phylogenetic systematics and molecular biology: A new age of exploration. In *The hierarchy of life* (ed. Fernholm, B., Bremer, K., and Jornvall, H.), pp. 43–61. Elsevier, New York.

Gregory, W. K. (1910). The orders of mammals. *Bulletin of the American Museum of Natural History*, **27**, 1–524.

Haeckel, E. (1866). *Generelle Morphologie der Organismen*. G. Reimer, Berlin.

Harland, W. B., Holland, C. H., House, M. R., Hughes, N. F., Reynolds, A. B., Rudwick, M. J. S., Satterthwaite, G. E., Tarlo, L. B. H., and Willey, E. C. (1967). The fossil record; a symposium with documentation. *Geological Society of London*, London, 827 pp.

Hecht, M. K. (1976). Phylogenetic inference and methodology as applied to the vertebrate record. *Evolutionary Biology*, **9**, 335–363.

Hennig, W. (1966). *Phylogenetic systematics*. University of Illinois Press, Urbana, 263 pp.

Hennig, W. (1981). *Insect phylogeny*. John Wiley, New York, 528 pp.

Hooker, J. J. (1989) Character polarities in early perissodactyls and their significance for *Hyracotherium* and infraordinal relationships. In *The evolution of perissodactyls* (ed. Prothero, D. R. and Schoch, R. M.), pp. 79–101. Clarendon Press, New York.

Hopson, J. A. (1991). Systematics of the nonmammalian Synapsida and implications for patterns of evolution in synapsids. In *Origins of the higher groups of tetrapods* (ed. Schultze, H.-P. and Trueb, L.), pp. 635–693. Cornell University Press, Ithaca, N.Y.

Hopson, J. A. and Barghusen, H. R. (1986). An analysis of therapsid relationships. In *The ecology and biology of mammal-like reptiles* (ed. Hotton III, N., MacLean, P. D., Roth, J. J., and Roth, E. C.), pp. 83–106. Smithsonian Institution Press, Washington, D.C.

Huelsenbeck, J. P. (1994). Comparing the stratigraphic record to estimates of phylogeny. *Paleobiology*, **20**, 470–483.

Jablonski, D. (1986). Causes and consequences of mass extinctions: a comparative approach. In *Dynamics of extinction* (ed. Elliott, D. K.), pp. 183–229. John Wiley & Sons, New York.

Jaeger, J.-J. (1988). Rodent phylogeny: new data and old problems. In *The phylogeny and classification of the tetrapods. Volume 2. Mammals* (ed. Benton, M. J.), pp. 177–199. Systematics Association Special Volume 35B. Clarendon Press, Oxford.

Janis, C. M. and Scott, K. M. (1988). The phylogeny of the Ruminantia (Artiodactyla, Mammalia). In *The phylogeny and classification of the tetrapods. Volume 2. Mammals* (ed. Benton, M. J.), pp. 273–282. Systematics Association Special Volume 35B. Clarendon Press, Oxford.

Kemp, T. S. (1982). *Mammal-like reptiles and the origin of mammals*. Academic Press, London, 363 pp.

Lauder, G. V. and Liem, K. F. (1983). The evolution and interrelationships of the actinopterygian fishes. *Bulletin of the Museum of Comparative Zoology*, **150**, 95–197.

Laurin, M. (1991). The osteology of a Lower Permian eosuchian from Texas and a review of diapsid phylogeny. *Zoological Journal of the Linnean Society*, **101**, 59–95.

Løvtrup, S. (1977). *The phylogeny of Vertebrata*. Wiley, London, 330 pp.

Mader, B. J. (1989). The Brontotheriidae: a systematic revision and preliminary phylogeny of North American genera. In *The evolution of perissodactyls* (ed. Prothero, D. R. and Schoch, R. M.), pp. 458–484. Clarendon Press, New York.

Marshall, C. R. (1990). Confidence intervals on stratigraphic ranges. *Paleobiology*, **16**, 1–10.

Marshall, C. R. (1994). Confidence intervals on stratigraphic ranges: partial relaxation of the assumption of randomly distributed fossil horizons. *Paleobiology*, **20**, 459–69.

Maxwell, W. D. and Benton, M. J. (1990). Historical tests of the absolute completeness of the fossil record of tetrapods. *Paleobiology*, **16**, 322–335.

McKenna, M. C. (1975). Toward a phylogenetic classification of the Mammalia. In *Phylogeny of the primates* (ed. Luckett, W. P. and Szalay, F. S.), 21–46. Plenum, New York.

Meehl, P. E. (1983). Consistency tests in estimating the completeness of the fossil record: a Neo-Popperian approach to statistical paleontology. *Minnesota Studies in the Philosophy of Science*, **10**, 413–473.

Milner, A. R. (1988). The relationships and origin of living amphibians. In *The phylogeny and classification of the tetrapods. Volume 1. Amphibians, reptiles, birds* (ed. Benton, M. J.), pp. 59–102. Systematics Association Special Volume, 35A. Clarendon Press, Oxford.

Miyamoto, M. M. and Goodman, M. (1986). Biomolecular systematics of eutherian mammals: phylogenetic patterns and classification. *Systematic Zoology*, **35**, 230–240.

Moody, R. T. J. (1984). The relative importance of cranial/ postcranial characters in the classification of sea turtles. In *Studia Palaeocheloniologica I* (ed. Broin, F. de and Jiménez-Fuentes, E.), pp. 205–213. Studia Geologica Salamanticensia, Volumen Especial 1. Universidad de Salamanca, Salamanca.

Nelson, G. J. (1969). Origin and diversification of teleostean fishes. *Annals of the New York Academy of Sciences*, **167**, 18–30.

Norell, M. A. (1993). Tree-based approaches to understanding history: comments on ranks, rules, and the quality of the fossil record. *American Journal of Science*, **293A**, 407–417.

Norell, M. A. and Novacek, M. J. (1992a). The fossil record and evolution: comparing cladistic and paleontologic evidence for vertebrate history. *Science*, **255**, 1690–1693.

Norell, M. A. and Novacek, M. J. (1992b). Congruence between superpositional and phylogenetic patterns: Comparing cladistic patterns with fossil records: *Cladistics*, **8**, 319–337.

Norman, D. B. (1984). A systematic reappraisal of the reptile order Ornithischia. In *Third symposium on Mesozoic terrestrial ecosystems, short papers* (ed. Reif, W.-E. and Westphal, F.), pp. 157–162. Attempto, Tbingen.

Novacek, M. J. (1982). Information for molecular studies from anatomical and fossil evidence on higher eutherian phylogeny. In *Macromolecular sequences in systematic and evolutionary biology* (ed. Goodman, M.), pp. 3–41. Plenum, New York.

Novacek, M. J. (1989). Higher mammal phylogeny. In *The hierarchy of life* (ed. Fernholm, B., Bremer, K., and Jornvall, H.), pp. 421–435. Elsevier, Amsterdam.

Novacek, M. J. (1992). Fossils as critical data for phylogeny. In *Extinction and phylogeny* (ed. Novacek, M. J. and Wheeler, Q. D.), pp. 46–88. Columbia University Press, New York.

Novacek, M. J. and Wyss, A. R. (1986). Higher-level relationships of the Recent eutherian orders: morphological evidence. *Cladistics*, **2**, 257–287.

Novacek, M. J., Wyss, A. R., and McKenna, M. C. (1988). The major groups of eutherian mammals. In *The phylogeny and classification of the tetrapods. Volume 2. Mammals* (ed. Benton, M. J.), pp. 31–71. Systematics Association Special Volume 35B. Clarendon Press, Oxford.

Panchen, A. L. and Smithson, T. R. (1987). Character diagnosis, fossils and the origin of tetrapods. *Biological Reviews*, **62**, 341–438.

Panchen, A. L. Smithson, T. R. (1988). The relationships of the earliest tetrapods. In *The phylogeny and classification of the tetrapods. Volume 1. Amphibians, reptiles, birds* (ed. Benton, M. J.), pp. 1–32. Systematics Association Special Volume 35A. Clarendon Press, Oxford.

Patterson, C. (1981). Significance of fossils in determining evolutionary relationships. *Annual Review of Ecology and Systematics*, **12**, 195–223.

Paul, C. R. C. (1982). The adequacy of the fossil record. In *Problems of phylogenetic reconstruction* (ed. Joysey, K. A. and Friday, A. E.), pp. 75–117. Academic Press, London.

Paul, C. R. C. (1990). Completeness of the fossil record. In *Palaeobiology; a synthesis* (ed. Briggs, D. E. G. and Crowther, P. R.), pp. 298–303. Blackwell Scientific, Oxford.

Platnick, N. I. (1979). Philosophy and the transformation of cladistics. *Systematic Zoology*, **28**, 537–546.

Prothero, D. R., Manning, E. M., and Fischer, M. (1988). The phylogeny of the ungulates. In *The phylogeny and classification of the tetrapods. Volume 2. Mammals* (ed. Benton, M. J.), pp. 201–234. Systematics Association Special Volume 35B. Clarendon Press, Oxford.

Prothero, D. R. and Schoch, R. M. (eds.) (1989). *The evolution of Perissodactyls.* Clarendon Press, New York.

Raup, D. M. (1972). Taxonomic diversity during the Phanerozoic. *Science*, **215**, 1065–1071.

Rieppel, O. (1988). The classification of the Squamata. In *The phylogeny and classification of the tetrapods. Volume 1. Amphibians, reptiles, birds.* (ed. Benton, M. J.), pp. 261–293. Systematics Association Special Volume, 35A. Clarendon Press, Oxford, 377 pp.

Romer, A.S. (1966). *Vertebrate Paleontology.* University of Chicago Press, Chicago.

Rosen, D. E., Forey, P. L., Gardiner, B. G., and Patterson, C. (1981). Lungfishes, tetrapods, paleontology, and plesiomorphy. Bulletin of the American Museum of *Natural History*, **167**, 159–276.

Rowe, T. (1988). Definition, diagnosis, and origin of Mammalia. *Journal of Vertebrate Paleontology*, **8**, 241–264.

Schaeffer, B., Hecht, M. K., and Eldredge, N. (1972). Phylogeny and paleontology. *Evolutionary Biology*, **6**, 31–46.

Schoch, R. M. (1986). *Phylogeny reconstruction in paleontology.* Van Nostrand Reinhold, New York, 353 pp.

Schoch, R. M. (1989). A review of the tapiroids. In *The evolution of perissodactyls* (ed. Prothero, D. R. Schoch, R. M.), pp. 298–321. Clarendon Press, New York.

Schultze, H.-P. (1987). Dipnoans as sarcopterygians. *Journal of Morphology, Supplement*, **1**, 39–74.

Schwenk, K. (1988). Comparative morphology of the lepidosaur tongue and its relevance to squamate phylogeny. In *Phylogenetic relationships of the lizard families. Essays commemorating Charles L. Camp* (ed. Estes, R. and Pregill, G.), 569–598. Stanford University Press, Stanford, 631 pp.

Sepkoski, J. J., Jr. (1982). A compendium of fossil marine families. *Milwaukee Public Museum, Contributions in Biology and Geology*, **51**, 1–125.

Sepkoski, J. J., Jr. (1992). A compendium of fossil marine animal families, 2d ed. *Milwaukee Public Museum, Contributions in Biology and Geology*, **83**, 1–156.

Sepkoski, J. J., Jr. (1993). Ten years in the library: how changes in taxonomic data bases affect perception of macroevolutionary pattern. *Paleobiology*, **19**, 43–51.

Sereno, P. C. (1984). The phylogeny of the Ornithischia: a reappraisal. In *Third symposium on Mesozoic terrestrial ecosystems, short papers* (ed. Reif, W.-E. and Westphal, F.), pp. 219–226. Attempto, Tbingen.

Sereno, P. C. (1986). Phylogeny of the bird-hipped dinosaurs (Order Ornithischia). *National Geographic Research*, **2**, 234–256.

Sheehan, P. M. (1977). Species diversity in the Phanerozoic: a reflection of labor by systematists? *Paleobiology*, **3**, 325–328.

Shoshani, J. (1986). Mammalian phylogeny: comparison of morphological and molecular results. *Molecular Biology and Evolution*, **3**, 222–242.

Signor, P. W. (1990). Patterns of diversification. In *Palaeobiology; a synthesis* (ed. Briggs, D. E. G. and Crowther, P. R.), pp. 130–135. Blackwell Scientific, Oxford.

Simpson, G. G. (1945). The principles of classification and a classification of mammals. *Bulletin of the American Museum of Natural History*, **85**, 1–350.

Smith, A.B. (1994). *Systematics and the fossil record*. Blackwell Scientific, Oxford.

Storrs, G. W. (1991). Anatomy and relationships of *Corosaurus alcovensis* (Diapsida: Sauropterygia) and the Triassic Alcova Limestone of Wyoming. *Bulletin of the Peabody Museum of Natural History*, **44**, 1–151.

Storrs, G. W. (1994). The quality of the Triassic sauropterygian fossil record. *Révue de Paléontologie*, **7**, 217–228.

Strauss, D. and Sadler, P. M. (1989). Classical confidence intervals and Bayesian probability estimates for ends of local taxon ranges. *Mathematical Geology*, **21**, 411–427.

Tassy, P. and Shoshani, J. (1988). The Tethytheria: elephants and their relatives. In *The phylogeny and classification of the tetrapods. Volume 2. Mammals* (ed. Benton, M. J.), pp. 283–315. Systematics Association Special Volume 35B. Clarendon Press, Oxford.

Trueb, L. and Cloutier, R. (1991). A phylogenetic investigation of the inter- and intrarelationships of the Lissamphibia (Amphibia: Temnospondyli). In *Origins of the higher groups of tetrapods* (ed. Schultze, H.-P. and Trueb, L. S.), pp. 223–313. Comstock Publishing Associates, Ithaca, New York.

Weishampel, D.B., Dodson, P. and Osmólska, H. (eds.) (1990). *The Dinosauria*. University of California Press, Berkeley.

Weishampel, D. B. and Heinrich, R. E. (1992). Systematics of Hypsilophodontidae and basal Iguanodontia (Dinosauria: Ornithopoda). *Historical Biology*, **6**, 159–184.

Weishampel, D. B. and Horner, J. R. (1990). Hadrosauridae. In *The Dinosauria* (ed. Weishampel, D. B., Dodson, P., and Osmólska, H.), pp. 534–561. University of California Press, Berkeley.

Wiley, E. O. (1981). *Phylogenetics: The theory and practice of phylogenetic systematics*. Wiley, New York.

2

What can phylogenetic patterns tell us about the evolutionary processes generating biodiversity?

Hervé Philippe and André Adoutte

Introduction

Although perhaps not fully appreciated by all ecologically oriented evolutionary biologists, there is a deep and necessary inter-relationship between the knowledge of **pattern** (i.e. phylogeny) and **process** (i.e. mechanisms) in all studies of comparative biology and especially in those devoted to the analysis of biodiversity and adaptation. This has been eloquently argued in several papers and manuals (Felsenstein 1985; Coddington 1988; Brooks and McLennan 1991; Harvey and Pagel 1991; Hillis and Bull 1991; Stearns 1992; Lauder *et al.* 1993; Avise 1994) and we do not intend to rephrase all the arguments here. Instead, after briefly summarizing these arguments, we will take the proposal one step further: we will ask whether, in some instances, a phylogenetic pattern cannot in itself be indicative of an underlying evolutionary process. We will reverse the usual procedure in which one seeks in the phylogeny a framework, a confirmation, or a control to an adaptive hypothesis, and we will focus instead on cases where the evolutionary hypothesis is actually revealed by the pattern. More specifically, we will focus upon the unresolved multifurcations which are found in an increasing number of molecular phylogenetic trees, and we will ask whether these correspond to explosive evolutionary radiations. If such is the case, then the analysis of the phylogenetic pattern will provide a major clue about the generation of biodiversity. The idea of extracting information on tempo and mode of evolution from molecular phylogenies is largely due to Harvey and co-workers (Nee *et al.* 1992, 1994). These authors have devised powerful methods to infer birth and death rates from the knowledge of the phylogeny of extant species (Harvey *et al.* 1994). Here, we will limit ourselves to the question of radiations as seen through molecular phylogenies, with special emphasis on the identification and elimination of possible artefacts leading to unresolved branching orders.

'Why worry about phylogeny?'

This is the title of the second chapter of Harvey and Pagel's book *The comparative method in evolutionary biology* (1991) from which we borrow some ideas. The following sections list possible answers to this question.

The pattern as a prerequisite

The first reason anybody interested in comparative biology and the study of adaptation requires a phylogeny is self-evident: how can one reason about a process without having a precise idea of the pattern that this process attempts to explain?

More explicitly, what we study in comparative biology are characters: we try to link character states to adaptative scenarios. For that, we need to place these character states in their historical order of emergence: we need to know the ancestral state of the character and when and in what direction it was modified *before* we can start thinking about the selective pressures at play. This is true at all levels of the taxonomic hierarchy, from within species to between phyla. In fact, it is probably fair to say that almost all adaptationist scenarios contain at least an implicitly formulated phylogenetic assumption.

The pattern as a basis to identify adaptations

In a slightly more subtle way, knowledge of the phylogeny allows the discrimination of character states that are identical because of shared ancestry from those that are identical because of independent evolution. Stearns (1992) dedicates a full chapter to 'lineage specific effects', starting with this quotation from Darwin:

It is generally acknowledged that all organic beings have been formed on two great laws—Unity of Type, and the Conditions of Existence. By unity of type is meant that fundamental agreement in structure which we see in organic beings of the same class, and which is quite independent of their habits of life. On my theory, unity of type is explained by unity of descent... Hence, in fact, [natural selection] is the higher law; as it includes, through the inheritance of former variations and adaptations, that of Unity of Type.

What Darwin, with his usual insight, is telling us is that any extant organism, as we see it today, is the product of both phylogenetic inertia and of adaptative pressure. One of the most interesting applications of the comparative method in phylogeny is indeed to delineate that part of the phenotype which is due to inertia from that due to adaptation. Harvey and Pagel's book is, in fact, largely devoted to discussing this question and indicating the best methods to approach it. The same can be said of the lucid chapter of Stearns. The message of both is that knowledge of the phylogeny is paramount.

To the examples described in Stearns' and in Harvey and Pagel's books, let us add one more concerning 'intracellular ecology', and in particular the question of the origin of chloroplasts and mitochondria. It is well recognized that the acquisitions of aerobic respiration and of photosynthesis have constituted major evolutionary advantages for eukaryotic cells. A long-standing question is whether acquisition of corresponding organelles, mitochondria and chloroplasts has occurred through intracellular differentiation without exogenous contribution (the 'autogenous' theory), or whether it was the result of the establishment of stable relationships with foreign, bacterial invaders (the 'endosymbiotic' theory). An ancilliary question is whether these processes occurred once or several times independently for each class of organelle.

Molecular phylogenies have clearly solved the main question in favour of the endosymbiotic theory (see Gray 1992 for review) by demonstrating that a number of mitochondrial and chloroplastic gene sequences have a much closer kinship to eubacterial homologues than to eukaryotic ones. That, of course, completely constrains all further analyses of mechanisms—what we must now attempt to understand is how the symbiosis originated, how some genes from the bacterial genomes became lost and others were transferred to the nucleus of the host cells, how proteins are addressed back to the organelles, etc. These questions are quite different from understanding how small parts of a genome may have become segregated within the cytoplasm of a proto-eukaryotic cell to become a mitochondrion or a chloroplast.

In addition, the question of single or multiple origin of both types of organelle is now being addressed, and a solution is in sight (Gray 1993). Thus, this example shows how a knowledge of the correct phylogeny constrains and orients our mechanistic explanations, and also how it enables us to discriminate shared ancestry (homology, i.e. common origin of each class of organelle) from convergence (analogy, i.e. independent acquisition).

Phylogeny as an account for biogeographic distribution

There are a number of instances in which knowledge of phylogeny helps explain a process that is not necessarily adaptive. One example is vicariance biogeography, where one attempts to fit a phylogenetic pattern to a geographic one and, if a good match is obtained, then diversification can be accounted for by the occurrence of allopatric speciation following a vicariant geographic event, and hence there is no need to resort to a specific proximal adaptationist scenario. In a sense, this is a counter example to those cited above—phylogenetic information is used to refute an adaptationist explanation.

Phylogeny and conservation

Finally, knowledge of phylogeny is increasingly used to elaborate conservation biology policies. Indeed, when (1) discussing the choice of individuals or sub-species of an endangered species to be protected (or reintroduced into the wild) or (2) making decisions about the relative importance of taxa to be protected, it is crucial to have a knowledge of the phylogenetic relationships and position of the taxa under consideration (for a full discussion of this point, see Avise 1994).

In the four aforementioned domains, molecular data have had a profound impact. There are several reasons for this, discussed by Avise (1994), but we would like to single out the one which, in the present context, appears to be the most important: it is the aptitude of molecular data for discriminating homology from homoplasy in phenotypic traits. As stated above, one of the key needs in comparative evolutionary biology is to be able to reconstruct 'adaptive scenarios' of character evolution. To this aim, phylogeny must obviously be independent of the characters whose adaptiveness is being assessed. Often, however, this is not the case since these characters have formed the basis of previous 'traditional' taxonomic and phylogenetic schemes. By providing characters located at the genetic level (and usually totally independent from the phenotypic traits under analysis), molecular markers allow one to bypass this difficulty and to assess phylogenetic relationships independently of adaptations.

Examples of unresolved nodes in molecular phylogenies

During the last few years, there has been an intense renewal in interest in phylogeny resulting from both the development of more rigorous methods of analysis of the data and access to the wealth of information contained in the genomes of extant organisms (Doolittle 1990; Swofford and Olsen 1990; Miyamoto and Cracaft 1991; Zimmer et al. 1993; Avise 1994). Many different portions of the genome have been put to use with more variable portions employed for closely related taxa and highly conserved portions for distantly related ones. Of the various genes or portions of genes analysed, the most useful for high-level phylogeny have so far proven to be ribosomal RNA in part, because there exists a large gene database. However, other genetic domains have also been widely used and among these the mitochondrial genome and several of its gene sequences have recently been intensively studied.

Through these studies, one recurrent result has emerged: while some taxa yield 'well resolved' patterns (i.e. portions of trees in which the branching order of the species analysed could be confidently established), other taxa yielded 'unresolved nodes' (i.e. points in the tree, from which several lineages emerge 'simultaneously' in an undetermined order). This lack of resolution is usually tightly correlated with the shortness of the corresponding internal branches, instability of the branching order, and low bootstrap values of the corresponding nodes. Such branching points can thus be likened to a 'bush' and are best represented as a multifurcation rather than as a series of bifurcations. Quite interestingly, they often seem to be correlated with a diversification in body plan organization of the ensuing lineages.

Are such patterns telling us something fundamental about accelerated rates of cladogenesis? Do they correspond to evolutionary bursts? In short, are they a reflection of explosive evolutionary radiations? Before answering these questions let us examine some examples.

We have chosen four cases, corresponding to diverse levels of the taxonomic hierarchy (albeit all fairly high) and to very distinct levels of organism complexity, since they range from protists to mammals. For three of them, palaeontological information is available. We will first document the phylogenetic pattern in a purely factual way by providing the trees and the bootstrap values, then we will discuss the question of whether such lack of resolution results from artefacts. Finally, using recently developed tools in conjunction with palaeontological data, we will show how an idea of the time interval covered by unresolved nodes can be obtained. In order to facilitate the presentation of the results, all the trees have been obtained using the neighbour-joining distance method (Saitou and Nei 1987) and the reliability of the nodes was ascertained using the bootstrap method (Felsenstein 1985) with 1000 replicates. All the handling and treatment of the data were achieved using the MUST computer package (Philippe 1993).

The 'terminal crown' of eukaryotes

Since the early days of rRNA sequence analysis, a rather striking pattern of branching has been recognized within the eukaryotic clade: a number of extremely distant, well-separated protist lineages emerged at the base of the eukaryotic tree, followed by a huge terminal 'bush' comprising all the multicellular kingdoms (metazoans, metaphytes and fungi) as well as several additional protistan groups, both photosynthetic and non-photosynthetic (chromophytes, rhodophytes, ciliates, dinoflagellates, apicomplexa = sporozoa, etc.) (Sogin et al. 1986; Perasso et al. 1989; Schlegel 1991; Sogin 1991; Cavalier-Smith 1993). Because diploblastic metazoans are very deeply split from triploblastic ones, even the monophyly of the Metazoa was not clearly supported in these trees (Field et al. 1988; Christen et al. 1991). A typical tree of this type is depicted in Fig. 2.1. Note the shortness of internal branches in the portion of the tree corresponding to the base of the 'bush' (indicated by an arrow) and the correspondingly low bootstrap values. Also note that the clades emerging from this bush correspond to remarkably different 'body plans' since they include plant cells, animal cells, fungal cells and a variety of additional distinct cell types.

Several attempts at improving the resolution of this terminal crown have been carried out more recently, some of which included analysis of protein coding genes (Gouy and Li 1989; Baldauf and Palmer 1993). The one relatively clear result emerging is that of a possible sister-group relationship of animals and fungi (Baldauf and Palmer 1993; Wainright et al. 1993). Even this result, however, remains only moderately supported by bootstrap resampling; it may, in fact, be due to an accelerated rate of evolution in both lineages.

In short, then, the terminal part of the eukaryotic tree has proven quite difficult to resolve and this has been taken to reflect a period of intense phyletic diversification. Possible environmental causes for this diversification have been sought, especially in the form of a sharp increase in atmospheric

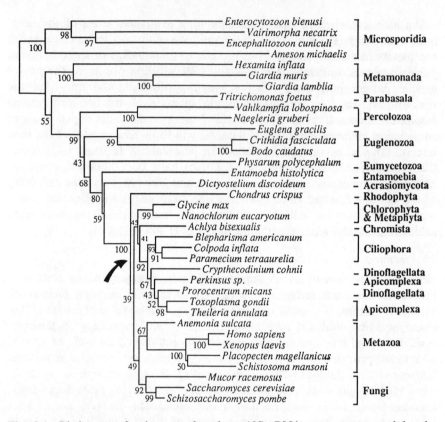

Fig. 2.1 Phylogeny of eukaryotes based on 18S rRNA sequences treated by the Neighbor-Joining distance method. Transversions only were taken into account. The numbers on branches refer to bootstrap proportions (1000 resamplings). The arrow points to the unresolved node of the 'eukaryotic crown'. Origin of the sequences, alignment, definition of boundaries are given in Philippe (in preparation) and are available on request for this dataset as well as for all the others. About 1000 nucleotides were reliably aligned. The tree was rooted using microsporida as an outgroup.

oxygen concentrations at the supposed time of the 'burst'. However, the fit among the palaeontological data on early eukaryotes, the environmental history and the molecular phylogeny is still far from perfect (Knoll 1992).

The diversification of the major classes of ciliates

A broad-scale molecular phylogeny of the ciliated protozoa has been recently reconstructed using 28S rRNA partial sequences and involving over 30 species representing the major classes within this phylum of highly elaborate single cells (Baroin-Tourancheau *et al.* 1992; Fleury *et al.* 1992). Interestingly, although the analysed species clustered according to traditional taxonomy, yielding seven of the expected classes, the **order of emergence** of these classes was quite unexpected. In particular, after early emergence of two lineages, a 'bush' comprising the five other classes was obtained. Further, it is striking that the diversification into each of the seven lineages was correlated with a given strategy of cytoskeletal organization of the cortex, and, in particular, the five lines emerging from the 'bush' could be grouped into three distinct types of sub-cortical cytoskeleton organization. Indeed it was found, through immunocytochemical and electron-microscopical analyses, that the cells strengthened their cortex either through (1) the use of longitudinal micro-tubules, (2) post-ciliary microtubules, (3) an epiplasm (a thick membrane skeleton), or (4) an ecto-endoplasmic boundary (made of thin filaments of Ca^{2+}-binding proteins); within each of the clades identified by the molecular phylogeny, only one of these four strategies was used. This, for a ciliate, corresponds to its 'body plan' since the major problem such a cell has to solve is the mode of anchoring its many cilia over its surface. Quite congruent results have been obtained using 18S rRNA (Bernhard et al. 1993) with better resolution of some of the basal nodes due to use of longer sequences.

Thus, in these unicellular organisms, some major morphogenetic 'choices' appear to have been made during a rapid evolutionary episode, and within each of the ensuing lineages, the cells later remained constrained by the choice initially made.

The radiation of the major triploblast metazoan phyla and the resolving power of rRNA

Amongst the most striking examples of evolutionary bursts is the simultan-eous appearance of the major invertebrate phyla in the fossil record dating from the early and middle Cambrian, as exemplified by the Burgess Shale fauna. Briefly, the present status of palaeontological evidence, as summarized by Conway Morris (1993), consists of recognizing three early episodes of animal life:

(1) the Precambrian Ediacara fauna (570–555 myr ago), probably mostly of a diploblastic 'grade' of organization;

(2) the early Cambrian fauna, largely dominated by triploblasts (~ 540 myr ago); and

(3) the explosive diversification of this fauna into most of the extant 35 triploblastic coelomate metazoan phyla in the middle Cambrian (~ 520 myr ago).

Thus, it appears that the major diversification of metazoans may have occurred in less than 20 myr. Further, Gould (1989) suggests that this diversification should be viewed as an early and rapid diversification into a large number of distinct phyla (greater than the 35 current ones) followed by decimation of a substantial number of the component lineages and elimina-tion of numerous phyla, leaving those within the pre-established 'bauplans'.

Very few if any new phyla would have emerged later; only diversification and 'tinkering' around the initially generated body plans would have occurred.

Does molecular phylogeny substantiate this view of a rapid and almost simultaneous origin of the major coelomate triploblast phyla in the form of a vast multifurcation of the corresponding lineages? Indeed, the results of Field *et al.* (1988) and our own more extended dataset (Adoutte and Philippe 1993), show that the rRNA-based phylogenies of metazoans fail to resolve the order of emergence of the major coelomate invertebrate phyla (Fig. 2.2): diploblasts are clearly segregated from triploblasts, and the acoelomate triploblasts, represented by the Platyhelminthes (flatworms), constitute a clear outgroup to the vast group of coelomate triploblasts; the coelomate triploblasts themselves constitute a solid monophyletic group (92% BP (bootstrap percentage)), but within them all the lineages emerge almost from the same point, and their deepest internal branches have very low bootstrap values, indicating low resolution. Thus, there is no support for the monophyly of the

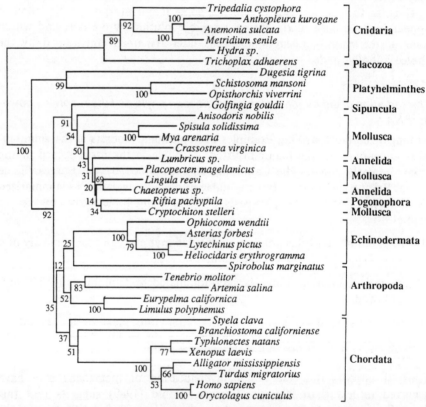

Fig. 2.2 Phylogeny of metazoa based on 18S rRNA sequences treated by the NJ method. The tree was rooted using the diploblasts (Cnidaria + Placozoa) as an outgroup. Sequences are the same as in Adoutte and Philippe (1993) with a few additions.

protostome and deuterostome lineages. In fact, some key phyla (e.g. arthropods, molluscs) are not even monophyletic and some others are strangely located (*Lumbricus*, an annelid, in the middle of molluscs, etc.).

In a separate paper (Philippe *et al.* 1994*a*), we analyse in greater detail these unresolved nodes of the metazoan phylogeny by employing a simple mathematical relation between the bootstrap value and the number of nucleotides used. Indeed, at each node, an 'experimental' curve of bootstrap percentage (*BP*) as a function of the number of nucleotides can be obtained. For nodes displaying average *BP*s, this allows us to extrapolate to high bootstrap values and calculate the sequencing effort that would be required to ascertain the reliability of that node (Lecointre *et al.* 1994).

The relationship is of the form:

$$BP = 100 \ (1 - e^{-bx})$$

where b is a parameter characteristic of the node and x is the number of nucleotides.

We further establish that b is directly correlated to the time elapsed between two cladogeneses (i.e. as demonstrated on the trees by the length of internal branches), provided that two conditions are fulfilled: (1) the NJ method is efficient at reflecting the true length of internal branches, and (2) a rough molecular clock exists in the domain of the tree under analysis. We can therefore estimate the ability of a given molecular dataset to resolve successive nodes in a tree on the scales of millions of years. Somewhat surprisingly, we find that the presently available dataset of full 18S rRNA cannot resolve cladogenetic events separated by less than 40 myr. Thus, in the case of the major metazoan coelomate phyla, the molecular data indeed point to a radiation if we consider that a time interval of 40 myr is 'rapid' in evolutionary terms.

To increase resolution, a considerable additional sequencing effort (involving the same type of gene) would be necessary. For example, to attain a resolving power of 1 myr, the equivalent of 40 times the length of the full 18S rRNA would have to be sequenced! This is due to the fact that the relationship between the time interval separating two cladogeneses and the number of nucleotides to be sequenced is a hyperbolic function (see below). Thus, for long time intervals, resolution is obtained with only short sequences while for short time intervals exceedingly long sequences are required.

It should be pointed out that a similar lack of resolution was observed at a later key stage of metazoan evolution—the split between the basal lines of vertebrates. Through analysis of a large database of 28S rRNA, Lê *et al.* (1993) found it impossible to establish confidently the branching order of the major lines of gnathostomes (chondrichthyans, actinopterygians, sarcopterygians). A similar result was obtained by Stock *et al.* (1991) using 18S rRNA and a somewhat different set of species. Here again the results were interpreted as probably reflecting rapid radiation.

The mammalian radiation and how to eliminate non-resolution of artefacts in molecular trees: a case study

There is probably no clearer case of evolutionary radiation than that corresponding to the emergence of the major orders of mammals (Simpson 1944, 1953). This evolutionary lineage diversified little for about 150 myr, then underwent a remarkable diversification during the 8 myr of the Palaeocene, yielding most of the major orders (insectivores, modern carnivores, dermopterans, bats, primates, perissodactyls, pangolins and rodents), the remaining orders arising during the following 22 myr of the Eocene (Benton 1990). It is not surprising that this rapid radiation has led to great difficulty in establishing the precise phylogenetic relationships between all these orders (reviewed in Novacek 1992).

How does molecular phylogeny represent these historical events? To anwser this, we employed a large database (EMBL, GenBank) of mammalian protein-coding genes, including cytochrome *b* to construct three types of trees:

(1) using transitions at the third position of codons (Fig. 2.3);

(2) using transversions at the third position of codons (Fig. 2.4); and

(3) using transversions at positions 1 and 2 (Fig. 2.5).

The first tree is totally unresolved, that is, it displays aberrant groupings, except for extremely close taxa, and the bootstrap values of internal branches are extremely low. This is the direct consequence of (1) the third base of codons being more free to mutate and (2) transitions being much more frequent than transversions in the mitrochondrial genome (Brown *et al.* 1982). Hence, over the time interval analysed, complete mutational saturation has occurred at this position. These processes preclude recovery of a correct topology and push all internal nodes to the same unresolved point.

Figure 2.6(a) shows how transitions are saturated with respect to transversions at the third position. When observed transitions are plotted over observed transversions, it is seen that the values of transitions rise very rapidly, reaching a plateau. Thus, as expected, transitions at the third position are much more saturated than transversions. When transversions at the third position are plotted against transversions at the first and second positions, some saturation is still seen (Fig. 2.6(b)). In sum, this approach allows one to identify transversions at first and second positions as the most reliable characters, but in using only these substitutions, the number of available informative sites is decreased.

An easy way to ascertain whether a set of sequences are mutationally saturated is to plot the number of observed differences between extant sequences over the number of substitutions inferred to have occurred between these two sequences during evolution (Philippe *et al.* 1994*b*). This is illustrated in Fig. 2.6(c) for first and second position transversions. The general distribution of points is quite different from that in Fig 2.6(a): there is an

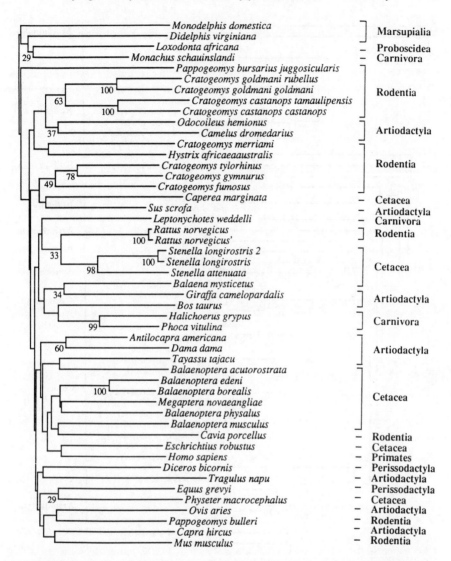

Fig. 2.3 Phylogeny of mammals based on third position transitions in cytochrome b (NJ method). Rooting is achieved using marsupials as an outgroup although they are not monophyletic. Bootstrap values below 25% have not been displayed.

almost linear relationship, indicating very little saturation in the transversions occurring in the first two bases.

The efficiency of this method is illustrated in Fig. 2.6(d), which shows saturation based on a plot of observed versus inferred transversions over the third position. The method thus allows the identification of third position saturation even for transversions. It also constitutes an internal test, which can be applied to all types of molecular data.

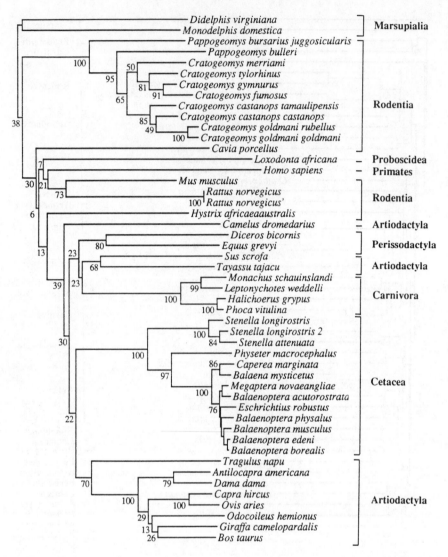

Fig. 2.4 Phylogeny of mammals based on third position transversions in cytochrome b (NJ method). Same rooting as in Fig. 2.3.

Why do these saturation problems occur? In Fig. 2.3, almost complete lack of resolution is observed except at very low sister taxonomic level (species or subspecies). When going from Fig. 2.4 to Fig. 2.5, improved resolution is obtained at the level of mammalian orders; however, within orders and suborders, resolution is better in Fig 2.4 (which uses 242 informative sites) than in Fig. 2.5 (which uses only 111 sites). In these two latter figures, two different effects are operating: in Fig. 2.4, it is mostly saturation that prevents the recovery of deep nodes, whereas in Fig. 2.5, it is the insufficient amount

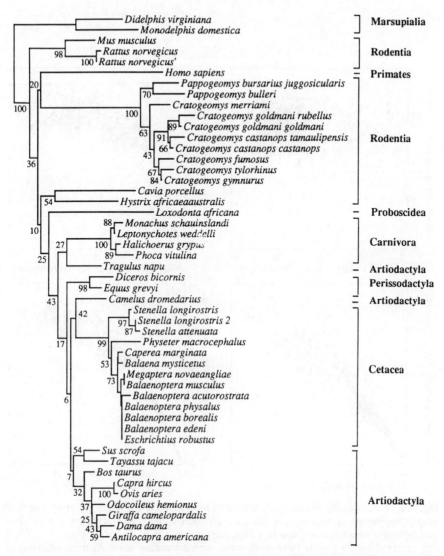

Fig. 2.5 Phylogeny of mammals based on first and second position transversions in cytochrome b (NJ method). Same rooting as in Fig. 2.3.

of information. A further manifestation of the saturation 'syndrome' is in the equality of branch lengths (Fig. 2.4), despite clear differences in evolutionary rates (Fig. 2.5) (for example in the case of Geomydae, see also De Walt *et al.* 1993).

The figures also show that the bootstrap values of most internal nodes are low and the corresponding branch lengths are short. This is in agreement with the notion of rapid radiation at the point of origin of the various orders. To see why is this the case, we have applied the procedure briefly outlined

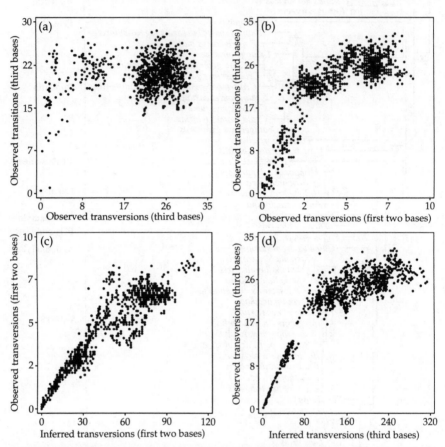

Fig. 2.6 Evidence for mutational saturation in cytochrome b. Each point on the four graphs represents species couples and the corresponding coordinates on the abscissa and ordinate represent evolutionary distances computed in different ways. In (a) and (b), crude distance values have been used, corresponding to characters evolving at different rates. This illustrates saturation of transitions at the third position of codons, with respect to transversions at these positions (a) and saturation of transversions at the third position of codons with transversions at positions 1 and 2 (b). In (c) and (d), observed distances were compared to inferred ones as described in Philippe *et al.* (1994*b*). Graph (d) displays the same shape as caption (b) illustrating the efficiency of the method. In constrast, (c) displays little saturation for transversions occurring at the first two positions of codons, demonstrating the usefulness of these characters.

above to evaluate the resolving power of the molecule under study (cyto-chrome *b*) as a function of the number of nucleotides used (Fig. 2.7). Two important points emerge from the resulting hyperbolic curve:

(1) the time interval between the two cladogeneses is of the order of 18 myr ('cyt *b*' on the ordinate). This is much longer than most of the expected time intervals separating the origin of mammalian orders as inferred from palaeontology; this molecule, by itself, is therefore insufficient to resolve the mammalian radiation, because of its low information content at the evolutionary level analysed here—first and second position transversions in cytochrome *b* provide only 111 informative sites.

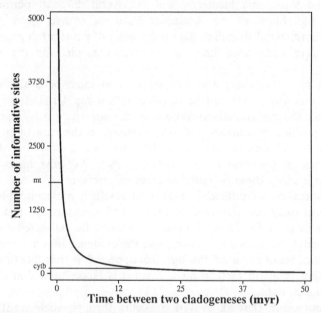

Figure 2.7 Graph relating the time between two cladogeneses as a function of the number of informative sites using a gene such as cytochrome b. The graph was computed using the data in Philippe and Douzery (1994) and the method of Philippe *et al.* (1994a). The function relating the number of informative sites, x, and the time, ΔT_c, is $x = 1800/\Delta T_c$. 'cytb' and 'mt' refer respectively to the number of informative sites of the cytochrome b gene and of the complete mitochondrial genome. The corresponding resolving power as read on the abscissa is respectively 18 and 1 myr. The four pairs of palaeontological divergence dates used to calibrate the curve are:
(1) *Mus/Rattus:* 15 myr (Catzeflis *et al.* 1992)
    (*Mus, Rattus*)/other rodents: 55 myr (Carroll 1988; Benton 1993)
(2) 1st Geomyidae in fossil record: 5 myr (Russel 1968)
    Geomyidae/other rodents: 55 myr (Carroll 1988; Benton 1993)
(3) Sus/Tayassu: 45 myr (Douzery, personal communication)
    Suiformes/other Artiodactyls: 60 myr (Carroll 1988; Benton 1993)
(4) 1st Cetaceans in fossil record: 40 myr (Carroll 1988; Benton 1993)
    Cetaceans/Artiodactyls: 60 myr (Carroll 1988; Benton 1993)

(2) an estimated 18000 nucleotides would be required to resolve time intervals of about 1 myr, that is, approximately the full size of the mitochondrial DNA ('mt' on the ordinate). The achievement of complete sequencing of mitochondrial DNA of representatives of several mammalian orders therefore presents good prospects for resolving the mammalian radiation.

Conclusion

The examples we have described were chosen to illustrate the potential of molecular phylogenies in revealing evolutionary radiations at many different levels of the taxonomic hierarchy and at several different periods in the history of life. Some of our examples have, of course, been previously inferred to correspond to radiations on the basis of palaeontological evidence and they have been used here as 'interval controls' for the molecular approach.

We have seen that there are two pitfalls that must be evaluated before unresolved multifurcations can be equated with a rapid radiation: they are, on one hand, the mutational saturation for the sequences used (such as seen with third position transitions and transversions in the case of mammalian cytochrome *b*) and, on the other hand, the insufficient amount of information used (as seen in the case of one ciliate clade). We now have the tools available to identify these potential sources of artefact.

After elimination of potential sources of artefact, molecular phylogenies do, in several cases, provide evidence for 'rapid' evolutionary radiations. A rough time interval for each of these events can be estimated by using a number of extrapolations and by assuming the existence of a molecular clock. One important result obtained through this approach is that this time interval is often rather large; in fact, in most cases it is larger than that defined by palaeontological data.

We do not claim that all evolution occurs through such events and we could cite a number of instances where major innovations in body plans occur in the form of well-separated transitions (the history of vertebrate classes is exemplary in this respect). We are struck, however, by the observation that, in many instances, key 'decisions' about the overall organization of the organisms, be it at the single cell or at the multicellular level, appear to have been taken during periods of rapid diversification. The mechanisms underlying this process are of course still open to discussion and we do not wish to enter into such speculation here (see Benton 1990; Jablonski and Bottjer 1990; Stanley 1990; Taylor and Larwood 1990). We wish to stress, however, that if the pattern we have identified is extended, if only in part, across all the lines of life, then a major process in the generation of biodiversity will have been illuminated. This is clearly an important insight to take into account both for understanding how diversity was (and is) generated in the living world and how we should define policy for its conservation.

Acknowledgements

This work was supported by the CNRS, the Université Paris-Sud and by a special grant from DRED (Direction de la Recherche et des Études Doctorales) for computer equipment. We thank our colleagues Hervé Le Guyader, Guillaume Lecointre, Guillaume Balavoine, Anne Fleury and Anne Baroin-Tourancheau for discussions throughout this work, Guillaume Lecointre for additional help with palaeontological information, Cécile Couanon for much help in the preparation of the manuscript, and Max Telford for critical reading of the manuscript.

References

Adoutte, A. and Philippe, H. (1993). The major lines of metazoan evolution: Summary of traditional evidence and lessons from ribosomal RNA sequence analysis. In *Comparative molecular neurobiology* (ed. Y. Pichon), pp. 1–30. Birkhäuser Verlag, Basel.

Avise, J. C. (1994). *Molecular markers, natural history and evolution*. Chapman and Hall, New York.

Baldauf, S. L. and Palmer, J. D. (1993). Animals and fungi are each other's closest relatives: congruent evidence from multiple proteins. *Proc. Natl. Acad. Sci. USA*, **90**, 11558–11562.

Baroin-Tourancheau, A., Delgado, P., Perasso, R. and Adoutte, A. (1992). A broad molecular phylogeny of ciliates: Identification of major evolutionary trends and radiations within the phylum. *Proc. Natl. Acad. Sci. USA*, **89**, 9764–9768.

Benton, M. J. (1990). *Vertebrate palaeontology*. Harper Collins Academic, London.

Benton, M. J. (1993) *The fossil Record 2*. Harper Collins Academic, London.

Bernhard, D., Leipe, D. and Schlegel, M. (1993). Phylogenetic relationship of ciliates inferred from small subunit ribosomal RNA sequence comparisons. *Abstract of IX Intern. Cong. Protozoology*, July 25–31, Berlin, p. 12.

Brooks, D. R. and McLennan, D. A. (1991). *Phylogeny, ecology, and behavior. A research program in comparative biology*. The University of Chicago Press.

Brown, W. M., Praeger, E. M., Wang, A., and Wilson, A. C. (1982). Mitochondrial DNA sequences of primates: Tempo and mode of evolution. *J. Mol. Evol.*, **18**, 225–239.

Carroll, R. L. (1988). *Vertebrate paleontology and evolution*. W.H. Freeman and Co., New York.

Catzeflis, F. M., Aguilar, J. P. and Jaeger, J. J. (1992). Muroid rodents: phylogeny and evolution. *TREE*, **7**, 122–126.

Cavalier-Smith, T. (1993). Kingdom protozoa and its 18 Phyla. *Microbiol. Rev.*, **57**, 953–994.

Christen, R., Ratto, A., Baroin, A., Perasso, R., Grell, K.G. and Adoutte, A. (1991). An analysis of the origin of metazoans, using comparisons of partial sequences of the 28S RNA, reveals an early emergence of triploblasts. *EMBO J.*, **10**, 499–503.

Coddington, J. A. (1988). Cladistic tests of adaptational hypotheses. *Cladistics*, **4**, 3–22.

Conway Morris, S (1993). The fossil record and the early evolution of the Metazoa. *Nature* **361**, 219–225.

DeWalt, T. S., Sudman, P. D., Hafner, M. S. and Davis, S. K. (1993). Phylogenetic relationships of pocket gophers (*Cratogeomys* and *Pappogeomys*) based on mitochondrial DNA cytochrome *b* sequences. *Mol. Phyl. Evol.*, **2**, 193–204.

Doolittle, R. F. (1990). Molecular evolution: Computer analysis of protein and nucleic acid sequences. *Methods in Enzymology*, Vol. 183, pp. 1–736, Academic Press Inc., San Diego, Ca.

Felsenstein, J. (1985). Confidence limits on phylogenies: an approach using the bootstrap. *Evolution*, **39**, 783–791.

Field, K.G., Olsen, G. J., Lane, D. J., Giovannoni, S. J., Ghiselin, M. T., Raff, E. C., Pace, N. R. and Raff, R. A. (1988). Molecular phylogeny of the animal kingdom. *Science*, **239**, 748–753.

Fleury, A., Delgado, P., Iftode, F. and Adoutte, A. (1992). Molecular phylogeny of Ciliates: What does it tell us about the evolution of the cytoskeleton and of developmental strategies? *Dev. Genetics*, **13**, 247–254.

Gould, S. J. (1989). *Wonderful life*. W.W. Norton & Company, New York, London.

Gouy, M. and Li, W. H. (1989). Molecular phylogeny of the kingdoms animalia, plantae, and fungi. *Mol. Biol. Evol.* **6**, 109–122.

Gray, M. W. (1992). The Endosymbiont hypothesis revisited. In *Mitochondrial genomes*. (eds. D.R. Wolstenholme and K.W. Jeon). *Intern. Rev. Cytol.*, Vol 141, pp. 233–357. Academic Press Inc., San Diego, Ca.

Gray, M. W. (1993). Origin and evolution of organelle genomes. *Curr. Opin. Genet. Develop.*, **3**, 884–890.

Harvey, P. H., May, R. M. and Nee, S. (1994). Phylogenies without fossils. *Evolution*, **48**, 523–529.

Harvey, P. H. and Pagel, M. D. (1991). *The comparative method in evolutionary biology*. Oxford University Press.

Hillis, D. M. and Bull, J. J. (1991). Of genes and genomes. *Science*, **254**, 528–558.

Jablonski, D. and Bottjer, D. J. (1990). The origin and diversification of major groups: environmental patterns and macroevolutinary lags. In *Major evolutionary radiations*. (eds. P. D. Taylor and G. P. Larwood) pp. 17–57. Clarendon Press, Oxford.

Knoll, A. H. (1992). The early evolution of eukaryotes: A geological perspective. *Science*, **256**, 622–627.

Lauder, G. V., Leroi, A. M. and Rose, M. R. (1993). Adaptations and history. *TREE*, **8**, 294–297.

Lê, H. L. V., Lecointre, G. and Perasso, R. (1993). A 28S rRNA-based phylogeny of the Gnathostomes: First steps in the analysis of conflict and congruence with morphologically based cladograms. *Mol. Phyl. Evol.*, **2**, 31–51.

Lecointre, G., Philippe, H., Lê, H. L. V. and Le Guyader, H. (1994). How many nucleotides are required to resolve a phylogenetic problem? The use of a new statistical method applicable to available sequences. *Mol. Phyl. Evol.*, **3**, 292–309.

Miyamoto, M. M. and Cracraft, J. (1991). *Phylogenetic analysis of DNA sequences*. Oxford University Press, New York.

Nee, S., May, R. M. and Harvey, P. H. (1994). The reconstructed evolutionary process. *Phil. Trans. Roy. Soc.*, **339**, 139–146.

Nee, S., Mooers, A. Ø. and Harvey, P. H. (1992). Tempo and mode of evolution revealed from molecular phylogenies. *Proc. Natl. Acad. Sci. USA*, **89**, 8322–8326.

Novacek, M. J. (1992). Mammalian phylogeny: shaking the tree. *Nature*, **356**, 121–125.

Perasso, R., Baroin, A., Qu, L. H., Bachellerie, J. P. and Adoutte, A. (1989). Origin of the algae. *Nature*, **339**, 142–144.

Philippe, H. (1993). MUST, a computer package of Management Utilities for Sequences and Trees. *Nucleic Acids Res.* **21,** 5264–5272.

Philippe, H. and Douzery, E. (1994). The pitfalls of molecular phylogeny based on four species as illustrated by the Cetacea/Artiodactyla relationships. *J. Mammalian Evol.,* **2,** 133–152.

Philippe, H., Chenuil, A. and Adoutte, A. (1994*a*). The Cambrian radiation: Can it be revealed through molecular phylogeny? *Development,* suppl., 15–25.

Philippe, H., Sörhannus, U., Baroin, A., Perasso, R., Gasse, F. and Adoutte, A. (1994*b*). Comparison of molecular and paleontological data in diatoms suggests a major gap in the fossil record. *J. Evol. Biol.,* **7,** 247–265.

Russell R. J. (1968). Evolution and classification of the pocket gophers of the subfamily Geomyinae. *Univ. Kans. Pub. Mus. Nat. Hist.* **16,** 473–597.

Saitou, N. and Nei, M. (1987). The neighbor-joining method: a new method for reconstructing phylogenetic trees. *Mol. Biol. Evol.* **5,** 406–425.

Schlegel, M. (1991). Protist evolution and phylogeny as discerned from small subunit ribosomal RNA sequence comparisons. *Europ. J. Protistol.* **27,** 207–219.

Simpson, G. G. (1944). *Tempo and mode in evolution.* Columbia University Press, New York.

Simpson, G. G. (1953). *The major features of evolution.* Columbia University Press, New York.

Sogin, M. L. (1991). Early evolution and the origin of eukaryotes. *Curr. Opin. Genet. Develop.,* **1,** 457–463.

Sogin, M. L., Elwood, H. J., and Gunderson, J. H. (1986). Evolutionary diversity of eukaryotic small-subunit rRNA genes. *Proc. Natl. Acad. Sci. USA,* **83,** 1383–1387.

Stanley, S. M. (1990). Adaptative radiation and macroevolution. In *Major evolutionary radiations.* (eds. P. D. Taylor and G. P. Larwood) pp. 1–16. Clarendon Press, Oxford.

Stearns, S. C. (1992). *The evolution of life histories.* Oxford University Press.

Stock, D. W., Gibbons, J. K., and Whitt, G. S. (1991). Strengths and limitations of molecular sequence comparisons for inferring the phylogeny of the major groups of fishes. *J. Fish. Biol.,* **39,** suppl A., 225–236.

Swofford, D. L. and Olsen, G. J. (1990). Phylogeny reconstruction. In *Molecular systematics* (eds. D.M. Hillis and C. Moritz), pp. 411–501. Sinauer Associates Inc., Sunderland, MA.

Taylor, P. D. and Larwood, G. P. (1990). *Major evolutionary radiations.* Clarendon Press, Oxford.

Wainright, P. O., Hinkle, G. and Sogin, M. L. (1993). Monophyletic origins of the Metazoa: An evolutionary link with fungi. *Science,* **260,** 340–342.

Zimmer, E. A., White, T. J., Cann, R. L., Wilson, A. C. (1993). Molecular evolution: Producing the biochemical data. *Methods in Enzymology,* Vol. 224, pp. 1–725, Academic Press Inc., San Diego.

3

New computer packages for analysing phylogenetic tree structure.

Paul H. Harvey, Andrew Rambaut and Sean Nee

Introduction

What can the structure of molecular phylogenies tell us about the evolutionary processes responsible for their genesis? Remarkably little attention has been paid to answering this question, despite the rapid accumulation of molecular sequence data which allows the construction of ever more precise phylogenetic trees. Our goal in this chapter is to provide research groups examining particular taxa with a set of routine procedures for extracting the footprints of the past that may lie hidden in their sequence data. Elsewhere, we have presented illustrative examples showing the types of conclusions that we believe can or cannot be derived from molecular phylogenies. It is not our intention to go over the same ground here, although we shall summarise some relevant points of this previous work insofar as they are relevant to our immediate aim.

The procedures

Phylogenetic trees based on molecular data can tell us when, on absolute or relative time scales, independently evolving lineages last shared common ancestry. The footprint of the past, which we wish to interpret, consists of the relationships among the times between the nodes in the tree (Fig. 3.1).

Our approach is statistical. We recognise that only two basic processes are responsible for the production of a phylogenetic tree: lineage splitting and lineage extinction. We now need to identify a null model and, fortunately, there is a natural one to use. It is a constant-rates Markov birth-death process in which, at any point in time, each lineage has the same probability of dividing and the same probability of going extinct; these probabilities are assumed constant (Raup *et al.* 1973). If the structure of our tree (i.e. the pattern of the internode intervals) fits Markov expectations, then we can estimate lineage birth and death rates (Nee *et al.* 1994*a,b*, 1995*a*); large trees with many branches give more accurate estimates of birth and death rates than smaller trees. On the other hand, if our tree deviates from Markov expectations, then we need to know how it deviates. For example, lineage

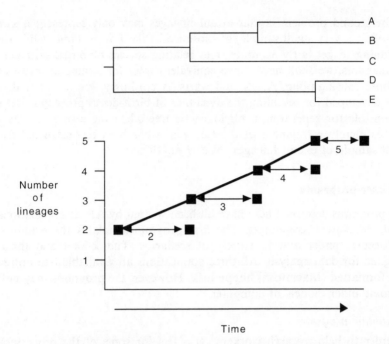

Fig. 3.1 A phylogenetic tree linking a sample of lineages (A to E). Time is measured increasing towards the present. Labelled internode intervals show the times between successive nodes, that is when 2, 3 and 4 lineages are represented. The interval labelled 5 shows the time between the present and the most recent node. The bold line, which is the line joining the dots in subsequent figures, links the points denoting the earliest historical times at which each number of lineages existed.

birth or death rates may change through time (Harvey *et al.* 1994*a,b*), or they may be higher in some branches than others (Harvey and Nee 1993; Nee *et al.* 1992). We need simple ways of deciding which patterns of rate changes could be responsible for the structure of the tree examined (Harvey and Nee 1994). Once an appropriate model is chosen (e.g., Nee *et al.* 1994*b*), we then need to estimate the relevant parameters.

Even if the relative or absolute timing of a set of nodes is known, two problems remain. First, it is virtually certain that details of many, or even most, extinct lineages will be unknown. For example, it is conceivable that we shall soon have a complete molecular phylogeny of all extant primate species based on so much base sequence data that tree structure and the timings of nodes will been estimated with uncanny accuracy. In contrast, it is not realistic to expect equivalent data on more than a few of the lineages that have gone extinct. Fortunately, accurate phylogenies reconstructed from all extant lineages can retain in their structure traces of the dynamical properties of the birth and death processes that produced them (Harvey *et al.* 1994*a,b*, Harvey and Nee 1994, Nee *et al.* 1994*a,b*, 1995*a*).

The second problem is that extant lineages may only represent a sample, and often a very small sample, of those available. For example, while we are beginning to get fairly accurate trees relating species of birds, primates and salamanders, we shall never have equivalent trees for strains of viruses such as those causing polio, AIDS, and warts. Accordingly, special methods have been developed for revealing the dynamics of birth-death processes that were responsible for generating a phylogenetic tree when the available data constitutes merely a fragment of the tree which has been reconstructed from a small sample of extant lineages (Nee *et al.* 1995*b*).

Software programs

The programs described here have all been written by AR as applications for Apple Macintosh computers. The first program simulates the evolution of phylogenies under a wide variety of scenarios. The second and third are designed for data analysis. All three applications are available free on receipt of a formatted (Macintosh) floppy disk. However, the programs may not run on some older models of computer.

Simulation program

In order to help research workers get a feel for some of the processes that may have produced their phylogenies, we have produced a computer program which simulates lineage splitting and extinction under a variety of scenarios. Once a model is selected and simulated, it then produces phylogenetic trees (1) from all the extant lineages and (2) from a random sample of extant lineages. The output is displayed in a graphical form as a series of lineages-through-time plots (see Fig. 3.1). The plots can be printed and the data are written to files for reading and manipulating in other graphical or statistical packages.

In its current incarnation, the simulation program incorporates up to eight consecutive time windows *within* each of which the lineage birth and death rates remain the same, but *among* which they differ according to user-specified instructions. Each time window finishes either after a specified amount of time or when a set number of lineages has been reached. At the end of each time period a mass extinction event of specified magnitude can occur. The simulation allows the evolution of very large numbers of lineages (it is only limited by the time taken to execute and the amount of disc space available on the computer). This then allows the sampling of a very small proportion of the actual extant lineages to produce the phylogeny.

Consider the following suite of examples. In a sample run we might specify the birth and death rates as 1.0 and 0.5 respectively, with the run to finish the first occasion 1000 extant lineages have been produced. We might also specify output based on phylogenies of 50%, 10%, and 2% of extant lineages (500, 100, and 20). A sample output is presented in Fig. 3.2. We might then ask how the picture would change if a mass extinction, eliminating 90% of the

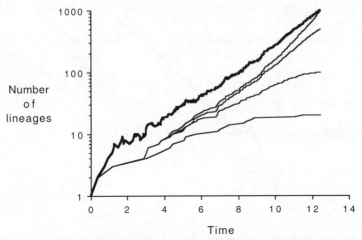

Fig. 3.2 Output from the simulation program for a birth-death process where birth and death rates (1.0 and 0.5, respectively) remained the same through time and in all branches of the tree until 1000 lineages existed. The top bold line shows the growth of the actual clade, in terms of the number of lineages that existed at each time (including lineages which will subsequently go extinct), while the lower light lines are the numbers of lineages in phylogenies derived from samples of extant taxa (100%, 50%, 20% and 2% respectively from top to bottom). These lower lines are equivalent to the ideal data available to molecular phylogeneticists; the aim is to use them to determine the form of the upper bold line. The 'reconstructed' lines (i.e. those showing the numbers of lineages in our phylogenies) all lie below that showing the actual number of lineages in the clade if extinctions have occurred. Note that the reconstructed line based on the full sample is roughly straight between 3 and 9 time units but curves up towards the present. The straight portion has a slope of the birth rate minus the death rate, whereas the most recent steep portion has a slope approaching the birth rate (Harvey *et al.* 1994*b*); as a consequence it is possible to estimate birth and death rates from the phylogenies based only on extant lineages, even in the absence of fossil records (Nee *et al.* 1994*a,b*, 1995*a*). However, as smaller samples of extant species are used, the reconstructed line flattens-off so that it is no longer possible to estimate the lineage birth rate from a steepening towards the present.

lineages, occurred the first time that the number of lineages reached 700. A sample output is given in Fig. 3.3. Finally, we might abandon the mass extinction event and instead lower the birth rate from 1.0 to 0.55 the first time that the number of lineages reached 700. A sample output is given in Fig. 3.4. We discuss the more general importance of the results presented in Figs 3.2–3.4 in the figure legends and in the conclusion. The program also allows the investigation of situations in which the number of lineages has remained constant, with every birth being matched by a death, for long periods of time. The application is user-friendly with plots being rescaled to fill the page or screen depending on the time interval and number of lineages produced in the run.

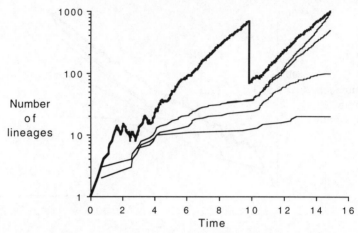

Fig. 3.3 Output from the simulation program with birth and death rates and final numbers set as in Fig. 3.2 (1.0, 0.5 and 1000 respectively), but with a mass extinction event removing a phylogenetically random sample of 90% of extant species the first time that 700 lineages were extant. The phylogenies based on 100%, 50%, 20% and 2% of extant species (as in Fig. 3.2) show a definite kink in lineage-through-time plots at the time of the mass extinction, but the kink is less pronounced as the sample becomes smaller and could not be statistically distinguished from random noise in the phylogeny based on the smallest sample.

Analysing tree structure

Lineages-through-time   The other programs are for the analysis of phylogenetic trees, produced from a researcher's own molecular data. Trees can be reconstructed using a variety of widely available packages, the most commonly used of which is PHYLIP (Felsenstein 1993). The method used to create the trees does not matter provided that the time of divergence of each of the internal nodes is known (which also means that the tips are contemporaneous). Current methods of molecular phylogenetic reconstruction producing such trees all assume the existence of a molecular clock. While many data sets may not conform to this assumption, statistical tests are available to test it (Kimura 1983, Tajima 1993).

The first of the analytical programs requires a tree from a PHYLIP output file, creating a new file which is then structured so that lineages-through-time plots can be readily produced and analysed from standard statistical packages. Lineages-through-time plots are a convenient visual representation of the internode intervals, which constitute the raw data. Although semi-logarithmic plots (as shown in Figs 3.2–3.4) provide a useful starting point for most analyses, other transformations may be necessary for revealing the past history of a population or clade when only a small random sample of extant lineages is available for analysis (Nee *et al.* 1995*b*). The appropriate application of these transformations can reveal whether the population, or clade, has been of constant size for a long period, if it has been growing exponentially at a

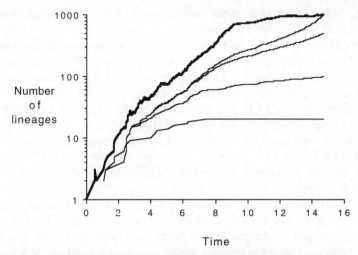

Fig. 3.4 Output from the simulation program with parameters and samples set as in Fig. 3.2 except that the per lineage birth rate drops from 1.0 to 0.55 the first time that 700 lineages were extant. There are no definite kinks in the reconstructed curves, but those based on the largest samples do differ from their equivalents in Fig. 3.2, the form taking on a sigmoid shape.

constant or changing rate, or if it has been declining. The output from the program provides appropriately transformed data so that plots can be easily produced from standard, commercially available graphical or statistical packages. Additionally, this program can produce likelihood surfaces for various functions of lineage birth and death rates.

Locating localised pattern in the tree The second analytical program also uses output from tree reconstruction packages (such as PHYLIP) to identify regions of a tree that are bushier or more comb-like than expected under the null model that all lineages are equivalent with respect to their probabilities of cladogenesis. The program is interactive in the sense that it produces a visual representation of the phylogenetic tree on the computer monitor and that when a node is 'clicked' the appropriate statistical comparison concerning tree balance is shown (Fig. 3.5). This allows the rapid statistical identification of evolutionary radiations which stand out against a background picture of birth and death rates that do not change significantly. Future versions of this program will allow the detection of departures from the null hypothesis in different directions.

Conclusion

Intuition often leads to the incorrect interpretation of molecular phylogenies. For example, one may interpret a flattening-off of lineage-through-time curves based on phylogenies reconstructed from small samples of lineages as a decrease in speciation rates in the recent past. As Fig. 3.2 demonstrates, such a flattening-off is actually expected under a constant-rate birth-death model. Similarly, again as shown in Fig. 3.2, the steepening towards the present in lineage-through-time plots based on a full sample of extant lineages, is expected under the constant rate model and does not imply a recent increase in per-lineage birth rates or a decrease in per-lineage death rates.

Guidelines and methods are now available for detailed interpretation of lineage-through-time plots (Harvey *et al.* 1994*b*, Nee *et al.* 1995*b*). For taxa as diverse as whales and viruses, such plots based on genetic similarity between individuals of a population or a species can reveal the recent history of population growth or decline. Indeed, by transforming the lineage axis from logarithmic to other defined scales, it is possible to determine whether populations have been growing faster or slower than exponentially, or whether they have been of constant size or decreasing. It is even possible to reveal changes in the pattern of population growth or the spread of a disease and, with an appropriately scaled time axis, to determine when those changes occurred. For example, we have been able to show that an epidemic of Hepatitis C occurred between 30 and 50 years ago, even though that epidemic is not recorded in history (indeed the virus was not identified until 1989).

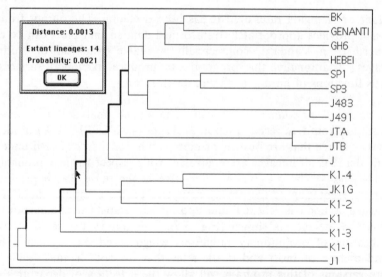

Fig. 3.5 An example display of the tree structure analysis program. This is a tree of Type 1b Hepatitis C Virus (NS-5 gene) isolates, constructed using the UPGMA method as implemented in Felsenstein (1993). The branches in bold are those giving rise to significantly more extant lineages (tips) than expected under a random model of lineage splitting and extinction (Nee *et al.* 1994*b*). The dialog box gives some details for the selected branch (next to the arrow): the distance or length of the branch, the number of tips arising from this branch, and the probability of having this many tips under the null model.

Not only can these methods be used to look at population processes, they can also be used to examine evolutionary processes. For example, using a published molecular phylogeny, we have shown that the rate of lineage diversification in birds has been decreasing since their origin, while two aberrant taxa (the Passeri and the Ciconiiformes) radiated at an unusually high rate (Harvey *et al.* 1991, Nee *et al.* 1992).

Analyses of the type described above will, we believe, become commonplace and the computer packages described in this chapter will help researchers to interpret better their own and other people's published data.

Acknowledgements

PHH thanks the Wellcome Foundation (38468), the BBSRC (H53655) and the NERC (8515) for research grants which supported much of the work summarised in this chapter. SN is supported by the BBSRC.

References

Felsenstein, J. (1993) *PHYLIP (Phylogeny Inference Package). Version 3.5c.* Distributed by Author at Department of Genetics, University of Washington, Seattle WA 98195, U.S.A.

Harvey, P. H. and Nee, S. (1993) New uses for new phylogenies. *European Review* **1**, 11–19.

Harvey, P. H. and Nee, S. (1994) Comparing real with expected patterns from molecular phylogenies. In *Phylogenetics and ecology* (ed. P. Eggleton and R. I. Vane-Wright) pp. 219–231. Academic Press, London.

Harvey, P. H., Nee, S., Mooers, A. Ø. and Partridge, L. (1991) These hierarchical views of life: phylogenies and metapopulations. In *Genes in Ecology* (ed. R. J. Berry and T. J. Crawford) pp. 123–137. Blackwell Scientific, Oxford.

Harvey, P. H., Holmes, E. C., Mooers, A. O. and Nee S. (1994a) Inferring evolutionary processes from molecular phylogenies. In *Models in phylogeny reconstruction* (ed. R.W. Scotland, D.J. Siebert and D.M. Williams) pp. 313–333. Systematics Association Special Volume Number 52, Oxford.

Harvey, P. H., May, R. M. and Nee, S. (1994b) Phylogenies without fossils: estimating lineage birth and death rates. *Evolution* **48**, 523–529.

Kimura, M. (1983) *The neutral theory of molecular evolution.* Cambridge University Press.

Nee, S., Mooers, A. Ø. and Harvey, P. H. (1992) The tempo and mode of evolution revealed from molecular phylogenies. *Proc. Natl. Acad. Sci. (USA)* **89**, 8322–8326.

Nee, S., Holmes, E. C., May, R. M. and Harvey, P. H. (1994a) Extinction rates can be estimated from molecular phylogenies. *Phil. Trans. Roy. Soc. (Lond.) B* **344**, 77–82.

Nee, S., May, R. M. and Harvey, P. H. (1994b) The reconstructed evolutionary process. *Phil. Trans. Roy. Soc. (Lond.) B* **344**, 305–311.

Nee, S., Holmes, E. C., May, R. M. and Harvey, P. H. (1995a) Estimating extinction from molecular phylogenies. In *Estimating extinction rates* (ed. J. L. Lawton & R. M. May), pp. 164–182. Oxford University Press.

Nee, S., Holmes, E. C. and Harvey, P. H. (1995b) Inferring population history from molecular phylogenies. *Phil. Trans. Roy. Soc. (Lond.) B* **349**, 33–40.

Raup, D. M., Gould, S. J., Schopf, T. J. M. and Simberloff, D. S. (1973) Stochastic models of phylogeny and the evolution of diversity. *J. Geol.* **81,** 525–542.
Tajima, F. (1993) Simple methods for testing the molecular evolutionary clock hypothesis. *Genetics* **135,** 599–607.

4

Is it ancient or modern history that we can read in the genes?

Richard A. Nichols and Mark A. Beaumont

Introduction

The impact of gene flow on the population structure of a species depends markedly on its population dynamics, and can also vary strikingly from locus to locus within the same individuals. The renaissance of the genealogical approach to population genetics has provided a series of simple and useful tools that enable us to use the distribution of allele frequencies to draw inferences about a species' history and demography. In particular, comparisons between loci with different rates and types of mutation will be valuable in conservation planning. Conversely, they can provide insight into strategies for maintaining genetic variation.

Some of the familiar results of population genetics follow from very simple rules of genealogy construction. Besides providing a fresh perspective on these well-known results, the approach can be extended to new problems. In particular, we will address the interactions between gene flow and colonisation and growth. These interactions are important when individuals (or gametes) disperse into sparsely populated areas and especially during the colonisation of virgin territory.

Range expansion and colonisation have been used to explain marked genetic differentiation between regions. Particularly convincing examples can be attributed to post-glacial range expansions of insects (Hewitt, 1989) and human neolithic migrations (Ammerman and Cavalli-Sforza 1984). The resulting genetic patterns appear to have persisted for thousands of years. It should be possible to explain why the subsequent gene flow has not obliterated these patterns and to evaluate whether extinction and colonisation on more local scales can generate similar persistent patterns.

The population ecology of actual species is rarely as simple as that found in genetical models. We will therefore draw on real examples to illustrate the use and the limitations of the theoretical results. Humans are chosen for their unique historical record, *Drosophila* because of the extensive genetic evidence, and grasshoppers provide a particularly clear case of an interaction with selection.

Gene genealogies

Much of the early progress in population genetics was based on the properties of gene genealogies (Fig. 4.1). The nodes of the tree represent the genes' common ancestors. Sewell Wright developed his inbreeding coefficients (F statistics) in terms of the proportion of genes inherited unmodified from these common ancestors (see Wright 1969). Malecot's (1948) investigation of isolation by distance was also founded in genealogical reasoning. Whilst this approach was never dropped from the theoreticians' armoury, much of the progress in the last two decades has drawn on the introduction of diffusion equations to describe distributions of allele frequencies following Kimura's seminal work (Kimura 1955).

Time to a common ancestor

Genealogical reasoning may be applied to the whole population, or a sample of its alleles. Recent advances have been reviewed by Ewens (1990) and Hudson (1990). Here we provide a brief sketch of the attributes of genealogies as a foundation for our subsequent analysis of population growth.

Figure 4.1 represents the genealogy of all the alleles in a population of constant size. If alleles are equally likely to be descended from each of the N alleles in the previous generation, then any pair share a common ancestor in that previous generation with probability $1/N$; the lineages are said to coa-

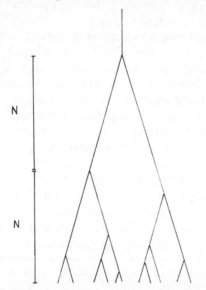

Fig. 4.1 An example of a gene genealogy. The open branches of the tree represent genes in a present-day population. The nodes represent common ancestors. The scale shows the expected number of generations to a common ancestor for the whole population: twice the haploid population size, ($2N$). For a period averaging N generations there were only two common ancestors.

lesce. Otherwise, with probability $(N - 1)/N$, their immediate ancestors were distinct and had probability $1/N$ of common ancestry in the grandparental generation, and so forth. This series, which gives the average time to a common ancestor, sums to N generations.

This result applies to a single pair of alleles. In a group of N alleles there are many possible pairs that could have an immediate common ancestor, specifically $\binom{N}{2}$. Hence the rate at which coalescent events occur is much higher—so much higher that the expected time to the common ancestor of the whole population is only $\approx 2N$ generations. Now, at some time in the past, there were only 2 common ancestors for the whole population (Fig. 4.1). None of the other alleles extant at that time left descendants. However, there is no reason to assume that the ancestry of the two successful alleles is atypical, so their expected time to coalescence is N generations. Hence, on average, the whole population spent half is history as two lineages.

The effect of this typical history can be seen in comparisons of the DNA sequence between alleles. Alleles will differ as a consequence of mutation events since the time of their common ancestor. Within the two main lineages, the alleles will be relatively similar but the two groups may be markedly different. Not all genealogies will have a two-branched structure, of course, but there will be a strong tendency for the more ancient branches to be longer. This means that pairwise comparisons of all pairs of DNA sequences can show the double-peaked pattern of differences typical of two-branched genealogies, although single or multiple peaks are not uncommon (Slatkin and Hudson 1991, Marjoram and Donnelly 1994).

It is not only DNA sequences for which we can judge allele divergence. Some loci show polymorphism in allele length (e.g. minisatellites and microsatellites), and mutation may change allele length only slightly (Jeffreys 1991, Jeffreys *et al.* 1991). Hence, alleles with similar lengths are more likely to have a recent common ancestor. Before using this line of reasoning to interpret patterns in human minisatellite allele distributions, we must consider more realistic population structures.

Subdivided populations

Most species do not consist of single randomly mating populations, but are typically divided into partially isolated subpopulations. If we look at a pair of alleles in one subpopulation, it is possible that their common and subsequent ancestors were all resident there. In this case, the coalescence time would be determined by the subpopulation size. However, if one of their ancestors was an immigrant, then we would expect that their common ancestor lived a very long time ago. For, in order to find the common ancestor, we must look back to a time when the ancestors were again in the same subpopulation. If the migration rate is low, or the number of subpopulations large, then this meeting is unlikely.

Despite these considerations, Strobeck (1987) showed that the expected time to a common ancestor of alleles taken from the same subpopulation may be unaffected by population subdivision. We can demonstrate the principle underlying this somewhat counter-intuitive result in a simple Monte-Carlo simulation. Fig. 4.2 shows the distribution of the number of generations back to coalescence for a pair of alleles in a simulated population of one million. If they are all in a single panmictic population, the coalescence times are distributed around the total population size (zero when plotted as ln(generations/N_{tot})). If the population is divided into 20 000 subpopulations with equal migration among them (i.e. an island model with 10% migration each generation), then most of the coalescence times are much faster: a time approximating the subpopulation size (N). However, this is balanced by the long time to coalescence with immigrant alleles, leaving the mean coalescence time unaltered.

It is these differences in coalescence times that are responsible for local genetic differentiation, because, if the mutation rate is low, those pairs in the left-hand category of Fig. 4.2 are likely to be identical by descent. The coalescence times of alleles in the right-hand category are the same as those taken from any two subpopulations. The larger the proportion in the first category, the more strongly differentiated the subpopulations will be. This

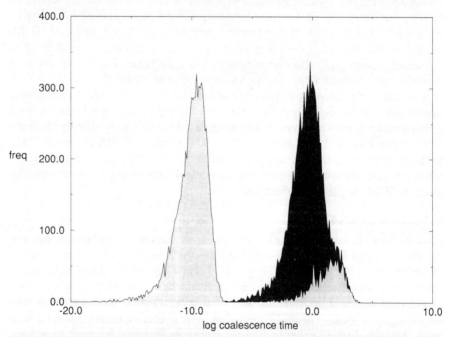

Fig. 4.2 A comparison of coalescence times in a single panmictic and a subdivided population. The distribution of coalescence times for pairs of lineages drawn from a simulated population of 1 000 000 individuals. Black distribution: a single panmictic population. Grey distribution: population subdivided into 10 000 demes with island model type migration at a rate of 0.1 per island, per generation. The time scale is ln(number of generations/1 000 000).

proportion can be estimated as follows. If we look back one generation, there is a probability $1/N$ of any two alleles having a common ancestor and, if the migration rate is m, a probability $2m$ that one or other parent is a newly arrived migrant. If neither event occurred, these probabilities remain unchanged for the preceding generations. Assuming simultaneous events are negligibly rare, and that one or the other of these events must occur if we look far enough back in time, the probability of a local common ancestor without subsequent migration, and hence the proportion in the left-hand group of Fig. 4.2, is

$$\frac{\frac{1}{N}}{\frac{1}{N} + 2m} = \frac{1}{1 + 2Nm}. \tag{4.1}$$

Wright's F_{ST}

This proportion has long been known as the expected value of F_{ST} when mutation rates are low (Wright 1951). For diploids it is convenient to use N to represent subpopulation size rather than the number of alleles, so $2Nm$ becomes $4Nm$. F_{ST} was conceived as the correlation between alleles in the same subpopulation; thus, if an allele drawn from it has frequency p in the population at large, the probability of drawing a second matching allele is

$$F_{ST} + (1 - F_{ST})p. \tag{4.2}$$

This can be interpreted as the sum of the probability of descent through a common ancestor in the subpopulation and the probability of matching otherwise.

Depending on the types of mutation involved, many genetic markers may only be distinguished by certain methods. Examples include allozyme loci with distinct alleles at the DNA level that are indistinguishable by electrophoresis and molecular markers which involve variable numbers of tandem repeats (VNTRs), wherein different mutations can arrive at the same number of repeats.

Alleles which are identical because the most recent mutation in each lineage produced indistinguishable alleles are defined as identical by state, rather than identical by descent.

Estimation If mutation rates are low ($\ll$ subpopulation size $^{-1}$) then alleles sharing a common ancestor in the subpopulation will be overwhelmingly identical by descent. The remaining allele pairs (the right-hand peak) have at least one ancestor that migrated into the subpopulation and hence usually have a much more distant common ancestor. They may be either identical by descent or, if there has been a mutation since that ancestor, identical by state. The probability of drawing a matching allele is then given by the population frequency p and hence equation (4.2) applies. (We will later investigate the effects of higher mutation rates.) Similar logic applies to heterozygosity which can be calculated from the probability of drawing a different allele from the same population $(1 - F_{ST})(1 - p)$, compared to $(1 - p)$ in the population at large. Nei's G_{ST} is a statistic that makes use of this reduced heterozygosity to estimate F_{ST} (Nei 1973). Slatkin (1993) has shown that with low mutation this is equivalent to estimating a ratio of the mean coalescence times for genes in the same (t_0) and different (t_1) subpopulations namely

$$\frac{t_1 - t_0}{t_1 + t_0} \tag{4.3}$$

for pairs of populations.

Non equilibrium populations

Equation (4.1), widely used to infer the rate of gene flow from studies of genetic variation, assumes an equilibrium between genetic drift and gene flow, even though it is well established that the assumption may be unsafe in some cases (Slatkin 1977, Wade and McCauley 1988, Boileau *et al.* 1992). In particular, the rate of approach to this equilibrium can be quite slow in many species.

Reassuringly, Slatkin (1991), has shown that the time to equilibrium is a function of subpopulation size, rather than total population size, concurring with Slatkin and Barton's (1989) simulations. This time can be interpreted as that required for common ancestry in the subpopulations to develop, and will be of the order $2N$ generations. Nevertheless, for species with large subpopulations, this result takes us back to before the last ice age. Other species have subpopulations which are not in equilibrium because they are particularly ephemeral. Can we describe the magnitude of genetic differentiation after colonisation, but before equilibrium in such populations?

Differentiation in Drosophila

The treatment of differentiation after colonisation presented here was inspired particularly by Singh and Rhomberg's (1987) remarkably extensive data set. They quantified genetic variation at 61 loci from *Drosophila* populations from 15 locations around the world. The histograms in Fig. 4.3 show the distribution of F_{ST} at these loci. Singh and Rhomberg identified two parts to the distribution: a peak around $F_{ST} = 0.1$ which they interpreted as indicating migration of 2 individuals per generation, and the more genetically structured loci. They postulated that the greater variation of the latter was due to geographically varying selection.

Local populations may reach enormous size, and the evidence suggests a relatively recent origin for most populations of *Drosophila melanogaster* (David and Capy 1988). David and Capy argue that the ancestral distribution was African, that Eurasian populations arose 10–15 000 years ago, and that other populations have arisen in the last few hundred years through their commensal relationship with humans.

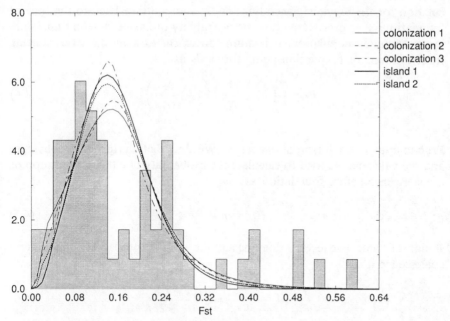

Fig. 4.3 Observed and expected distributions of F_{ST}. The histograms show the observed distribution of F_{ST} at different loci within the same sample of *Drosophila melanogaster* (Singh and Rhomberg 1987). The curves show the expected distributions from five very different models of population structure described in the text.

Applying the genealogical approach   To encapsulate this type of history, we consider a large ancestral population from which a series of new populations are produced, each with N_0 founders, foundation being followed by a period of exponential growth at rate r. We find that during this period genetic differentiation reaches a quasi-equilibrium. Once the populations have reached a large size, the effects of genetic drift are minimal, and so this distribution is frozen in place until it is eroded by gene flow. However, as we have already argued, this would take a time comparable to the subpopulation size; in *Drosophila* this may be hundreds of thousands of generations. Migrants arrive from the ancestral population at a rate of M individuals per generation. In early generations, when population sizes are small, these migrants can have a large impact on the genetics. The effect in later generations will be negligible, as will the effect of back-migration to the large ancestral population.

The specification of migration is slightly different from our previous treatment. Thus, the probability of any allele being a newly arrived migrant in generation t is M/N_t, where N_t is the population size in generation t. The probability of neither lineage being immigrant is $(1 - (M/N_t))^2$. The probability that the lineages then remain distinct (do not coalesce) is $(1 - (1/N_{t-1}))$. If the lineages are distinct, then we can then repeat the same calculation for the previous generation, $t - 2$, and continue back to the time of foundation. This procedure gives the probability of having different ancestors in the founding population. It is more convenient to write the series starting from the time of foundation going forwards as

$$P_f = \prod_{t=0}^{\infty} \left(1 - \frac{1}{N_t}\right)\left(1 - \frac{M}{N_{t+1}}\right)^2. \tag{4.4}$$

Probabilities of each type of ancestry   We find that the product converges and the value can be used to calculate the probability of either a migration or a coalescence before foundation as

$$P_{cm} = 1 - P_f. \tag{4.5}$$

If one of these two events does occur, the relative probability that it is a coalescence is

$$\frac{P_c}{P_{cm}} = \frac{\frac{1}{N_{t-1}}}{\frac{1}{N_{t-1}} + \frac{2M}{N_t}} = \frac{1+r}{1+r+2M}, \tag{4.6}$$

(since $N_t = N_{t-1}(1 + r)$).

P_f in equation (4.4) is approximated by $e^{(-r + 2M + 1)/N_0 r}$; hence, combining equations (4.4) and (4.5) we obtain the probability of coalescence in the subpopulation

$$P_c = \frac{(1 - e^{(-r+2M+1)/N_0 r})(1 + r)}{1 + r + 2M}. \tag{4.7}$$

We have also found a reasonable and robust approximation for P_c,

$$\frac{1+r}{1 + N_0 r + 2M}. \tag{4.8}$$

Expectations of F_{ST} In this case, P_c is not exactly equivalent to F_{ST} unless the number of subpopulations is quite large, but we can relate the two values using the same logic as Slatkin (1991). He used the relationship between coalescent time (t) of two alleles and probability of identity (f). Identity indicates that there have been no mutations since the common ancestor. For low mutation rate (μ) Slatkin uses $f = 1 - 2\mu t$. F_{ST} can be interpreted as

$$\frac{f_0 - \bar{f}}{1 - \bar{f}}, \tag{4.9}$$

where f_0 is the identity of alleles drawn from the same subpopulation and $\bar{f}$ of alleles drawn at random from the whole population. Where there are $s - 1$ populations derived from one ancestral population, it can be shown that

$$F_{ST} = \frac{P_c(s-1)}{s - P_c + \frac{s}{(s-1)}}. \tag{4.10}$$

Distributions of F_{ST} These equations give expectations of F_{ST}, but not the distributions around them. In order to generate the distribution, we ran stochastic simulations of genealogy construction (see Hudson 1990 for methodology) with the same probabilities used to derive the equations. Mutations were added to the genealogy as an infinite allele process with a rate consistent with the observed heterozygosity. G_{ST} (Nei 1973), as an estimator of F_{ST}, was then calculated from the simulated samples.

Because they are nearly independent in diploids, we can evaluate both the mean, and the distribution of F_{ST} from different loci in the same individuals. There is a 50% probability that two alleles at different loci are inherited from the same parent, simply because there are two possible parents. For grand-parents the probability is 25%, and so forth. After a handful of generations back in time, they are effectively independent.

It is clear from equations (4.7) and (4.8) that a particular mean value of F_{ST} could be generated by a whole range of reasonable values for r, N_0 and M. We used the simulations to see if the distributions of F_{ST} around these means were informative. We quantified the fit between the observed and expected distributions using the Kolmogoroff–Smirnoff statistic (Steel and Torrie 1981). We fixed two of the three parameters (r, N_0 and M) and varied the other to maximise the fit.

The results are shown in Fig. 4.3. The distributions are nearly identical for three radically different assumptions about colonisation; the growth rates can differ ten-fold and migration can be absent, yet the results are effectively indistinguishable (Case 1: $r = 0.1$, $N_0 = 4$ and M fitted; Case 2: $r = 1$, $N_0 = 4$ and M fitted; Case 3: $M = 0$, $N_0 = 50$ and r fitted). Furthermore, simulations of the standard island model give an equally convincing fit, whether they consist of 15 or 60 islands (the only parameter, M, is fitted).

Interpreting genetic variation

These calculations illustrate how, as populations are founded and then grow, the genetic differentiation from the ancestral population may stabilise. This variation will be misinterpreted if it is assumed to be due to a balance between gene flow and genetic drift, as specified by the standard island model. Instead of depending solely on the number of migrants (N_m), the variation can be explained by a range of combinations of population growth rate (r), number of founders (N_0) and number of immigrants per generation (M) (equation 4.8).

Difficulties in distinguishing population structure   By studying the genetic variation alone, even very different histories may be indistinguishable. In the case of recent colonisation, the gene lineages can be divided into the proportion that coalesce quickly in the founded subpopulation and those that are derived from founders or immigrants. Lineages in the latter two categories extend separately back to the ancestral population, where they coalesce much more slowly at a rate that is the inverse of the ancestral population size $(1/N_a)$ per pair. A very similar pattern of coalescence times can be generated if the population structure resembles an island model. In that case, a proportion of lineages ($\approx F_{ST}$) coalesce very quickly in the subpopulations (Fig. 4.2), but the rest coalesce at a rate determined by the migration rate (m), the chances of being in the same subpopulation (the inverse of the number of subpopulations: $1/s$), and the probability of coalescing whilst they are in the same subpopulation ($\approx F_{ST}$). The rate per pair is $\approx 2mF_{ST}/s$. With the appropriate choice of s and N_a the rates could be the same. A biologist would need to draw on knowledge of the species' ecology to decide which is the most plausible history.

The imprecision of single locus estimates   The enormously broad distributions shown in Fig. 4.3 illustrate the uncertainty of deductions about gene flow founded on small numbers of loci. It underlies Slatkin and Barton's (1989) observation that, due to their genealogical history, variation in estimates of F_{ST} among loci can far outweigh sampling variation. Indeed, in pilot simulations we found that as long as sample size was greater than 30 (= 15 diploid individuals) per subpopulation, sampling error had negligible effect on the estimated distribution. For the same reason, it is difficult to attribute genetic variation at any one locus to the action of selection without corroborating evidence.

The effects of mutation with population subdivision

So far, we have considered loci with low mutation rates. Many widely used molecular markers, particularly those with variable numbers of tandem repeats (VNTRs), have very high mutation rates and correspondingly high heterozygosities. We will use the example of human VNTR distributions to show that the most variable loci are not necessarily the most informative about the extent of population subdivision. Our extensive knowledge of human demography also allows us to interpret more readily genetic differentiation over different geographical scales.

Heterozygosity

Heterozygosity within a population is a consequence of the relative rates of coalescence and mutation. A mutation occurs in one or the other lineage at twice the mutation rate: 2μ. We have seen that a pair of lineages coalesce at a rate $1/N$ per generation. The ratio is the familiar expression derived by Kimura and Crow (1964)

$$\frac{\frac{1}{N}}{\frac{1}{N} + 2\mu} = \frac{1}{1 + 2N\mu}. \tag{4.11}$$

If N is the diploid population size then $2N\mu$ is replaced by $4N\mu$. If all mutations are unique then this is the expected heterozygosity.

With other mutation processes and population structures, the picture becomes a little more complex. If mutant alleles are drawn from the same distribution as immigrants, then equation (4.1) takes the form

$$\frac{1}{1 + 4N(m + \mu)}. \tag{4.12}$$

A mutation process of this type appears to explain different types of allele frequency distributions seen at human VNTR loci MS1 and MS43a. Large databases of London population frequencies have been compiled because of their use in forensic science (Buffery *et al.* 1991). Individuals have been classified into broad ethnic groups. The groups are not well defined as they are based on information available to police, which is often limited to physical appearance. For our purposes, it is sufficient that there is clear genetic differentiation between the groups. In addition, there will be substantial differentiation within the groups, which we will consider later.

Allele length distributions

The data are often presented as smoothed distributions, which can be misleading. In Fig. 4.4 the database is presented to a resolution of 10bp, which emphasises that MS43a frequencies show significant differentiation between people of Afro-Caribbean descent and the other two categories. The allele measurements have an error (with variance σ^2) which increases with length (Buffery *et al.* 1991). A crude estimate of F_{ST} was obtained by dividing the distribution into sections of 3σ and then calculating G_{ST}. This will tend to be an underestimate because many alleles are not distinguished, and, in addition, measurement error will have led to the misclassification of some alleles into the wrong section. Nevertheless, G_{ST} is 0.11, a value comparable with the variation typically seen between ethnic groups at other loci (Lewontin 1972).

In stark contrast, the distribution of MS1 in the same individuals appears identical in all three ethnic groups ($G_{ST} \approx 0.0$). MS1 has a mutation rate approaching 0.1 (Jeffreys 1991). The MS1 distribution can be explained as the stationary distribution produced by the mutation process, whereby the height of the distribution is proportional to the expected time that a lineage has alleles of each length. The difference from MS43a illustrates how loci with different mutation rates can be used to investigate different periods of a species history. Common ancestry within these human groups is sufficiently ancient for there to have been multiple MS1 mutations, which have obliterated any traces of genealogy. MS43a has a lower mutation rate, and we can therefore use it to look further back into history.

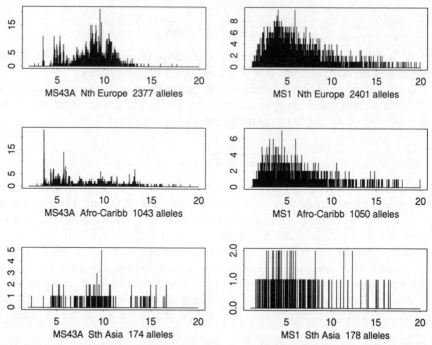

Fig. 4.4 The distribution of VNTR allele lengths in Londoners. Estimated allele lengths are shown in kilobases. The distributions are shown for two loci (MS43a and MS1) and three ethnic groups (North European, Afro-Caribbean and South Asian). The classification into ethnic groups was done by officers on the basis of appearance.

At MS43a, there are distinct peaks that characterise each ethnic group. Most striking is the peak around 4 kb in the Afro-Caribbean distribution, which makes up nearly 10% of the whole distribution. There is a peak in a similar position in the North European distribution, but there is low measurement error for such small alleles and the peaks' locations are significantly different; this fact is obscured if the distributions are smoothed. It is tempting to interpret a single large peak as a result of a founder event, or similar dramatic episode in population history. However, Roe (1993) has shown that such peaks arise as a simple consequence of genealogy shape in stable populations.

To see why peaks arise, consider Fig. 4.1 again. Our earlier description drew attention to the considerable length of the first branches of a typical genealogy. Following this period in which there were only a few common ancestors, the genealogy tends to branch profusely and rapidly. During the initial long period, mutations can accumulate so that the ancestral alleles become atypically long or short. Thereafter, mutations also occur on the branches issuing from the common ancestors, but there is less time for them to accumulate. As mutations typically cause small changes in allele length, the recent mutants will be clumped around the ancestral length, producing a peak.

Geographic scale and estimates of F_{ST} Human population substructuring is relevant to the use of VNTRs in forensic science (Nichols and Balding 1991, Balding and Nichols 1993). Disagreement about the magnitude of the effects has highlighted some disparities in the calculation of F_{ST}s and the inferences that are drawn from them. Most F_{ST}s reported for humans are much less than 0.01, indicating very little differentiation within the major ethnic groups (Morton 1992). However, in their survey of human genetic diversity, Cavalli-Sforza and Piazza (1993) present values exceeding 0.05 within Africa and within Asia.

Slatkin and Voelm (1991) present a simple model that can generate such discrepancies between the variation at regional and finer levels. They consider regions consisting of groups of interconnected subpopulations, which are relatively isolated from each other. They show that the genetic differentiation both within and between the groups is simply given by equation (4.1). However, for the between-group variation the parameters have a different meaning: m refers to the migration between groups, rather than the more familiar definition of migration between subpopulations. Similarly, N represents the total size of a group rather than a component subpopulation. This example clearly illustrates that genetic variation cannot be characterised by a single value. In real populations there are further considerations: the population structures are not static, and involve multiple levels.

Humans, like many species, are distributed among different regions, with reduced migration between them. Slatkin and Voelm's (1991) results indicate that these regional groups have the potential to differentiate quite markedly through drift. However, if the regional population size is large, this may take an inordinate time. Instead, the differentiation between regions appears to date from the time of population foundation. Ammerman and Cavalli-Sforza (1984) argue that the clines in allele frequency throughout Europe date from the Neolithic migrations that accompanied the spread of agriculture. Subsequent population growth has inured the regional frequencies to the impact of gene flow because the immigrant genes are diluted by the resident populations. At such times of population foundation and range expansion, there are important interactions between population structure and selection. Some of the best evidence for this type of process comes from studies of insect chromosomes.

Selection and history

The clearest example of interactions with selection are those of underdominant alleles, where heterozygotes have lower fitness than the homozygotes. Periods of range expansion are particularly important in their spread; at other times their increase from low frequencies is inhibited because at low frequency they tend to occur in the unfit heterozygotes. Even if locally common, they are unlikely to invade existing populations in surrounding regions because they will arrive at low frequency. Many species have alleles whose geographical distributions can be explained in this way (Barton and Hewitt 1985).

Grasshopper hybrid zones

The alpine grasshopper *Podisma pedestris* has a well-studied hybrid zone running from east to west along the main ridge of the Alpes Maritimes (Hewitt and John 1972, Hewitt 1975). To the south of the zone the grass-hoppers carry a chromosome that appears to have been produced by fusion between the ancestral X-chromosome and an autosome. To the north, the X is unfused. Chromosomal hybrids are only found in a narrow zone 800 m wide, stretching from the Italian border 200 km into France. It does not appear to run along a habitat transition (Nichols and Hewitt 1988), and the simplest explanation of the zone is selection against the chromosomal hybrids. This might be expected if there is non-disjunction of the X-autosome trivalent at hybrid meiosis.

The narrow width of the zone can be attributed to surprisingly weak selection. Barton (1979) has shown that the width (w) of the zone resulting from the balance between dispersal (with parent–offspring variance, σ^2) and selection against the hybrids (s) is given by

$$w = \sqrt{\frac{8\sigma}{s}}. \tag{4.13}$$

From the width of the zone, and estimates of grasshopper dispersal, he estimated the selection against hybrids as 0.05%. For much of its length the *Podisma* zone is of approximately constant width, 800 m, as would be expected if selection and dispersal were essentially constant.

A discrepant region   Nichols *et al.* (1990) described an area in the same region which contrasts sharply with the pattern described above. The fre-quency of the fused chromosome actually rises significantly in a 1 km transect running north away from the zone near the Chabanon ski lodge (Fig. 4.5(a)). They tentatively attributed this rise to the interaction between gene flow and small effective population size. Much of the northern region was previously wooded, and would have only maintained sparse populations of *Podisma*, many of them in ephemeral clearings. In such regions of very low effective population size, underdominant alleles can become established through genetic drift. If an immigrant allele becomes established in a sufficiently large area, then it will tend to persist because it is locally in the majority and hence favoured by selection. In this way a hybrid zone can break down into a mosaic of patches fixed for alternative alleles, so that the transition is much wider than predicted by equation (4.13) (Nichols 1989).

In the early 1970s, ski runs were built through the woods, thus providing the sunny conditions needed for the grasshopper to thrive. On these *pistes* it

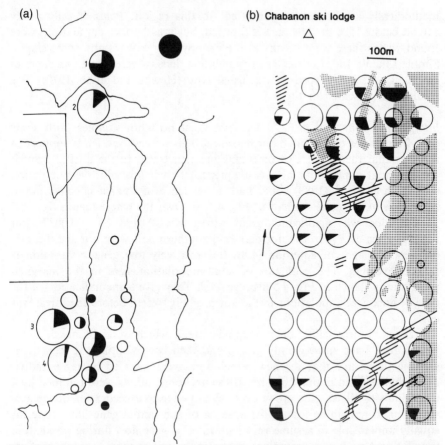

Fig. 4.5 (a) A map of the distribution of the fused chromosome in the marginal habitat around the Chabanon region. The black portion of the pie charts shows the frequency of the fused chromosome. The size of each circle is proportional to sample size (maximum 20). The area of the inset (b) is delineated. The contours are, from the east to west 1700 m, 1600 m, 1500 m. Estimated population sizes were 51 and 62 for the lower piste populations marked 1 and 2; 129 and 328 for the highest populations, 3 and 4 which were around the tree line. There was no significant heterogeneity in the other estimates (they each have a very large error), and their average estimated size was 14.8.

Inset (b). A map of the distribution of the fused chromosome near Chabanon ski lodge. The maximum sample size is 12. The stippled areas are forested, and the shaded areas are scree.

proved possible to collect reasonably large and evenly spaced samples. It is much more difficult to survey the surrounding areas because there are fewer open areas and populations are very sparse. Nevertheless, we considered it worthwhile in order to test the prediction that there are areas where the fused X-chromosome is locally at high frequency. The area roughly bounded by the 1700 m and 1500 m contours around the previous study area was searched as methodically as the terrain permitted. In this region, *Podisma* only rarely extend below 1500 m. The area is forested, but grasshopper populations were found where there were breaks in the canopy near ravines and scree slopes. Simple Lincoln Index estimates of population density were made and samples collected and karyotyped in the usual way (Hewitt and John 1972). The results are presented in Fig. 4.5.

Interpreting estimates of F_{ST} The two most northern samples both show high frequencies of the fused chromosome. This observation lends support to the idea that the fused X-chromosome has penetrated to the north, through the marginal populations as a consequence of their low effective population size. Nichols *et al.* (1989) could find no evidence of smaller effective population size in the northern half of Fig. 4.5(b). Over the whole region G_{ST} at 7 allozyme loci was significantly greater than zero (0.0138, SE = 0.0037), but the estimates of F_{ST} in the northern and southern half were not significantly different. In light of equation (4.8), this relatively low genetic variation is rather revealing. The estimates of adult population sizes in the marginal populations shown in Fig. 4.5 average 14.8. They give some indication of the type of ancestral populations in the north of Fig. 4.5(b) before the forest was cut.

Presumably tree felling opened up new areas, which were colonised. These populations have subsequently grown and then been sampled. It seems reasonable to assume that the new populations were founded by fewer individuals than the old population size. If the new populations were founded by 4 individuals (8 alleles), the growth rate would have to exceed 8 for the genetic variation to remain so low in the absence of substantial gene flow. It seems equally implausible to assume very high rates of gene flow during population foundation (> 18 immigrants per generation), so it appears that we must re-evaluate our view of the history of population structure and believe that there were already large long-established populations connected by high rates of gene flow before the ski runs were cut.

Conclusion

Genetic variation within species is one of the most valuable and yet neglected components of biological diversity. One might hope to manage it effectively and to use the genetic patterns as a source of information about the species' past and present. Both these objectives require some insight into the genesis and maintenance of that diversity. Our three examples illustrate the fundamental principle that the interpretation of genetic patterns requires some independent knowledge of a species' history.

Disentangling the effects of ancient and modern history

The analysis of *Drosophila* variation demonstrated that a recent episode of colonisation and growth can produce genealogies effectively indistinguishable from those in a long-established subdivided population. We know that these flies are capable of explosive growth and that the current populations are very large. This, together with our beliefs about the species' history, leads us to be suspicious of explanations that assume the genetic processes are in equilibrium. Furthermore, measurements are needed of genetic differentiation from many loci in order to draw inferences about species' history and gene flow. These observations are of wider relevance, particularly in the planning of genetic surveys of population structure. At present, practical considerations may mean that biologists must choose between scoring the DNA sequence of a few loci, or the allozyme/VNTR variation of many. It is often more worthwhile to collect low-resolution data from a large number of loci than detailed information from a few loci.

Just as the informativeness of DNA sequence data can prove illusory, the value of highly polymorphic loci may be overestimated. The example of human VNTR variation demonstrated that the high mutation rates producing the most highly polymorphic loci can actually obscure the population structure we wish to detect. Mutation rates will vary between the loci available, and the appropriate choice will depend on the time scale over which the genetic differentiation developed. The distinct patterns at different loci are consistent with our existing knowledge of human history. In other species we may make use of the differences between loci to reconstruct history.

We have focused on the interpretation of patterns created by genetic drift and gene flow rather than by selection. Slatkin and Barton (1989) have shown that the broad conclusions are robust to some forms of weak selection, particularly balancing selection. Nevertheless, other types of selection cannot be overlooked. The example of grasshopper chromosomes shows that even weak underdominance (selection = 0.5%) can generate dramatically different patterns. Despite this selection, other loci in the same individuals can still be used to draw strong inferences about the species' past. Hence, as with each of our examples, the genetic evidence alone was ambiguous and liable to misinterpretation. If we make use of our knowledge of the species' biology and the loci involved, it is possible to avoid these pitfalls and draw firm conclusions. We suggest that the pursuit of understanding biological diversity inevitably involves integrating our knowledge from genetic and non-genetic sources. In this task the simple rules that govern genealogy construction are particularly effective guides.

Acknowledgements

We are particularly grateful to the members of the Metropolitan Police Forensic Science Laboratory and to Simon Astbury for practical help and discussion. This work was supported in part by SERC and NERC grants to RAN and a Wellcome fellowship to MAB.

References

Ammerman, A. J. and Cavalli-Sforza, L. L. (1984). *The neolithic transition and the genetics of populations in Europe*. Princeton University Press.

Balding, D. J. and Nichols, R. A. (1993). DNA profile match probability calculation how to allow for population stratification relatedness, database selection and single bands. *Forensic Science International*, **64**, 125–140.

Barton, N. H. (1979). The dynamics of hybrid zones. *Heredity*, **43**, 341–359.

Barton, N. H. and Hewitt, G. M. (1985). Analysis of hybrid zones. *Annual Review of Ecology and Systematics*, **16**, 113–148.

Boileau, M. G., Herbert, P. D. M. and Schwartz, S. S. (1992). Non-equilibrium gene frequency divergence: persistent founder events in natural populations. *Journal of Evolutionary Biology*, **5**, 25–39.

Buffery, C., Burridge, F., Greenhalgh, M., Jones, S. and Willot, G. (1991). Allele frequency distributions of four variable number tandem repeat (VNTR) loci in the London area. *Forensic Science International*, **52**, 53–64.

Cavalli-Sforza, L. L. and Piazza, A. (1993). Human genomic diversity in Europe: a summary of recent research and prospects for the future. *European Journal of Human Genetics*, **1**, 3–18.

David, J. R. and Capy, P. (1988). Genetic variation of *Drosophila melanogaster* natural populations. *Trends in Genetics*, **4**, 106–111.

Ewens, W. J. (1990). Population genetics theory—the past and future. In *Mathematical and statistical developments of evolutionary theory*, pp. 177–227 (ed. S. Lessard). Kluwer, Amsterdam.

Hewitt, G. M. (1975). A sex-chromosome hybrid zone in the grasshopper *Podisma pedestris* (Orthoptera: Acrididae). *Heredity*, **35**, 375–385.

Hewitt, G. M. (1989). The subdivision of species in hybrid zones. In *Speciation and its consequences*, pp. 85–110 (eds. D. Otte and J. A. Endler). Sinaur, Sunderland.

Hewitt, G. M. and John, B. (1972). Interpopulation sex chromosome polymorphism in the grasshopper *Podisma pedestris* II: population parameters. *Chromosoma*, **37**, 23–42.

Hudson, R. R. (1990). Gene genealogies and the coalescent process. In *Oxford surveys in evolutionary biology*, **7**, 1–44 (eds. D. J. Futuyma and J. Antonovics) Oxford University Press.

Jeffreys, A. J. (1991). Variation and mutation in human minisatellites. *American Journal of Human Genetics*, **49**, 64.

Jeffreys, A. J., MacCloed, A., Tamaki, T., Neil, D. L. and Monckton, D. G. (1991). Minisatellite repeat coding as a digital approach to DNA typing. *Nature*, **354**, 204–209.

Kimura, M. (1955). Solution of a process of random genetic drift with a continuous model. *Proceedings of the National Academy of Sciences* (USA), **41**, 144–150.

Kimura, M. and Crow, J. F. (1964). The number of alleles that can be maintained in a finite population. *Genetics*, **49**, 725–738.

Lewontin, R. C. (1972). The apportionment of human diversity. *Evolutionary Biology*, **6**, 381–398.

Malecot, G. (1948). *Les mathematiques de l'heredité*. Masson et Cie, Paris.

Marjoram, P. and Donnelly, P. (1994). Pairwise comparisons of mitochondrial DNA sequences in subdivided populations and implications for early human evolution. *Genetics*, **136**, 673–683.

Morton, N. E. (1992). Genetic structure of forensic populations. *Proceedings of the National Academy of Sciences* (USA), **89**, 2556–2560.

Nei, M. (1973). Analysis of gene diversity in subdivided populations. *Proceedings of the National Academy of Sciences* (USA), **70**, 3321–3323.

Nichols, R. A. (1989). The breakdown of hybrid zones at small effective population size. *American Naturalist*, **134**, 969–977.

Nichols, R. A. and Balding, D. J. (1991). Effects of population structure on DNA fingerprint analysis in forensic science. *Heredity*, **66**, 297–302.

Nichols, R. A. and Hewitt, G. M. (1988). Genetical and ecological differentiation across a hybrid zone. *Ecological Entomology*, **13**, 39–49.

Nichols, R. A., Humpage, E. A. and Hewitt, G. M. (1990). Gene flow and the distribution of karyotypes in the alpine grasshopper *Podisma pedestris* (L.) (Orthoptera: Acrididae). *Boletin de Sanidad Vegetal* (Feura de serie), **20**, 373–379.

Roe, A. M. P. (1993). *Correlations and interactions in random walks and population genetics*. PhD Thesis. QMW, University of London.

Singh, R. S. and Rhomberg, L. R. (1987). A comprehensive study of the genetic variation in natural populations of *Drosophila melanogaster* II: estimates of heterozygosity and patterns of geographic differentiation. *Genetics*, **117**, 255–271.

Slatkin, M. (1977). Gene flow and genetic drift in a species subject to frequent local extinctions. *Thoeretical Population Biology*, **12**, 253–262.

Slatkin, M. (1991). Inbreeding coefficients and coalescence times. *Genetic Research* (Camb), **58**, 167–175.

Slatkin, M. (1993). Isolation by distance in equilibrium and nonequilibrium populations. *Evolution*, **47**, 264–279.

Slatkin, M. and Barton, N. H. (1989). A comparison of three indirect methods for estimating average levels of gene flow. *Evolution*, **43**, 1349–1368.

Slatkin, M. and Hudson, R. R. (1991). Pairwise comparisons of mitochondrial DMA sequences in stable and exponentially growing populations. *Genetics*, **129**, 555–562.

Slatkin, M. and Voelm, L. (1991). F_{ST} in a hierarchical island model. *Genetics*, **127**, 627–629.

Steel, R. G. D. and Torrie, J. H. (1981). *Principles and Procedures of Statistics*. McGraw-Hill International, London.

Strobeck, C. (1987). The average number of nucleotide differences in a sample from a single subpopulation: a test for population subdivision. *Genetics*, **117**, 149–153.

Wade, M. J. and McCauley, D. E. (1988). Extinction and recolonization: their effects on the genetic differentiation of local populations. *Evolution*, **42**, 995–1005.

Wright, S. (1951). The genetical structure of populations. *Annals of Eugenics*, **15**, 323–354.

Wright, S. (1969). *Evolution and the genetics of populations II: the theory of gene frequencies*. Chicago University Press.

5

Evolution of adaptive polymorphism in spatially heterogeneous environments

Thierry de Meeûs and François Renaud

Introduction

The genesis and maintenance of biological diversity can be studied at many levels from molecular biology to community ecology. One of these different, but complementary, levels is genetic diversity. Genetic polymorphism for protein structure has been found to be more extensive than previously thought (e.g. Lewontin and Hubby 1966) and, as a consequence, evolutionary biologists have debated several important questions concerning the maintenance of genetic polymorphism:

(1) How can selection maintain polymorphism?

(2) What is the role played by environmental heterogeneity?

(3) What role can habitat selection play?

(4) What can we infer from polymorphism to understand the process of speciation?

Two major arguments for molecular polymorphism are widely recognised. The 'selectionist' argument is the selective advantage of heterozygous individuals; this prediction, however, is not in accord with the high genetic load that would emerge from the multilocus polymorphism of proteins (e.g. Wallace 1975). Conversely, 'neutralists' consider that the balance between mutation, drift and population structure is enough to maintain most of the molecular polymorphism observed (e.g. Kimura 1983). Indeed, in a single finite population of effective size N_e, with a mutation rate μ, it can be shown that the expected number of alleles maintained per locus at equilibrium will be $n = 4N_e\mu + 1$ (e.g. Hartl and Clark 1989). For structured populations, much more polymorphism can be maintained providing there are sufficiently small rates of migration between demes (e.g. Latter 1973). Interestingly, in such theoretical approaches, geographic clines of allele frequencies, often used as proof of the progressive action of selective factors, can be found as consequences of random genetic drift and migration alone (e.g. Hartl and Clark 1989).

A corollary to the selectionist view (and that focused on hereafter) is the action of selection in heterogeneous environments (e.g. Felsenstein 1976). Despite extensive work over the last 40 years, it does not appear that this kind of mechanism fully explains observed molecular polymorphisms (Barton and Clark 1990), despite the fact that this research domain is directly linked with those on speciation and the maintenance of ecological diversity (e.g. de Meeûs et al. 1993).

It is known that temporal heterogeneity leads less easily to adaptive polymorphism than spatial heterogeneity: for example, adaptive polymorphism is possible for haploids in spatially heterogeneous environments while this is not the case in temporally varying environments (e.g. Felsenstein 1976). Consequently, more work has been conducted on spatial heterogeneity, and thus we treat this subject more extensively below.

Selection in heterogeneous environments

Here, and in the following sections, populations are assumed to exhibit discrete and non-overlapping generations. We begin by considering basic models.

Basic models

In models incorporating environmental heterogeneity it is important to distinguish between 'soft' and 'hard' selection (Wallace 1968). Wallace (1975) defined selection as soft when it is both frequency and density *dependent* and as hard when it is both frequency and density *independent*.

With regard to genetic polymorphism in a variable environment, the soft and hard selection concepts were developed by Levene (1953) and Dempster (1955), respectively (schematically described in Fig. 5.1). In Levene's model, population regulation occurs independently within each habitat (constant number of individuals produced by each habitat in each generation) (Fig. 5.1(a)). Selection coefficients neither directly depend on density nor on frequency, while the resulting mean fitness of each genotype does (see Fig. 5.1(a)) (Barton and Clark 1990). In Dempster's model, population regulation occurs before habitat colonisation (constant number of colonisers), leading to both density and frequency independent selection (despite the fact that regulation indeed occurs) (Fig 5.1(b)). Because in hard-regulated populations habitat productivity depends on the number of surviving colonisers, group selection may act in favour of genes coming from the most favourable habitats. It is worth noting that Levene's and Dempster's frameworks can be applied to the study of phenotypic and/or species diversity, with fairly different assumptions from those found in demographic approaches of the Lotka–Volterra form.

In Dempster's model (hard selection) the conditions for the maintenance of polymorphism are the same as those in a homogeneous environment (i.e. overdominance relative to the arithmetic mean). In other words, in

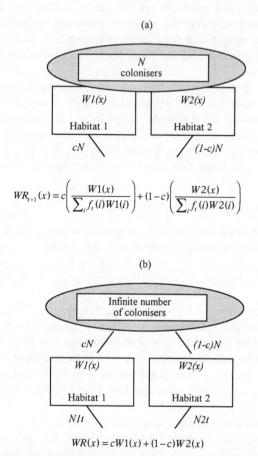

Fig. 5.1 Schematic representations of (a) Levene's and (b) Dempster's models. These models assume the random colonisation of each habitat and a differential selection affects each genotype (or species) in each habitat (e.g. $W1(x)$ and $W2(x)$ as selective coefficients). In each model, regulation maintains population size at a constant level N. In Levene's model this regulation occurs within each habitat (i.e. coupled with selection) so that the contribution of each habitat remains constant in each generation (e.g. c and $(1-c)$), independent of the number of surviving colonisers. Because these surviving colonisers must share the productivity of the habitat in which they settle, their resulting fitness (WR) becomes frequency- ($f_t(i)$) and density- ($\sum_i f_t(i)W1(i)$ or $\sum_i f_t(i)W2(i)$) dependent. In Dempster's model, regulation occurs outside of the habitat, so that the number of colonisers in each habitat remains constant at each generation (cN and $(1-c)N$). Consequently, the contribution of each habitat to the next generation will depend on the number of surviving colonisers found in it ($N1_t$ or $N2_t$). The resulting fitness of each genotype is thus frequency and density independent.

environments where resources vary in quality, if regulation is such that there is no resource limitation, then environmental heterogeneity has no effect on adaptive heterogeneity. In Levene's (soft selection) model the maintenance conditions are much broader (i.e. overdominance relative to the harmonic mean). In particular, for cases of haploids or two species, adaptive poly-morphism cannot be maintained in Dempster's model when one allele is dominant, whereas it can in Levene's (Maynard-Smith 1962). In the latter framework, adaptive diversity is maintained providing the existence of an inverse ranking of the harmonic and arithmetic means of relative fitness (Prout 1968) (see the appendix to this chapter). Interestingly, because habitat productivity is constant, soft selection cannot produce a genetic load whereas hard selection can.

The conditions for the maintenance of polymorphism in soft selection models are very stringent (Maynard-Smith 1962, 1966; Maynard-Smith and Hoekstra 1980). A stable adaptive polymorphism 'requires either that the selective advantages are large, or that they are nicely adjusted to the niche size' (Maynard-Smith 1966). Consequently, considerable effort has been made to modify the basic models in order to explain the maintenance of adaptive diversity, incorporating, for example, direct frequency dependence, environ-mental fragmentation (coarse-grained environment), habitat selection and deviations from random mating (see below for references).

Direct frequency dependence

When habitat productivity is a decreasing function of the number of indi-viduals present (Wilson and Turelli 1986), the conditions for the maintenance of adaptive polymorphism are considerably widened, and even allow for stable underdominance. This scenario of productivity is likely to occur in nature, for instance, for pathogens exploiting different host species. However, the degrees of density and frequency dependence required for this model to function appear to apply only for 'heavy consumers', like plant pests and epidemic pathogens (e.g. myxomatosis), and not in more general situations.

Habitat fragmentation

In this kind of model the population is subdivided into N subpopulations which are connected by a constant rate of migration (m) per generation. Here, hard selection must be re-defined, because as m decreases the different demes become increasingly independent, and the population cannot be fully regulated outside the different habitats, as in Dempster's model. Thus, within-deme regulation (i.e. density dependence) must intervene even in this hard selection framework. At one extreme, for m close to 0, hard and soft selection models behave in a similar way (Christiansen 1975). A very different result is observed under habitat selection (see next section). Moreover, because habitat productivity depends on the relative number of surviving colonisers, hard selection allows for demic selection to occur in a coarse-grained environment. As a consequence, in fragmented habitats, the conditions for protected polymorphism when selection is hard may be much less stringent (Walsh 1984).

When an adaptive polymorphism is selected in such situations, any mutation-reducing migration capabilities will be at an advantage (Balkau and Feldman 1973), a mechanism that can easily lead to speciation. Paradoxically, this mechanism rarely attracts attention in speciation theories (but see Felsenstein 1981). Attention is usually devoted to the effects of habitat selection in fine-grained environments.

Habitat selection

Two categories of habitat selection have been identified in the literature: environmentally and genetically induced.

Environmentally induced habitat selection was first introduced by Maynard-Smith (1966) and developed by Hoekstra et al. (1985). Here, only soft selection has been explored. In such models, females tend to lay their eggs in the types of demes where they themselves have developed. As more females carrying genes adapted to niche i lay their eggs in that niche, there is some degree of correlation between the habitat selected and the fitness of the offspring within it. Conditions for stable polymorphism are widened, but stay restricted to large selective advantages (Hoekstra et al. 1985) (Fig. 5.2).

If habitat selection is genetically induced, a correlation must exist between the quality of the selected habitat and the fitness expected within the habitat. Indeed, alleles for choosing the wrong habitat can never be maintained. This correlation can be obtained by two means: pleiotropy and epistasis.

If habitat selection is determined by pleiotropy, the same genes determine adaptation and habitat selection. Templeton and Rothman (1981), Rausher (1984), Garcia–Dorado (1986, 1987) and Hedrick (1990a) have demonstrated that the conditions required for the maintenance of polymorphism are considerably broadened if selection is soft. In certain cases, no selective advantage is necessary. However, no clear assumption is made concerning the origin and evolution of such habitat selection and the case of hard regulated populations has not been explored.

An alternative explanation can be found in de Meeûs et al. (1993). Consider two polymorphic loci, one coding for the adaptation to a particular habitat (one allele per habitat) and the other coding either for no habitat selection (one allele) or for habitat preference (a second allele). Assuming that the adaptive locus has an epistatic influence on the habitat preference allele, an individual carrying the habitat choice allele should prefer to settle on the most favourable habitat according to its genotype on the adaptive locus. Indeed, its within-habitat survival depends on the genotype displayed at the adaptive locus. de Meeûs et al. (1993) showed for both hard and soft regulation that there is an apparent conflict between the evolution of habitat specialisation and the maintenance of adaptive diversity. In addition, polymorphism can only be maintained if the population is soft regulated. If so, and if the habitat preference allele is initially fixed, polymorphism is always

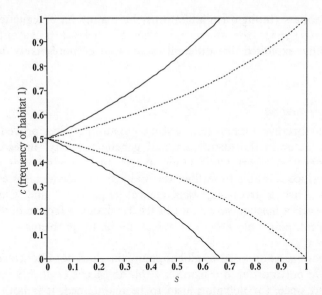

Fig. 5.2 Range of habitat sizes, c, for which polymorphism is stable as a function of the selection coefficient, s, for a one-locus, two-allele and two-habitat model (from Hoekstra *et al.* 1985). In habitat 1, with frequency c, fitnesses are 1, $1-ks$ and $1-s$ for genotypes A_1A_1, A_1A_2 and A_2A_2 respectively. In habitat 2, with frequency $(1-c)$ these fitness are $1-s$, $1-(1-k)s$ and 1 (in this example $k = 0.5$). Polymorphism is maintained between the two dotted lines without habitat selection and between the solid lines when the rate of habitat selection is 0.5.

maintained even for minimum efficiencies of habitat preference. However, habitat preference is always selected against if no polymorphism exists on the adaptive locus. In the absence of competition between adaptive alleles, refusing to exploit an empty habitat is always unfavourable. On the other hand, if the population is hard regulated, the evolution of habitat preference is easy, providing the existence of reasonably low dispersal costs. In such cases the choosy allele becomes fixed and does so with at most one allele adapted to just one habitat (monomorphic equilibria). It is worth noting here that the other habitat, although available, remains empty. As predicted by Futuyma and Moreno (1988), hard selection thus appears better at enhancing the evolution of habitat specialisation. Using a model that does not allow for adaptive divergence (one habitat less favourable or similar to the other), Rausher (1993) reached a similar conclusion.

Thus, soft selection remains the *sine qua non* condition allowing adaptive diversity. Adaptive diversity, in turn, is the *sine qua non* condition for the evolution of habitat specialisation under soft selection.

Nevertheless, the differences underlying the hard and soft selection models can prove useful for the interpretation of the patterns found in natural populations (de Meeûs *et al.* 1993). Specialised species not exploiting all the suitable habitats (though available) must be mainly hard regulated. On the other hand, species for which no available habitat is vacant and are sympatrically exploited by different entities, are more likely to be soft regulated.

Non-random mating

Negative assortative mating (e.g. self-incompatibility) has proved to be an important factor in the maintenance of genetic diversity. Indeed, for high levels of negative assortative mating, a rare allele is always at a selective advantage since it is able to fertilise the majority of individuals present in the population (strong frequency dependence) (e.g. Hartl and Clark 1988). However, such a mechanism cannot in itself enhance adaptive diversity from an ecological point of view and refers more to problems in molecular genetics.

Local mating, as opposed to global panmixia, does not alter the conditions for adaptive polymorphism (Strobeck 1974, de Meeûs et al. 1993). This may be expected since, for polymorphism to be maintained, it is both necessary and sufficient that any allele increases when rare (Strobeck 1974). A heterozygote for a very rare allele will mate with the homozygote for the common allele, irrespective of where it mates.

This might appear to contradict the conclusions of some authors (e.g. Hedrick 1990b) who have based their conclusions from Hoekstra et al. (1985). In this model, only females can habitat-select, and they tend to lay eggs in the habitat where they were born (Maynard-Smith 1966). Thus, in the case of global panmixia occurring outside the habitats in consideration, only the genes of females are involved in the process of habitat selection (genes of males can randomly come from any habitat). But when mating occurs within each habitat, the eggs carried by the females (which are habitat selectors) contain genes from males and females that have been submitted to the same kind of selection. Thus, when females lay their eggs, female and male genes are both involved in the homing process. Here, local mating widens the conditions for protected polymorphism because it leads more genes to select for the habitat. In this latter case, the situation actually corresponds to models involving a coarse-grained environment with limited migration (e.g. Christiansen 1975) (see above).

Moreover, deviation from panmixia can, at best, only be detrimental for the maintenance of polymorphism since it lowers overdominance effects (less heterozygotes) (e.g. Hartl and Clark 1989). At one extreme, totally selfing populations tend to fix homozygosity, and thus behave as haploids. In a varying environment, conditions for an adaptive polymorphism in haploids are just the same as those for diploids in the dominant case (Gliddon and Strobeck 1974). Interestingly, such findings make all these models appropriate for the theory of speciation and of maintenance species diversity (see below).

Sympatric speciation

The controversy surrounding sympatric mechanisms of speciation as an important generator of diversity (e.g. Mayr 1963, 1982, Maynard-Smith 1966, Bush 1975, Rosenzweig 1978, Futuyma and Mayer 1980, Felsenstein 1981, Tauber and Tauber 1989, Diehl and Bush 1989, Bush 1993) is intimately related to our arguments on habitat selection. The most convincing population genetics models leading to the sympatric evolution of two species must involve a previous stable polymorphism and/or the pre-existence of some kind of habitat selection (e.g. Maynard-Smith 1966, Rice 1984, 1987).

All the models leading to sympatric species diversification explicitly assume soft regulated populations. These same model approaches and results can be connected to studies of adaptive polymorphism. Indeed, a particular habitat favouring a particular allele against others (which are favoured elsewhere) can be viewed as the ecological niche of this variant. However, if adaptation depended on more than one locus, then local mating with habitat selection (or assortative mating) would certainly represent an advantage, because it would allow for the conservation of the fittest allelic combinations. In the case of global random mating, the production of less fit combinations at each generation would enhance the evolution of true assortative mating. This kind of process would appear very close to what Rosenzweig (1978) calls 'competitive speciation'. But a prerequisite for this phenomenon to occur is the existence of an adaptive polymorphism for each of the involved loci. Thus, the study of ecological niche diversification encounters the same problems as those described for genetic polymorphism.

Conclusion

Future study should focus on less caricatural assumptions than those defined for hard and soft selection. Alternatively, models with some hard selection (i.e. more favourable for the evolution of habitat preference) and some soft selection (where adaptive polymorphisms are more likely to be selected) could provide a more convincing theory of ecological diversification (de Meeûs *et al.* 1993).

Hard and soft selection assumptions explicitly refer to population regulation, that is demographic parameters that are hidden by the use of constant population sizes (within demes in soft selection, globally in hard selection). The analysis of these models with explicit demography should shed light on the role played by such parameters and provide much more powerful theories in terms of prediction and testability.

Finally, the different topics discussed here are likely to apply to many different kinds of organisms. Indeed, many if not most organisms experience sufficient amounts of habitat heterogeneity to display some kind of habitat selection. We believe that highly specialised organisms, such as symbionts (e.g. parasites and mutualists) provide the best illustrations of theoretical principles in evolution, such as adaptive polymorphism, specialisation and speciation.

Acknowledgements

The authors thank two anonymous referees for their comments which helped improve this paper.

References

Balkau, B. J. and Feldman, M. W. (1973). Selection for migration modification. *Genetics*, **74**, 171–174.

Barton, N. and Clark, A. (1990). Population structure and processes in evolution. In *Population biology: ecological and evolutionary viewpoints* (eds. K. Wöhrmann and S. K. Jain.), pp 115–73. Springer-Verlag, Berlin.

Bush, G. L. (1975). Sympatric speciation in phytophagous parasitic insects. In *Evolutionary strategies of parasitic insects and mites* (ed. P. W. Price), pp 187–204. Plenum Press, New York.

Bush, G. L. (1993). A reaffirmation of Santa Rosalia, or why are there so many kinds of small animals? In *Evolutionary patterns and processes* (eds. D. R. Lees and D. Edwards), pp 230–249, Academic Press, London.

Christiansen, F. B. (1975). Hard and soft selection in a subdivided population. *American Naturalist*, **109**, 11–16.

de Meeûs, T., Michalakis, Y., Renaud, F. and Olivieri, I. (1993). Polymorphism in heterogeneous environments, habitat selection and sympatric speciation: soft and hard selection models. *Evolutionary Ecology*, **7**, 175–198.

Dempster, E. R. (1955). Maintenance of genetic heterogeneity. *Cold Spring Harbour Symposia of Quantitative Biological Sciences*, **20**, 25–32.

Diehl, S. R. and Bush, G. L. (1989). The role of habitat preference in adaptation and speciation. In *Speciation and its consequences* (eds. D. Otte and J. A. Endler), pp 345–365. Sinauer Associates Inc., Sunderland, MA.

Felsenstein, J. (1976). The theoretical population genetics of variable selection and migration. *Annual Review of Genetics*, **10**, 253–280.

Felsenstein, J. (1981). Skepticism towards Santa Rosalia, or why are there so few kinds of animals? *Evolution*, **35**, 124–138.

Futuyma, D. J. and Mayer, G. C. (1980). Non-allopatric speciation in animals. *Systematic Zoology*, **29**, 254–271.

Futuyma, D. J. and Moreno, G. (1988). The evolution of ecological specialization. *Annual Review of Ecology and Systematics*, **19**, 207–233.

Garcia-Dorado, A. (1986). The effect of niche preference on polymorphism protection in a heterogeneous environment. *Evolution*, **40**, 936–945.

Garcia-Dorado, A. (1987). Polymorphism from environmental heterogeneity: some features of genetically induced niche preference. *Theoretical Population Biology*, **32**, 66–75.

Gliddon, C. and Strobeck, C. (1974). Necessary and sufficient conditions for multi-niche polymorphism in haploids. *American Naturalist*, **109**, 233–235.

Hartl, D. L. and Clark, A. G. (1989). *Principles of population genetics*. Sinauer Associates Inc., Sunderland, MA.

Hedrick, P. W. (1990a). Genotypic habitat selection: a new model and its application. *Heredity*, **65**, 145–149.

Hedrick, P. W. (1990b). Theoretical analysis of habitat selection and the maintenance of genetic variation. In *Ecological and evolutionary genetics of Drosophila* (eds. J. S. F. Baeker, W. T. Starmer and R. J. McIntyre), pp 209–227. Plenum Press, New York.

Hoekstra, R. F., Bijlsma, R. and Dolman, A. J. (1985). Polymorphism from environmental heterogeneity: models are only robust if the heterozygote is close in fitness to the favoured homozygote in each environment. *Genetical Research Cambridge*, **45**, 299–314.

Kimura, M. (1983). *The neutral theory of molecular evolution*. Cambridge University Press.

Latter, B. D. H. (1973). The island model of population differentiation: a general solution. *Genetics*, **73**, 147–157.

Levene, H. (1953). Genetic equilibrium when more than one ecological niche is available. *American Naturalist*, **87**, 331–333.

Lewontin, R. C. and Hubby, J. L. (1966). A molecular approach to the study of genetic heterozygosity in natural populations. II. Amount of variation and degree of heterozygosity in natural populations of *Drosophila pseudobscura*. *Genetics*, **54**, 595–609.

Maynard-Smith, J. (1962). Disruptive selection, polymorphism and sympatric speciation. *Nature*, **195**, 60–62.

Maynard-Smith, J. (1966). Sympatric speciation. *American Naturalist*, **100**, 637–649.

Maynard-Smith, J. and Hoekstra, R. F. (1980). Polymorphism in a varied environment: how robust are the models? *Genetical Research Cambridge*, **35**, 45–57.

Mayr, E. (1963). *Populations, species and evolution*. Harvard University Press, Cambridge, MA.

Mayr, E. (1982). Speciation and macroevolution. *Evolution*, **36**, 1119–32.

Prout, T. (1968). Sufficient conditions for multiple niche polymorphism. *American Naturalist*, **102**, 493–496.

Rausher, M. D. (1984). The evolution of habitat preference in subdivided populations. *Evolution*, **38**, 596–608.

Rausher, M. D. (1993). The evolution of habitat preference: avoidance and adaptation. In *Evolution of insect pests: patterns of variation* (eds. K. C. Kim and B. A. McPherson), pp 259–283, John Wiley and Sons Inc., New York.

Rice, W. R. (1984). Disruptive selection on habitat preference and the evolution of reproductive isolation: a simulation study. *Evolution*, **38**, 1251–1260.

Rice, W. R. (1987). Speciation via habitat specialization: the evolution of reproductive isolation as a correlated character. *Evolutionary Ecology*, **1**, 301–314.

Rosenzweig, M. L. (1978). Competitive speciation. *Biological Journal of the Linnean Society*, **10**, 275–289.

Strobeck, C. (1974). Sufficient conditions for polymorphism with N niches and M mating groups. *American Naturalist*, **108**, 152–156.

Tauber, C. A. and Tauber, M. J. (1989). Sympatric speciation in insects: perception and perspective. In *Speciation and its consequences* (eds. D. Otte and J. A. Endler), pp 307–44. Sinauer Associates Inc., Sunderland, MA.

Templeton, A. R. and Rothman, E. D. (1981). Evolution in fine grained environments. II. Habitat selection as a homeostatic mechanism. *Theoretical Population Biology*, **19**, 326–340.

Wallace, B. (1968). Polymorphism, population size, and genetic load. In *Population biology and evolution* (ed. R. C. Lewontin), pp 87–108. Syracuse University Press, NY.

Wallace, B. (1975). Hard and soft selection revisited. *Evolution*, **29**, 465–73.

Walsh, J. B. (1984). Hard lessons for soft selection. *American Naturalist*, **124**, 518–526.

Wilson, D. S. and Turelli, M. (1986). Stable underdominance and the evolutionary invasion of empty niches. *American Naturalist*, **127**, 835–850

Appendix

Arithmetic and harmonic means of relative fitness

Suppose we have two habitats with productivity c and $(1-c)$ respectively. At one polymorphic locus (A/a) the within-habitat fitnesses are $W_1(A)$ and $W_2(A)$ for AA and Aa genotypes in habitat 1 and 2, respectively, and $W_1(a)$ and $W_2(a)$ for the aa genotype. Working on fitness relative to one genotype (e.g. AA) we have:

$$WR_i(A) = \frac{W_i(A)}{W_i(A)} = 1 \text{ and } WR_i(a) = \frac{W_i(a)}{W_i(A)}$$

where i is the habitat type. The arithmetic means are thus:

$$ARI(A) = 1 \text{ and } ARI(a) = cWR_1(a) + (1-c)WR_2(a)$$

for A and a respectively.

The harmonic means are:

$$HAR(A) = 1 \text{ and } HAR(a) = \frac{1}{c\frac{1}{WR_1(a)} + (1-c)\frac{1}{WR_2(a)}}$$

Note that the harmonic mean is always less than the arithmetic mean (e.g. Hartl and Clark 1989).

Part 2 Ecology: from populations to communities to ecosystems

Overview

Jean Clobert, Chris Gliddon, Bradford A. Hawkins and Michael E. Hochberg

With the increasing impact of humanity on the ecological processes of this planet, we find ourselves in a position, much like Noah, of attempting to save some semblance of the existing biodiversity of the earth. Human activities are modifying a large proportion of Earth's ecosystems by both reducing their geographical size and altering their functioning. Noah's solution of saving one pair of each species, albeit pragmatic, is inconsistent with current theories of population biology and, even if possible today, would simply be expected to fail as a global conservation strategy. Given the limited resources we have to maintain biodiversity, how should we choose which species or assemblages to conserve and, having made this first decision, how should we put our conservation aims into practice?

To begin to derive strategies aimed at conserving species and communities of species we must first understand how diversity is manifested at the ecosystem level. The problem is that the utter complexity of ecosystems impedes our progress in understanding. It is evident that we must break down the problem into smaller, more tractable parts, while remembering that the sum of the parts do not necessarily equal the whole. Part 1 of the book dealt with evolutionary patterns and processes. This part simplifies approaches to understanding the ecological processes governing the diversity of communities and ecosystems.

Drosophila are not simply species which keep geneticists busy. Their diversity and their large geographical distributions make them good model systems for studying processes acting on communities. Resource partitioning is one of the dominant explanations for the coexistence of species at local spatial scales; simply put, species must differ in the resources they use and/or in the way they use them. Indeed, Shorrocks (Chapter 6) presents evidence that species of *Drosophila* show some degree of specialisation in the type of decaying fruit used and in the preferred degree of microbial infection displayed by the fruit. However, the degree to which species niches may overlap in these systems without entailing competitive exclusion is unknown and mechanisms reducing competition such as spatial heterogeneity may equally well explain the observed patterns. Shorrocks uses a classification of diversity theories made by Cornell and Lawton to evaluate their respective roles in structuring *Drosophila* guilds. Models of local coexistence can be classified according to three axes: niche heterogeneity, spatial heterogeneity and the degree to which the niche space is used. Resource partitioning implies considerable niche heterogeneity, strong density-dependent species interactions, but no spatial heterogeneity. A competing hypothesis, the aggregation model, is based on strong spatial heterogeneity, strong intra-specific density dependence, but no resource heterogeneity. To discriminate between these two possible mechanisms, Shorrocks uses data sets of egg number of different *Drosophila* species deposited on a variety of breeding sites, patchily distributed and relatively ephemeral. Distribution patterns are then examined using Ives' recently derived index of the relative strengths of intraspecific and interspecific aggregation. The frequency of intraspecific aggregation is found to overwhelm that of interspecific clumping, suggesting that this type of exploitation of a shared resource base is at least partially responsible for species coexistence in drosophilid guilds.

There are many reasons why individuals of a same species aggregate, such as a consequence of patchy resource distributions or individuals using one another as cues for assessing the suitability of the resource. But individuals can also escape consumption though heterogeneity in their risk of being found and attacked by consumers. This is the central theme of Jones' and Hassell's chapter (Chapter 7), which focuses on the mechanisms of insect parasitoid–host coexistence and what is implied for the population dynamics. There are many ways of modelling host–multi-parasitoid systems depending on the sequence of attack by the two parasitoids, the outcome of the interspecific competition within the host, the respective number of host species used, and the capacity of each parasitoid to explore new patches. Jones and Hassell consider three different key points in parasitoid coexistence. First, they show how coexistence in systems of two monophagous parasitoids may largely be dependent on the trade-off between the time of attack by the two parasitoids (or their competitive ability within the host, if the order of arrival is the same) and their respective searching efficiencies. In their basic model, the introduction of a second species into a system containing but a single parasitoid species must lower host density. However, when stage-structured models are considered (parasitoids attacking different host stages, for example eggs and larvae), this conclusion does not hold in all cases. When some heterogeneity in the time that host individuals spend in each stage is introduced, coexistence is made easier. Secondly, Jones and Hassell examine models where one parasitoid is a generalist (its attack rate will depend in part on current host abundance), and the other is a specialist. The introduction of a specialist into an existing host–generalist interaction can lead to stable three-species systems, whereas the primitive two-species system was unstable and therefore the host was liable to go extinct. In brief, the three-species system shows a range of dynamics which is not revealed by either of the two species couplings in isolation. Thirdly, by introducing explicit spatial heterogeneity the conclusions derived above are further supported—the introduction of a second parasitoid in one host–one parasitoid system not only increases the range of conditions for coexistence (thus promoting bio-diversity), but also increases the range of possible types of population dynamics. The extent to which this wide range of population dynamics can be found in nature is, however, unknown.

Indeed, hosts of insect parasitoids also have predators and competitors, and they also feed on other living species. In their introduction, Jones and Hassell remark that most insects are attacked by only a few parasitoid species, suggesting that some constraint exists on the number of interactions with other species that a given animal can have. By representing all the species in an ecosystem in terms of feeding relationships Warren (Chapter 8) investigates how such constraints act on food web assembly. Food webs show some regularities in chain length, in the proportion of basal, intermediate and top-level species, and the number of feeding links a species has with other species with respect to web size. Simple models based on simple assumptions do well in predicting such regularities, except for connectance. There are several hypotheses to explain patterns observed in connectance and in the evolution of the number of feeding links with web size. Biological constraints such as feeding specialisation and the ratio of predator to non-predator species are just two of these. Warren defines interactions between species in terms of feeding niche overlap with the addition of a few other hypotheses (feeding always occurs on species of smaller size; predicted proportions of basal, non-basal and predator species must fit the observed ones). He finds that the model does surprisingly well in predicting food-chain length, linkage density and the number of links per species. His emphasis is therefore clearly on understanding the processes responsible for structuring food webs and not simply on deriving indicators for food web function. This approach suggests that biological constraints probably impose strong pressures on web structure and, most likely, on their temporal and spatial stability.

In a comprehensive review of the limnological literature, Lacroix, Lescher-Moutoué and Pourriot (Chapter 9) follow Warren in considering how trophic interactions influence food web structure. They present three decisive steps which have led to our present understanding:

(1) the emphasis of the role of abiotic factors and nutrient availability as contributors to the bottom-up control of freshwater communities;

(2) the theory of niche partitioning, and the roles of competition and pre-dation in freshwater community structure; and in the last decade,

(3) a more synthetic view where both top-down and bottom-up forces act simultaneously.

However, which of these latter two actually dominate particular interactions is highly both model- and experiment-dependent, especially with regards to assumptions on the type of predation and the use of nutrients. These contrasting findings are explainable in that most of the models and experiments considered only restricted parts of the ecosystem, and neglected complex indirect interactions such as cascading effects, facilitations, and bottlenecks. Using their own data, Lacroix *et al.* investigate the relative importance of bottom-up and top-down processes, concluding that strong hierarchical effects in body size appear to be pivotal in structuring freshwater food web assembly. In reporting on a variety of the interactions among trophic levels, they present a detailed yet clear picture of how freshwater food webs function.

It is appropriate in some respects to complete this section with the processes that 'tie the knot' in ecosystems, that is, the decomposition and re-entry of nutrients into communities. Grover and Loreau (Chapter 10) begin by pointing out that in neglecting constraints such as matter and energy balances, decomposition processes, and nutrient cycling we may be overestimating the frequency of top-down control (i.e. by predators and parasites) as compared to bottom-up driven processes (nutrient and energy limitation). Indeed, the first models employed to take into account nutrient cycling showed that its inclusion increased the probability that food webs were locally stable at a steady state. In focusing mainly on consumers, Grover and Loreau envisage several ways of modelling matter and energy transfer between the different trophic levels. They find that an increase in consumption rate results in an increase of the nutrient pool, an increase in consumer biomass, and a decrease in producer biomass (even when predators are added). Consumers seem to play, at least theoretically, a more important role than decomposers in maximising productivity. However, consumption rate will depend on the response of producers to grazing which, in turn, will depend on their efficiency to use nutrients, particularly in nutrient-limited ecosystems. To study this, functional groups are divided into categories which may represent species, genotypes or clones. Grover and Loreau present the case of the competition between two producers and contrast their predictions with those made by other similar models. In general, consumers promote coexistence between producer species when the latter have different positions on the trade-off between resistance to herbivory and efficacity of nutrient exploitation. Selection will favour species which persist with the smallest available nutrient pool or species which sustain the highest herbivore population size, depending on the size of the nutrient supply. However, the authors recognise that most of these results are based on very simple scenarios, and that data are badly needed.

So, can we describe ecosystems through a restricted number of key elements (species or nutrients) and use the dynamics of the latter as predictors of ecosystem evolution? Is ecosystem function strongly homeostatic and constrained by simple general rules? Are there generic answers to system stability? Are species interchangeable for a given functional group, and how may this be related to biodiversity? Some preliminary answers to these questions are provided in the chapters of this section, but we have some way to go before a comprehensive understanding of community and ecosystem structure and function will be within our grasp. By comparing different situations and combining features of these different approaches, a first corpus of theory will be developed for use in understanding large-scale diversity patterns and their conservation.

6

Local diversity: a problem with too many solutions

Bryan Shorrocks

Introduction

Population ecologists have increasingly become interested in multi-species ecology. This is clearly shown in an analysis of papers published in the *Journal of Animal Ecology* (JAE), between 1932 (the first volume) and 1992 (Shorrocks 1993). In the early days, population ecologists were predominantly interested in understanding the dynamics of single species systems. There were papers in JAE on multispecies communities, but they were simply descriptive. Following the golden age of the 'bottle experiment' (Kareiva 1989) when theoretical ideas about two-species interactions were rigorously scrutinised in the laboratory (Gause 1934; Crombie 1946; Park 1948; Utida 1953, 1957) interest in the dynamics of multi-species systems grew. This dialogue between theory and experiment laid the foundations for what was to become the dominant explanation for local biodiversity (MacArthur 1972)—resource partitioning. However, there are many levels of species diversity, and there are other possible mechanisms which may help to explain the local coexistence of species. In this chapter I will examine species diversity in a model group of insects—*Drosophila*.

Levels of diversity in *Drosophila*

Drosophila are primarily consumers of the yeasts and bacteria associated with the fermentation and decay of plant material. Without the living bacteria and yeasts these substrates are probably not sufficient for *Drosophila* growth and survival (Kearney 1979). Within a local area, groups of *Drosophila* (guilds) appear to coexist within relatively few types of larval substrate. Traditionally, these have been grouped into five main categories (Carson 1971; Kimura *et al.* 1977; Begon and Shorrocks 1978; Shorrocks 1981). These are flowers (restricted to the tropics), fermenting fruits, sap fluxes, fungi, and decaying plant material such as leaves, stems and roots.

This separation of *Drosophila* into larval feeding guilds is well illustrated by the emergence records reported by Kimura *et al.* (1977). The area surveyed by these Japanese workers is near Sapporo, on the northern island of Hokkaido, and the data represent five years of collecting effort. What is apparent is the very high degree of specificity actually showed by the 35 species of *Drosophila*. Twenty-one species (60%) only emerged from a single substrate type. A further five species had over 90% of their emergence from a single substrate, and even a relatively unspecialised species like *D. testacea* showed 73% emergence from its preferred fungal resource.

Shorrocks and Rosewell (1986, 1987) used this type of information to construct a distribution of guild sizes for drosophilids, which is shown in Fig. 6.1. They estimated guild size twice. The 'simple' estimate was obtained by just counting the species recorded locally from a particular resource type. The 'complex' estimate was obtained by examining, where possible, the detailed breeding records and determining if the simple groups could be split or reduced because species had never actually been recorded sympatrically. Also, records for species represented by single individuals only were removed. Both distributions are in fact rather similar, with modal guild size about seven in both cases. Estimation of the guild size for the different substrates used by the larvae fails to suggest any difference in average size. Mean guild size ($\pm$ SE) for the 'complex' data gives, flowers (5.6 $\pm$ 0.9), fungi (6.3 $\pm$ 0.7), sap (6.7 $\pm$ 0.9), fruit (7.5 $\pm$ 1.2), and rotten vegetation, mainly leaves (8.5 $\pm$ 1.4). It is not possible on the basis of this data to suggest any difference in guild size between temperate, tropical or Hawaiian drosophilids ('simple' guilds, $\chi^2 = 1.27$, df = 4, $P = 0.867$; 'complex' guilds, $\chi^2 = 6.22$, df = 4, $P = 0.182$). However, there are more species of *Drosophila* in Hawaii (366) than in Europe (49) for example, and we will return to this difference later. What other patterns of diversity are seen in *Drosophila*? Below I discuss very briefly some examples from our studies at Leeds.

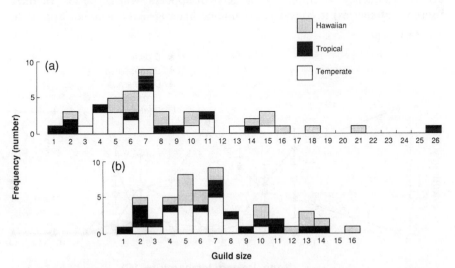

Fig. 6.1 Frequency distribution of natural guild sizes in drosophilids using (a) 'simple' and (b) 'complex' classifications (see text for explanation of these terms) (from Shorrocks and Rosewell 1986).

Clinal diversity in the European fungal-breeding guild.

Figure 6.2 shows the latitudinal change in frequency seen in the five most
frequent species of the eight-species fungal-breeding group of *Drosophila* in
western Europe (Shorrocks 1977). Three of the five species appear to replace
each other in a regular way from southern to northern Europe. *Drosophila
cameraria* has an Iberian distribution with a northern Atlantic extension into
France and the British Isles. *Drosophila transversa* is the dominant fungal
species in northern Scandinavia and *D. phalerata* lies between the two, with a
central European distribution. In parentheses it might be noted that the two
species *D. phalerata* and *D. transversa* are examples of quite different patterns
of 'central' and 'marginal' distribution. While the ecological margin for *D.
phalerata* is small compared to the central area, marginal populations of *D.
transversa* extend across most of Europe. Like many other species therefore,
Drosophila show latitudinal changes in β diversity (Harrison *et al.* 1992;
Whittaker 1960) and this may be directly related to temperature. In the
northern island of Japan (Hokkaido) all eight of the European fungivorous
Drosophila are found, with the exception of *D. phalerata* and *D. limbata*.
Drosophila transversa and *D. testacea* are the commonest species (Makino
and Takeharu 1951; Makino *et al.* 1956, 1958; Takada 1956; Kaneko and
Shima 1960), although this island is equivalent in latitude to the Iberian
Peninsula. However, the mean January temperature is comparable to Switzerland and southern Finland (-3 to -7°C). In Japan, therefore, these two
species occupy a similar climatic zone to their counterparts in western
Europe. Finally, it should be noted that this cline in species diversity (looking
at the frequency component), largely disappears when species richness
(number of species) is considered. Climatic heterogeneity does not appear to

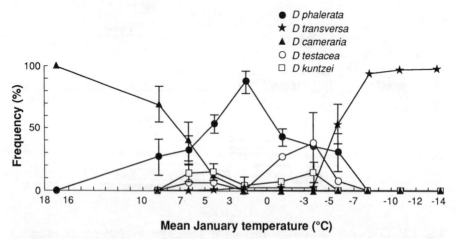

Fig. 6.2 Mean frequency ($\pm$ SE) of the five abundant fungal-breeding *Drosophila* in
Europe, plotted against mean January temperature ($^\circ$C) (from Shorrocks 1977).

affect species richness (apart from northern Scandinavia), and over most of the cline, the β component is small.

Local spatial–temporal diversity in the fungal-breeding guild

Shorrocks and Charlesworth (1980) collected fungi from a number of British woodlands to the north of Leeds. Five species of fungivorous *Drosophila* (*D. phalerata*, *D. transversa*, *D. cameraria*, *D. confusa*, and the generalist *D. subobscura*) were reared from 125 species of fungi over a three-year period (1974–76). One method of presenting the resulting diversity is shown in Fig. 6.3. Diversity was measured using the Shannon and Weaver (1949) formula,

$$H' = -\sum_{i-1}^{n} p_i \ln p_i$$

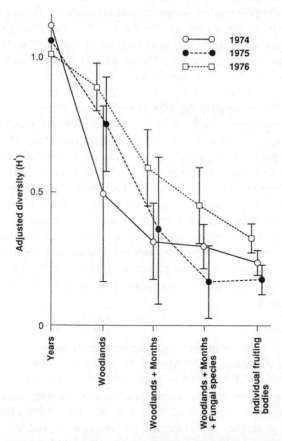

Fig. 6.3 Spatio-temporal hierarchy of local diversity in fungal-breeding *Drosophila* in northern England (from Shorrocks and Charlesworth 1980). Values of *H'* are adjusted to account for the smaller size of samples to the right of the figure.

where p_i is the frequency of species i. The drosophilid diversity emerging from fungi has been arranged into a nested hierarchy of spatio-temporal collections, with the largest collection, for the whole year, to the left and the smallest collections, individual fungal bodies, to the right. There appears to be a cumulative component to local diversity, rather similar to what has been termed β diversity when considering regional (or γ) diversity. This 'fractal' nature of diversity leads to some confusion in the literature, since both α and γ diversity can be partitioned. Interestingly, as with the European cline above, this β-like component largely disappears when 'species richness' is considered. The local spatio-temporal heterogeneity affects the frequencies of species but not their presence and absence.

Resource selection within the domestic Drosophila

One of the most studied groups of *Drosophila* are those frequenting 'domestic habitats' such as markets, breweries and vineyards (Atkinson 1979). Atkinson and Shorrocks (1977) studied seven species (*D. busckii*, *D. funebris*, *D. hydei*, *D. immigrans*, *D. melanogaster*, *D. simulans*, and *D. subobscura*) in a fruit and vegetable market 4 km southeast of the centre of Leeds. Breeding site species were compared to emerging *Drosophila* species using Raabe's percentage similarity (Southwood 1966) and showed clear differences in species composition between the different fruits and vegetables. In fact, two main groups of breeding sites were revealed by this analysis, a group of 'vegetables' and a group containing mainly fruits. Within the second group, some taxonomically related breeding sites were found to be closely associated. The three *Prunus* species, plum, peach, and apricot are very similar in their *Drosophila* fauna as are the three *Citrus* species, lemon, orange, and grapefruit. The flies may well be differentiating sites on the basis of their microflora (yeasts or bacteria) determined by the pH of the site. Fruits have a low pH that favours the growth of yeasts rather than bacteria that are the common spoilage agents of vegetables (Jay 1970). Pears are the only fruits that commonly undergo bacterial spoilage, and intriguingly they are found with the vegetables in the analysis. Heterogeneity between resource patches, of the type often linked with 'resource partitioning' as an explanation for species diversity, is clearly seen in *Drosophila*.

Fruit and Penicillium

Citrus fruits are unusual in that their decay, usually associated with infections of *Penicillium digitatum* Saccardo or *P. italicum* Wehmer (Fawcett 1936; Raper and Thom 1949), is rather predictable. In the market collections discussed in the previous section, three species of citrus were found (grapefruit, *C. paradisi*; lemon, *C. limon*; and orange, *C. sinensis*) from which five species of *Drosophila* emerged. Two species (*D. immigrans* and *D. melanogaster*) emerged in greater numbers from citrus compared to non-citrus fruits, while three species (*D. simulans*, *D. hydei*, and *D. subobscura*) emerged in greater numbers from non-citrus fruits (Atkinson 1981). The mean numbers of *D.*

immigrans and *D. melanogaster* emerging from citrus fruits with different proportions of the surface initially covered with *Penicillium* conidia are significantly different ($\chi_3^2 = 858.7$, $P < 0.001$), with numbers of *D. immigrans* increasing and numbers of *D. melanogaster* decreasing, with levels of infection. In a laboratory experiment Atkinson (1981) placed 10 larvae of either *D. immigrans* or *D. melanogaster* onto slices of lemon which either had or had not been inoculated with *Penicillium*. In a subsequent analysis of variance on larval survival, there was a significant effect of treatment (*Penicillium* or not; $F = 1072.6$, $P < 0.05$) and a significant interaction between *Penicillium* treatment and species ($F = 2620.7$, $P < 0.001$), suggesting that again the two species of *Drosophila* showed a different response to the microbial microhabitat. Resource partitioning on a fine scale?

A problem with too many solutions

Population ecology and population genetics share at least two problems: they both have many theoretical explanations that *could* explain their respective diversities, and the proponents of one explanation tend to dismiss the possibility that the proponents of another model have anything sensible to say on the matter. This latter problem has been called 'cognitive particularism' by Travis and Collins (1991) and the 'lunar green cheese' alternative by Turner (1992).

Many of these explanations involve some type of environmental heterogeneity—either present in the environment before the organism arrives, for example resource partitioning (MacArthur 1972) or imposed by the organism when it arrives, such as the aggregation model and the priority model (Shorrocks 1990). Many explanations are also similar in both ecology and genetics; examples include resource partitioning in ecology and multiple niche polymorphism (Hoekstra *et al.* 1985) in genetics, and density/frequency dependence in both (Begon 1992; Clarke and Beaumont 1992). Not all the mechanisms are mutually exclusive.

One attempt to classify the many explanations for diversity in ecology is that of Cornell and Lawton (1992), and their scheme is reproduced in Fig. 6.4. The explanations chosen are not a complete list but are representative of the range of existing models. There are three dimensions: niche heterogeneity, spatial heterogeneity, and the fullness of the niche space (saturation level). Traditional explanations such as resource partitioning (MacArthur 1972) involve considerable niche heterogeneity, no spatial heterogeneity and assume that interaction between species are intense and density dependent. They are therefore to the top/right of Fig. 6.4. Alternatively, models that invoke spatial heterogeneity, such as the aggregation model (Atkinson and Shorrocks 1981, Hanski 1981), are interactive with strong density dependence, require no resource partitioning, and are to the top/left of Fig. 6.4. All these mechanisms *can* theoretically maintain species diversity, and there is considerable evidence, from laboratory experiments, that many of them *could*

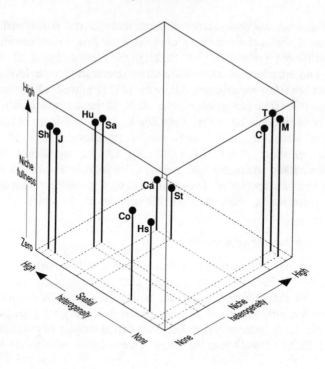

C = Chesson & Huntley 1989   Ca = Caswell 1976
Co = Connell 1978   Hs = Huston 1979
Hu = Hubbell & Foster 1986   J = Janzen 1970, Armstrong 1989
M = MacArthur 1972   T = Tilman 1986
Sa = Sale 1977   Sh = Shorrocks & Rosewell 1986
St = Strong, Lawton & Southwood 1984

Fig. 6.4 A three-way classification of community models in terms of niche heterogeneity, spatial heterogeneity, and the fullness of the niche space (saturation) (modified from Cornell and Lawton 1992).

maintain species diversity. There is some evidence, from field experiments or field observations, that these mechanisms *might* maintain species diversity. However, there is little feel for how the theoretical (and field) possibilities map on to the real world. All the proposed explanations of Fig. 6.4 can explain diversity, but how much of the variance do they each account for? In other words, does the real world span the whole box or simply occupy one corner?

I cannot answer this question for all the mechanisms in Fig. 6.4, but I will try to compare the top/right and top/left corners using *Drosophila*. Models that occupy these opposing corners are resource partitioning and aggregation, and in the remainder of this section I will briefly outline their respective positions. The question of whether all *Drosophila* communities are interactive and saturated (the third dimension in Fig. 6.4) cannot be answered at the moment, although there is evidence of competition in the field (Atkinson, 1979; Sevenster 1992) and invasion of alien species (*D. subobscura*) eliminating native Chilean species (Brncic and Budnik 1987), suggesting that they are towards the saturated (top) end of this dimension. However, Sevenster (1992) reports that an Asian species, *D. malerkotliana*, has recently colonised the neotropics with amazing success, despite its large resource overlaps with many resident species.

Resource partitioning

Resource partitioning has been the dominant explanation of local species diversity for nearly 60 years (Gause 1934; MacArthur 1972; Cody and Diamond 1975; Hutchinson 1978; Arthur 1987), and I do not intend to give a detailed account of the theory here. In simple descriptive terms it suggests that the reason for species coexistence (local diversity) is the partitioning of available resources, so that heterospecific individuals meet less frequently than conspecific individuals. The use of resource utilisation functions and measures of niche breadth and niche overlap have attempted to quantify this explanation (May and MacArthur 1972; Arthur 1987; Sevenster 1992). One difficulty with the theory is that the degree of partitioning required for coexistence (limiting similarity) is not known, although there have been attempts to quantify this (MacArthur and Levins 1967; Abrams 1975).

As we have seen in the previous section, there are many examples of species distributions and resource associations that could be examples of resource partitioning in *Drosophila*. The separation of species into guilds, based upon the type of substrate (fruit, fungi, decaying vegetation, sap and flowers) used by the larvae, is a clear example. Correlated with this partitioning of resources are morphological differences in larval mouth-parts and female ovipositors (Okada 1963) and nutritional differences on the associated yeasts (Kearney 1982). Similarly, the examples of work carried out at Leeds show what could be resource partitioning in *Drosophila*, on a spatio-temporal scale that ranges from woodlands/months to fruits with different amounts of *Penicillium*. Clearly, interspecific aggregation (resource partitioning) could promote species diversity in drosophilids, but does it?

Aggregation

As we have seen, many insects, including *Drosophila*, exploit patchy resources, consisting of small, separate units and which are ephemeral in the sense that they persist for only one or two generations. For *Drosophila,* such resources include fruit (Atkinson 1981, 1985; Sevenster 1992), fungi (Hackman and Meinander 1979; Shorrocks and Charlesworth 1980; Grimaldi and Jaenike 1984), sap-flows (Toda 1984), decaying leaves (Toda *et al.* 1984) and flowers (Pipkin *et al.* 1966), but for other insects may also include dung, carrion, seeds, dead wood, and small bodies of water held in parts of terrestrial plants (phytotelmata) (Shorrocks 1990). This general view of insect ecology inspired the 'aggregation model' of competition (Shorrocks *et al.*

1979, 1984; Atkinson and Shorrocks 1981) that allows a competitively inferior species to survive in probability refuges. These are patches of resource with no or a few superior competitors, that arise because the competing stages (usually larvae) have an aggregated distribution across the patches. An independent model proposed by Hanski (1981, 1983) uses different 'mechanics' to incorporate spatial variance into the environment and similarly promotes coexistence. These probability refuges are a permanent feature of such systems because the patches are ephemeral and aggregation increases mean crowding (Lloyd 1967). Mean crowding (m^*) is the mean density experienced by randomly chosen individuals and is equal to $m + V/m - 1$, where m and V are the mean and variance of the distribution over patches. With a random distribution $V = m$ and $m^* = m$, but with an aggregated distribution $V \gg m$ and $m^* \gg m$. Thus global population density is limited by strong intraspecific competition in patches with high local density while low-density patches still exist. Coexistence is promoted because aggregation of the superior species increases its intraspecific competition and reduces interspecific competition.

In the aggregation model, the eggs of both insect species are independently distributed over the patches according to a negative binomial distribution, which has an exponent, k, inversely related to the degree of intraspecific aggregation. The use of the negative binomial and the assumption of independence have been justified for drosophilids (Rosewell et al. 1990; Shorrocks et al. 1990). In the first version of the model (Atkinson and Shorrocks 1981) the parameter k was constant and independent of density. This is not valid for real populations, but relaxing this assumption does not prevent coexistence (Rosewell et al. 1990). Within each patch, competition is modelled by the equation of Hassell and Comins (1976):

$$N_i(t + 1) = \lambda_i N(t)[1 + a_i(N_i(t) + \alpha_{ij}N_j(t))]^{-bi} \qquad i \neq j$$

where $N_i(t)$ and $N_j(t)$ are the numbers of each species in a patch at time t, λ_i is the net reproductive rate, α_{ij} is a competition coefficient and a_i and b_i are constants. The parameter a_i (equal to $(\lambda_i^{1/bi} - 1)/N_i^*$, where N^* is the equilibrium density or 'carrying capacity' of a patch) is related to the population size at which density dependence starts to act and b_i describes the type of competition. At $b = 1$, competition is contest with density dependence exactly compensating. With increasing values of b, competition becomes increasingly scramble with density dependence being overcompensating (Nicholson 1954, Hassell 1975). This type of difference equation has been shown to model very well insect competition in general (Stubbs 1977, Bellows 1981) and drosophilid competition in particular (Gilpin and Ayala 1973). The predictions of the aggregation model are shown in Fig. 6.5 for $\lambda = 5$, $b = 1$ and $N^* = 10$, values that are quite representative of drosophilids (Shorrocks and Rosewell 1987). Also shown are the distributions of α and k collected by Shorrocks and Rosewell (1987). Clearly, intraspecific aggregation could promote species diversity in drosophilids, but does it?

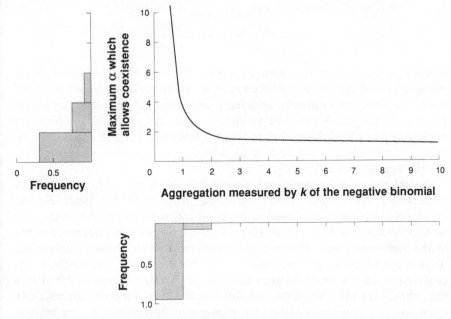

Fig. 6.5 Relationship between aggregation (k) and the maximum value for the competition coefficient (α) of the 'superior' species which allows coexistence in the aggregation model. Also shown are the distributions of α and k values from *Drosophila* laboratory and field experiments.

Intraspecific and interspecific aggregation

Following on from the ideas of Lloyd (1967) about mean crowding, Ives (1988, 1991) has suggested a measure of intraspecific aggregation, J, that is simply the proportionate increase in the number of conspecific competitors experienced by a random individual relative to a random distribution:

$$J_1 = \frac{\sum_{i=1}^{p}\dfrac{n_{1i}(n_{1i}-1)}{m_1 P} - m_1}{m_1} = \frac{\dfrac{V_1}{m_1}-1}{m_1}$$

where n_{1i}, m_1 and V_1 are the numbers in patch i, mean numbers, and the variance in numbers of species 1. Note that the index J is equal to $1/k$ of the negative binomial mentioned earlier. A value of $J = 0$ ($k = \infty$) indicates that individuals are randomly distributed, while a value of $J = 0.5$ ($k = 2$) indicates a 50% increase in the number of conspecifics expected in a patch compared to the random situation, that is an increase in mean crowding of 50%.

To measure interspecific aggregation, Ives (1988, 1991) derived a similar index, C, which measures the proportionate increase in the number of heterospecific competitors relative to a random association:

$$C_{12} = \frac{\displaystyle\sum_{i=1}^{p} \frac{n_{1i}n_{2i}}{m_1 P} - m_2}{m_2} = \frac{Cov_{12}}{m_1 m_2}$$

where Cov is the covariance between a pair of species, the subscripts indicate the species and the other symbols are as in the formula for J. When $C = 0$ the two species are randomly associated with respect to one another across patches. When $C > 0$ they are positively associated, when $C < 0$ they are negatively associated. Note that for each species pair there are two different values of intraspecific aggregation (J_1 and J_2) but only one value of inter-specific aggregation ($C_{12} = C_{21}$).

Several people have previously commented upon the relative strength of intraspecific aggregation (based upon the distribution parameter k) and interspecific aggregation (based upon statistical association) (Worthen and McGuire 1988, Shorrocks et al. 1990). However, these two measures are not in the same units and therefore not comparable when trying to assess the importance of intra- and interspecific mechanisms in their contribution to contact (= competition) between individuals. The significance of J and C is that they measure the same quantity within and between species. The reduction in competition caused by intraspecific aggregation or the relative strength of intraspecific versus interspecific aggregation, in a pair of species (1 and 2), can therefore be assessed with the quantity:

$$A_{12} = \frac{(J_1 + 1)(J_2 + 1)}{(C_{12} + 1)^2}$$

following Ives (1991), Jaenike and James (1991) and Sevenster (1992).

Only four drosophilid data sets are available to be analysed using this method. Unfortunately, although breeding sites have been extensively collected all over the world, records for individual patches are usually not available.

The first data-set is from Shorrocks and Charlesworth (1980) (see Fig. 6.3), a study of fungal-breeding species in woodlands to the north of Leeds, in the UK. Only the 1974 data have been analysed, based on fungi collected from four woodlands, from June to September, and involved a total of 168 patches and 4 Drosophila species. Values of A_{12} have been calculated for all pairs of species within each woodland and the distribution of values are shown in Fig. 6.6(a).

The second data-set is from Atkinson (1985) for diptera (including five Drosophila) breeding on fallen fruit, collected from subtropical rainforest sites in eastern Australia. Collections were made from three sites (one site collected twice in separate years) and the four collections involved a total of 289 patches and 10 species. Values of A_{12} have again been calculated for all pairs of species within each collection and the distribution of values is shown in Fig. 6.6(b).

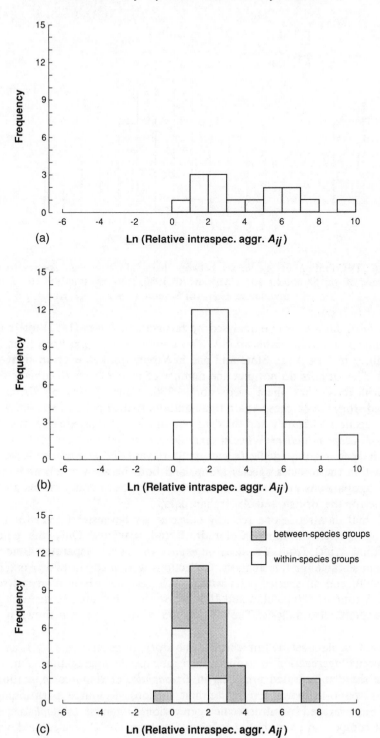

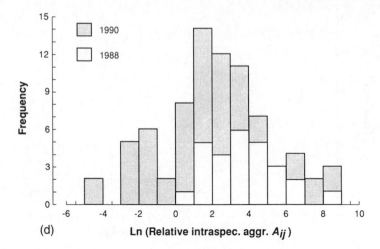

(d)

Ln (Relative intraspec. aggr. A_{ij})

Fig. 6.6 Distribution of A_{ij} values between pairs of *Drosophila* species for the collections of: (a) Shorrocks and Charlesworth 1980; (b) Atkinson 1985; (c) Jaenike and James 1991; (d) Sevenster 1992.

The third data-set is one analysed by Jaenike and James (1991) for fungal-breeding species in the eastern USA. This comprised six collections from four sites (three in New York State and one in Virginia) taken over an eight-year period. The authors do not give the number of patches, but I estimate, from statistical tests, that there were about 136. Four species of *Drosophila* emerged from the patches. Again these authors treated pairs of species within each collection separately and the distribution of A_{12} values are shown in Fig. 6.6(c). *Drosophila falleni* Wheeler and *D. recens* Wheeler are both in the quinaria species group, while *D. putrida* Sturtevant and *D. testacea* Roser are both within the testacea species group, and both between group and within group comparisons are also shown in Fig. 6.6(c). Such comparisons are not possible for the British and Australian data.

The final data-set is one recently collected by Sevenster (1992) for fruit-breeding species on Barro Colorado Island, Panama. Only the patches (individual fruit) from *Spondias mombin* were kept separately and are therefore suitable for this analysis. Collections were made in two years (1988 and 1990), and all species pairs within each year have been analysed separately. A total of 420 patches and 16 insect species, including 10 drosophilids, were available for analysis. The distribution of A_{12} values are shown in Fig 6.6(d).

These four data-sets tell much the same story, in general A_{ij} is > 1, so that intraspecific aggregation is stronger than interspecific aggregation. This is so despite the demonstrated presence, in *Drosophila*, of resource separation at various spatio-temporal levels (see first section). Resource partitioning is there, but swamped by intraspecific aggregation. Of course *Drosophila* species do not compete in pairs, but even when Sevenster (1992) recalculated his A_{ij}

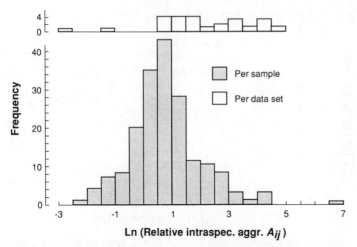

Fig. 6.7 Distribution of A_{ij} values between each species and all other species in the collections of Sevenster 1992.

values between each species and all his other species combined (Fig. 6.7) the dominance of intraspecific aggregation is still paramount.

Conclusion

It is now possible to draw a more precise, but more restricted, view of the world than that shown in Fig. 6.4., with the analysis from the four *Drosophila* data-sets superimposed (Fig. 6.8). The dimensions of this world are J and C (note the different scales because C only varies between 0 and $+1$ in these data), the comparable measures of intraspecific and interspecific aggregation, which have replaced spatial heterogeneity and niche heterogeneity in Fig. 6.4. Notice however, that parameter space within this world is not equally likely for *Drosophila*, since the points are very clustered. A world in which inter-individual contact was dominated by resource partitioning would lie to the bottom, but this is not where the points are! Intraspecific aggregation dominates this world and is the prime agent responsible for species diversity at this level. However, resource partitioning does come into play at a different level, between those species that utilise different resource types such as fruit, fungi, sap, decaying vegetation, or flowers.

In *Drosophila* local diversity is a product of these two forces, with the number of resource types perhaps expanding in the tropics to explain the higher diversity. For example, in the northern island of Hokkaido, Japan (and in Europe), decaying leaves are a substrate type while in Hawaii decaying *Cheirodendron* leaves contain an identifiable group of *Drosophila* (Shorrocks and Rosewell 1986). Within these substrate groups, coexistence is dominated by intraspecific aggregation and guild size is approximately seven (Fig. 6.1).

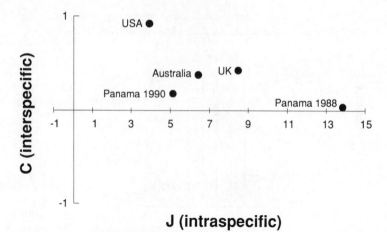

Fig. 6.8 The scheme of Cornell and Lawton (1992) redrawn without 'fullness of the niche space'. Spatial heterogeneity and niche heterogeneity are replaced by intraspecific and interspecific aggregation.

For *Drosophila*, therefore, I suggest that there are two important aspects to local diversity: (1) interspecific aggregation (resource partitioning) between categories of substrate (fruits, fungi, decaying vegetation, sap, flowers), and (2) intraspecific aggregation across patches. If for convenience we use a value of 7 for guild size in drosophilids, then local diversity in temperate regions is approximately $4 \times 7 = 28$. For the tropics and Hawaii the number of substrate types is larger, so that the equation may be more like $10 \times 7 = 70$. We have a good idea what causes the 7, but we have less idea what causes the 4, and we don't even know if the 10 is correct!

Acknowledgements

I would like to thank all those postgraduate students, postdoctoral fellows and technicians that have contributed to the *Drosophila* Population Biology Programme in the Department of Pure and Applied Biology over the last 20 years (Will Atkinson, Mike Begon, Marc Bingley, Roger Catchpole, Paul Charlesworth, Roland Cooper, Andrew Davis, Kevin Dooher, Calvin Dytham, Kathy Edwards, Linda Jenkinson, John Kearney, Sheena McNamee, Julian Miller, Jon Rosewell, Andrew Watson). I would also like to thank Jan Sevenster and two anonymous referees for their comments on a previous version of this manuscript.

References

Abrams, P. (1975). Limiting similarity and the form of the competition coefficient. *Theoretical Population Biology*, **8**, 356–375.

Armstrong, R. A. (1989). Competition, seed predation, and species coexistence. *Journal of Theoretical Biology*, **141**, 191–194.

Arthur, W. (1987). *The Niche in Competition and Evolution*. Wiley, New York.

Atkinson, W. D. (1979). A field investigation of larval competition in domestic *Drosophila*. *Journal of Animal Ecology*, **48**, 91–102.

Atkinson, W. D. (1981). An ecological interaction between citrus fruit, *Penicillium* moulds and *Drosophila immigrans* Sturtevant (Diptera: Drosophilidae). *Ecological Entomology*, **6**, 339–344.

Atkinson, W. D. (1985). Coexistence of Australian rainforest Diptera breeding in fallen fruit. *Journal of Animal Ecology*, **54**, 507–518.

Atkinson, W. D. and Shorrocks, B. (1977). Breeding site specificity in the domestic species of *Drosophila*. *Oecologia*, **29**, 223–232.

Atkinson, W. D. and Shorrocks, B. (1981). Competition on a divided and ephemeral resource: a simulation model. *Journal of Animal Ecology*, **50**, 461–471.

Begon, M. (1992). Density and frequency dependence in ecology: messages for genetics? *Genes in Ecology* (eds R. J. Berry, T. J. Crawford and G. M. Hewitt). Symposium of the British Ecological Society No. 33, pp. 335–352, Blackwell Scientific Publications, Oxford.

Begon, M. and Shorrocks, B. (1978). The feeding and breeding sites of *Drosophila obscura* Fallen and *D. subobscura* Collin. *Journal of Natural History*, **12**, 137–151.

Bellows, T. S. (1981). The descriptive properties of some models for density dependence. *Journal of Animal Ecology*, **50**, 139–156.

Brncic, D. and Budnik, M. (1987). Some interactions of the colonising species *Drosophila subobscura* with local *Drosophila* in Chile. *Genetica Iberica*, **39**, 249–267.

Carson, H. L. (1971). The ecology of *Drosophila* breeding sites. Harold L. Lyon Arboretum Lecture Number Two. University of Hawaii, Honolulu. (Unpublished.)

Caswell, H. (1976). Community structure: a neutral model analysis. *Ecological Monographs*, **46**, 327–354.

Chesson, P. L. and Huntly, N. (1989). Short-term instabilities and long-term community dynamics. *Trends in Ecology and Evolution*, **4**, 293–298.

Clarke, B. C. and Beaumont, M. A. (1992). Density and frequency dependence: a genetical view. *Genes in Ecology* (eds R. J. Berry, T. J. Crawford and G. M. Hewitt). Symposium of the British Ecological Society No. 33, pp. 353–364, Blackwell Scientific Publications, Oxford.

Cody, M. L. and Diamond, J. M. (1975). *Ecology and Evolution of Communities*. Harvard University Press, Cambridge, Mass.

Connell, J. H. (1978). Diversity in tropical rain forests and coral reefs. *Science*, **199**, 1302–1310.

Cornell, H. V. and Lawton, J. H. (1992). Species interactions, local and regional processes, and limits to the richness of ecological communities: a theoretical perspective. *Journal of Animal Ecology*, **61**, 1–12.

Crombie, A. (1946). On competition between different species of graminivorous insects. *Proceedings of the Royal Society of London (B)*, **132**, 362–395.

Fawcett, H. S. (1936). *Citrus Diseases and their Control*. McGraw-Hill, New York.

Gause, G. F. (1934). *The Struggle for Existence*. Williams and Wilkins, Baltimore.

Gilpin, M. E. and Ayala, F. J. (1973). Global models of growth and competition. *Proceedings of the National Academy of Sciences*, **70**, 3590–3593.

Grimaldi, D. and Jaenike, J. (1984). Competition in natural populations of mycophagous *Drosophila*. *Ecology*, **65**, 1113–1120.

Hackman, W. and Meinander, M. (1979). Diptera feeding as larvae on macrofungi in Finland. *Annales Zoologici Fenneci*, **16**, 50–83.

Hanski, I. (1981). Coexistence of competitors in patchy environments with and without predation. *Oikos*, **37**, 306–312.

Hanski, I (1983). Coexistence of competitors in patchy environments. *Ecology*, **64**, 493–500.

Harrison, S., Ross, S. J. and Lawton, J. H. (1992). Beta diversity on geographic gradients in Britain. *Journal of Animal Ecology*, **61**, 151–158.

Hassell, M. P. (1975). Density dependence in single species populations. *Journal of Animal Ecology*, **42**, 693–726.

Hassell, M. P. and Comins, H. N. (1976). Discrete time models for two-species competition. *Theoretical Population Biology*, **9**, 202–221.

Hoekstra, R. F., Bijlsma, R. and Dolman, A. J. (1985). Polymorphism from environmental heterogeneity: models are only robust if the heterozygote is close in fitness to the favoured homozygote in each environment. *Genetical Research*, **45**, 299–314.

Hubbell, S. P. and Foster, R. B. (1986). Biology, chance, and history and the structure of tropical rain forest communities. *Community Ecology* (eds J. Diamond and T. J. Case), pp. 314–329, Harper Row, New York.

Huston, M. (1979). A general hypothesis of species diversity. *American Naturalist*, **113**, 81–101.

Hutchinson, G. E. (1978). *An Introduction to Population Ecology*. Yale University Press, New Haven.

Ives, A. R. (1988). Aggregation and the coexistence of competitors. *Annales Zoologici Fennici*, **25**, 75–88.

Ives, A. R. (1991). Aggregation and coexistence in a carrion fly community. *Ecological Monographs*, **61**, 75–94.

Jaenike, J. and James, A. C. (1991). Aggregation and the coexistence of mycophagous *Drosophila*. *Journal of Animal Ecology*, **60**, 913–928.

Janzen, D. H. (1970). Herbivores and the number of three species in tropical rain forests. *American Naturalist*, **104**, 501–528.

Jay, J. M. (1970). *Modern Food Microbiology*. Van Nostrand Reinhold.

Kaneko, A. and Shima, T. (1960). Drosophilid species collected in Okoppe and its neighbouring districts in east-northern Hokkaido. *Drosophila Information Service*, **34**, 87.

Kareiva, P. (1989). Renewing the dialogue between theory and experiments in population ecology. *Perspectives in Ecological Theory* (eds J. Roughgarden, R. M. May and S. A. Levin), pp. 68–88, Princeton University Press, Princeton.

Kearney, J. (1979). *The Breeding Site Ecology of Three Species of Woodland Drosophila (Diptera, Drosophilidae)*. Unpublished PhD thesis, University of Leeds.

Kearney, J. (1982). Oviposition and feeding on naturally occurring yeasts by *Drosophila* spp., using natural substrates. *Oikos*, **39**, 103–112.

Kimura, M. T., Toda, M. J., Beppu, K. and Watabe, H. (1977). Breeding sites of drosophilid flies in and near Sapporo, Northern Japan, with supplementary notes on adult feeding habits. *Kontyu, Tokyo*, **45**, 571–582.

Lloyd, M. (1967). Mean crowding. *Journal of Animal Ecology*, **36**, 1–30.

MacArthur, R. H. (1972). *Geographical Ecology*. Harper and Row, New York.

MacArthur, R. H. and Levins, R. (1967). The limiting similarity, convergence and divergence of coexisting species. *American Naturalist*, **101**, 377–385.

Makino, S., Momma, E. and Tokada, H. (1958). *Drosophila* survey in Shiretoko Peninsula, Hokkaido. *Drosophila Information Service*, **32**, 137.

Makino, S., Momma, E. and Wakahama, K. (1956). Fluctuations of predominant species of *Drosophila* at Sapporo. *Drosophila Information Service*, **30**, 133.

Makino, S. and Takeharu, K. (1951). A monthly survey of *Drosophila* in the city of Sapporo, Hokkaido. *Drosophila Information Service*, **25**, 111.

May, R. M. and MacArthur, R. H. (1972). Niche overlap as a function of environmental variability. *Proceedings of the National Academy of Science of the USA*, **69**, 1109-1113.

Nicholson, A. J. (1954). An outline of the population dynamics of natural populations. *Australian Journal of Zoology*, **2**, 9–65.

Okada, T. (1963). Caenogenetic differentiation of mouthhooks in drosophilid larvae. *Evolution*, **17**, 84–98.

Park, T. (1948). Experimental studies of interspecific competition: 1. Competition between populations of the flour beetles *Tribolium confusum* and *Tribolium castaneum*. *Ecological Monographs*, **18**, 265–307.

Pipkin, S. B., Rodriguez, R. L. and Leon, J. (1966). Plant host specificity among flower-feeding neotropical *Drosophila* (Diptera: Drosophilidae). *American Naturalist*, **100**, 135–156.

Raper, K. B. and Thom, C. (1949). *A Manual of the Penicillia*. Baillière, Tindall and Cox, London.

Rosewell, J., Shorrocks, B. and Edwards, K. (1990). Competition on a divided and ephemeral resource: testing the assumptions. I. Aggregation. *Journal of Animal Ecology*, **59**, 977–1001.

Sale, P. F. (1977). Maintenance of high diversity in coral reef fish communities. *American Naturalist*, **111**, 337–359.

Sevenster, J. G. (1992). *The community ecology of frugivorous* Drosophila *in a neotropical forest*. Unpublished PhD Thesis, University of Leiden.

Shannon, C. E. and Weaver, W. (1949). *The Mathematical Theory of Communications*. Urbana: University of Illinois Press, Urbana, IL.

Shorrocks, B. (1977). An ecological classification of European *Drosophila* species. *Oecologia*, **26**, 335–345.

Shorrocks, B. (1981). The breeding sites of temperate woodland *Drosophila*. *The Genetics and Biology of* Drosophila (eds M. Ashburner, H. L. Carson and J. N. Thompson jr), pp. 385–428, Academic Press, London.

Shorrocks, B. (1990). Coexistence in a patchy environment. *Living in a Patchy Environment* (eds B. Shorrocks and I. Swingland), pp. 91–106, Oxford Science Publications.

Shorrocks, B. (1993). Trends in the *Journal of Animal Ecology*: 1932–92. *Journal of Animal Ecology*, **62**, 599–605.

Shorrocks, B. and Charlesworth, P. (1980). The distribution and abundance of the British fungal-breeding *Drosophila*. *Ecological Entomology*, **5**, 61–78.

Shorrocks, B. and Rosewell, J. (1986). Guild size in drosophilids: a simulation model. *Journal of Animal Ecology*, **55**, 527–541.

Shorrocks, B. and Rosewell, J. (1987). Spatial patchiness and community structure: coexistence and guild size of drosophilids on ephemeral resources. *Organisation of Communities: Past and Present* (ed. by J. H. R. Gee and P. S. Giller), pp. 29–51. Blackwell Scientific Publications, Oxford.

Shorrocks, B., Atkinson, W. D. and Charlesworth, P. (1979). Competition on a divided and ephemeral resource. *Journal of Animal Ecology*, **48**, 899–908.

Shorrocks, B., Rosewell, J., Edwards, K. and Atkinson, W. D. (1984). Interspecific competition is not a major organising force in many insect communities *Nature*, **310**, 310–312.

Shorrocks, B., Rosewell, J. and Edwards, K. (1990). Competition on a divided and ephemeral resource: testing the assumptions. II. Association. *Journal of Animal Ecology*, **59**, 1003–1017.

Southwood, T. R. E. (1966). *Ecological Methods*. Methuen, London.

Strong, D. R., Lawton, J. H. and Southwood, T. R. E. (1984). *Insects on Plants: Community Patterns and Mechanisms*. Blackwell Scientific Publications, Oxford.

Stubbs, M. (1977). Density dependence in the life-cycles of animals and its importance in k- and r-strategies. *Journal of Animal Ecology*, **46**, 677–688.

Takada, H. (1956). Notes on the vertical distribution of Drosophilidae. *Drosophila Information Service*, **30**, 154.

Tilman, D. (1986). Evolution and differentiation in terrestrial plant communities: the importance of the soil resource: light gradient. *Community Ecology* (eds J. Diamond and T. J. Case), pp. 359–380, Harper and Row, New York.

Toda, M. J. (1984). Guild structure and its comparison between two local drosophilid communities. *Physiology and Ecology (Japan)*, **21**, 131–172.

Toda, M. J., Kimura, M. T. and Enomoto, O. (1984). Bionomics of Drosophilidae (Diptera) in Hokkaido. VI. Decayed herbage feeders, with special reference to their reproductive strategies. *Japanese Journal of Ecology*, **34**, 253–270.

Travis, G. D. L. and Collins, H. M. (1991). New light on old boys: cognitive and institutional particularism in the peer review system. *Science, Technology and Human Values*, **16**, 322–341.

Turner, J. R. G. (1992). Stochastic processes in populations: the horse behind the cart? *Genes in Ecology* (eds R. J. Berry, T. J. Crawford and G. M. Hewitt). Symposium of the British Ecological Society No. 33, pp. 29–53, Blackwell Scientific Publications, Oxford.

Utida, S. (1953). Interspecific competition in two species of bean weevil. *Ecology*, **34**, 301–307.

Utida, S. (1957). Population fluctuation: an experimental and theoretical approach. *Cold Spring Harbour Symposium in Quantitative Biology*, **22**, 139–151.

Whittaker, R. M. (1960). Vegetation of the Siskiyou Mountains, Oregon and California. *Ecological Monographs*, **30**, 279–338.

Worthen, W. B. and McGuire, T. R. (1988). A criticism of the aggregation model of coexistence: non-independent distribution of dipteran species on ephemeral resources. *American Naturalist*, **131**, 453–458.

7

The population dynamics of single host–multiparasitoid interactions

T. Hefin Jones and Michael P. Hassell

Introduction

Insect parasitoids and their hosts are an integral part of the complex web of multispecies interactions in which insects exist. Although some insect species are host to a complex of 20 or more parasitoid species, a large number are attacked by only one or two species (Askew and Shaw 1986; Hawkins and Lawton 1987, 1988). Many parasitoid species are polyphagous to some degree, and their populations are thus buffered from fluctuations in abundance of any one of their host species (Southwood and Comins 1976).

A detailed quantitative understanding of the dynamics of species within such complex communities is rarely feasible. Instead, in an attempt to determine what factors are promoting the persistence of interacting species, the search has been more for broad patterns of community organisation (May 1981). This has frequently involved developing models that are either tailored to specific questions, or attempt to describe particular systems (Kareiva 1989; Jones *et al.* 1994). It is possible 'that interactions between relatively few keystone members of complex communities determine the overall state of the system, with the dynamics of many of the other species influenced by, but not influencing the keystone species' (Memmott and Godfray 1994).

Although 'there is no comfortable theorem assuring that increasing diversity and complexity beget enhanced community stability' (May 1973), natural systems do endure! By concentrating on host–parasitoid interactions in this chapter, we explore some of the mechanisms which may promote the persistence of multispecies systems. By going beyond the dynamics of single parasitoid–single host interactions, which have been the focus of so much empirical and theoretical work (Hassell 1978 and references therein), we illustrate how the addition of a second parasitoid can introduce quite different dynamics.

The chapter is divided into three main sections. First, we explore two general models developed to describe a single host–multiparasitoid system. The first, by May and Hassell (1981), is based on the general Nicholson and

Bailey (1935) framework, so frequently used for coupled single host–single parasitoid interactions with discrete generations, where

$$N_{t+1} = \lambda N_t f(P_t,N_t)$$

$$P_{t+1} = cN_t(1 - f(P_t,N_t)).$$

Here, N_t and P_t represent the number of hosts and parasitoids, respectively, in generation t; λ is the *per capita* net rate of increase of the host population; c is the average number of adult female parasitoids emerging from each parasitised host; and the function $f(P_t,N_t)$ defines the fraction of hosts that survive parasitism. We consider some more recent work by Briggs (1993) who uses a stage-structured model to describe the outcome when two parasitoid species attack different developmental stages of a single host species. We next explore systems in which the host is attacked by both a generalist and a specialist parasitoid species. A general treatment of this has been given by Hassell and May (1986), and more recently, a case study by Jones *et al.* (1993) investigates the effects of a generalist and a specialist parasitoid on the dynamics of the cabbage root fly *Delia radicum* (L.) (Diptera: Anthomyiidae). Finally, we review how the simple process of 'diffusive' dispersal in a patchy environment may promote the persistence of more complex multispecies systems and what underlying mechanisms allow such coexistence to occur (Hassell *et al.* 1994).

Single host–two parasitoids interactions

One of several different outcomes may result when a parasitised host individual is encountered by a second parasitoid species. Some species are able to detect that a host has been attacked by a different species and to refrain from further parasitism (Mackauer 1990). In the majority of species, however, interspecific discrimination does not occur (van Alphen and Visser 1990; Godfray 1994) and frequently it is the larvae themselves that directly fight each other (van Strien-van Liempt 1983). In some systems, larval size determines the outcome, with the first arriving (and larger) species killing or outdeveloping its competitor. In other systems, the reverse is true, especially when the earlier arriving species is an endoparasitoid and the later arriving species is an ectoparasitoid (Mackauer 1990).

The model described by May and Hassell (1981) is applicable to two types of interactions. First, it is appropriate where one parasitoid acts first, followed by another which only acts on the surviving hosts—a situation that sometimes arises when hosts with fairly discrete generations are attacked in different developmental stages by a range of parasitoid species. Secondly, it is appropriate when both parasitoids act on the same host stage but one species is superior in competition regardless of the order of arrival within the host.

A single vulnerable host stage

By modifying the general framework given above, May and Hassell (1981) developed a two parasitoid–single host system:

$$N_{t+1} = \lambda N_t f_1(P_t) f_2(Q_t)$$

$$P_{t+1} = N_t(1 - f_1(P_t))$$

$$Q_{t+1} = N_t f_1(P_t)\,(1 - f_2(Q_t))$$

where $f_1(P_t)$ and $f_2(Q_t)$ are the probabilities of a host (N_t) not being found by P_t or Q_t parasitoids, respectively. The probabilities are based on the zero term of the inherently stabilizing negative binomial distribution (May 1978):

$$f_1(P_t) = \left(1 + \frac{a_1 P_t}{k_1}\right)^{-k_1} \qquad f_2(Q_t) = \left(1 + \frac{a_2 Q_t}{k_2}\right)^{-k_2}$$

where a_1 and a_2 are the searching efficiencies of the two parasitoids, and k_1 and k_2 describe the degree of clumping in the parasitoid attacks. Handling time has been omitted for the sake of simplicity and because its effects are unlikely to be qualitatively important to the dynamics of the interaction (Hassell and May 1973).

Numerical examples of this model (Fig. 7.1) allow examination of the conditions necessary for coexistence of P and Q in a locally stable, three-species interaction. In general, when parasitoids attack a host species in temporal sequence, there are more possibilities for coexistence if the later acting species (Q) has the higher searching efficiency. If, however, the parasitoids attack the same host stage, then it is the species with the inferior larval competitor (again Q in May and Hassell (1981)) which should have the higher searching efficiency.

This model was further explored by Hogarth and Diamond (1984) who considered cases where the outcome of multiparasitism depended on the order of arrival within the host. By assuming that when multiparasitism occurs a proportion s of P and $(1 - s)$ of Q survived, Hogarth and Diamond proposed the model:

$$N_{t+1} = \lambda N_t f_1(P_t) f_2(Q_t)$$

$$P_{t+1} = N_t(1 - f_1(P_t))\,(s + (1 - s)f_2(Q_t))$$

$$Q_{t+1} = N_t(1 - s + s f_1(P_t))\,(1 - f_2(Q_t)).$$

If $s = 1$ or 0, the model is reduced to the two parasitoid - one host system of May and Hassell (1981), with either P (when $s = 1$) or Q $(s = 0)$ precluding the other by attacking first, or prevailing in cases of multiparasitism. Hogarth and Diamond (1984) found that regardless of whether the outcome of intra-larval competition was a 'one-way street' (as in May and Hassell (1981)) or intermediate $(0 < s < 1)$, a tendency towards parasitoid aggregation increased

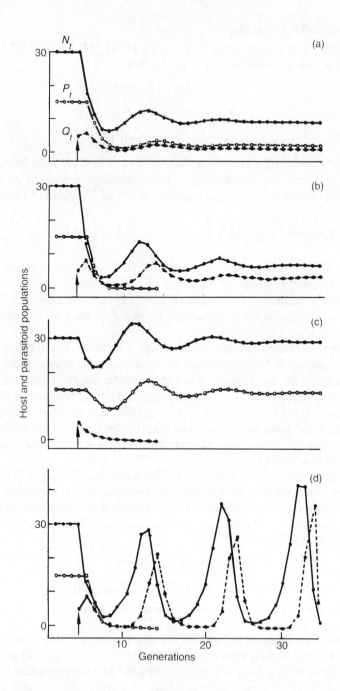

the likelihood of a stable, multispecies interaction, without the competitive exclusion of a parasitoid species.

Host age-structure

Using a similar three-species system, Briggs (1993) developed an age-structured model to investigate the role of competition among parasitoid species on a stage-structured host and its effects on host suppression. Here, it is the presence of an invulnerable adult stage that is stabilising the equilibrium rather than the density-dependent parasitoid attack rate assumed in May and Hassell (1981) and Hogarth and Diamond (1984). In Briggs's study the two parasitoid species attack different development stages of the host and two versions of the model are developed, differing in their assumptions on the outcome of competition between larvae of the two parasitoid species. Both versions of the model make the following general assumptions. The host has three developmental stages: an egg stage (E), a larval stage (L), and an adult stage (A), of duration T_E, T_L and T_A days, respectively. There are two solitary parasitoid species, P and Q, each having an immature and an adult stage, with developmental periods of T_{JP} and T_{JQ} days, respectively. Parasitoid P attacks the egg stage of the host, and parasitoid Q attacks the larval stage of the host. Both species exhibit Type I functional responses, with constant attack rates a_P and a_Q that are independent of host or parasitoid densities. The adult host stage is invulnerable to parasitism. In the first version of the model, which is similar to the first situation described by May and Hassell (1981) above, the larval parasitoid will attack only those hosts not previously parasitized by the egg parasitoid. In the second version, hosts that are parasitised by the egg parasitoid can later be attacked by the larval parasitoid which invariably succeeds.

When the egg parasitoid wins in larval competition, Briggs found that the two species are not able to coexist. In the majority of cases, the egg parasitoid gains a competitive advantage through attacking the earlier host stage. However, the species that wins in competition is not necessarily the one that would produce the lowest adult host density (Fig. 7.2(a)). The larval parasitoid (Q) can invade the system only if it has an attack rate much higher than that of P (Fig. 7.2(b)). This is necessary for successful invasion because the host eggs are a far more abundant resource and the egg parasitoid also

---

Legend to facing page

Fig. 7.1 Numerical simulations illustrating four of the possible outcomes following the introduction of a further parasitoid species (Q) to a stable host–parasitoid interaction. N and P commence at their equilibrium values and Q is introduced in generation 4. (a) $a_1 = 0.25$, $a_2 = 0.35$, $k_1 = k_2 = 0.25$. (b) $a_1 = 0.1$, $a_2 = 0.4$, $k_1 = k_2 = 0.5$. (c) $a_1 = 0.1$, $a_2 = 0.05$, $k_1 = k_2 = 0.5$. (d) $a_1 = 0.1$, $a_2 = 0.25$, $k_1 = 0.5$, $k_2 = 1.1$. (After Hassell 1978.)

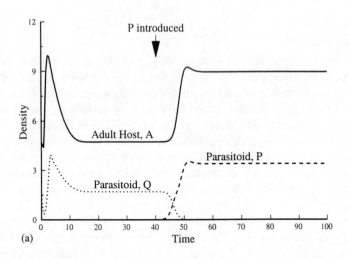

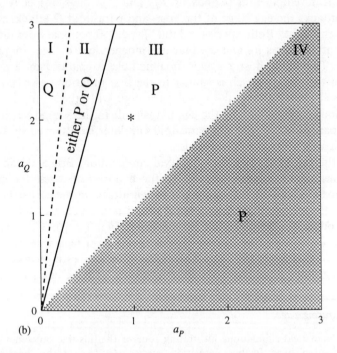

attacks before the host has been subjected to the various other sources of mortality.

In the second version of Briggs's model, hosts that are parasitised by the egg parasitoid can later be successfully attacked by the larval parasitoid. This results in an adult of the larval parasitoid species emerging. In this case, the two species can coexist (Fig. 7.3); hosts previously parasitised by the egg parasitoid now serve as an additional resource for the larval parasitoid. When the two species do coexist, the density of each of the stages of the host is intermediate between the densities obtained when each of the two parasitoid species is present alone.

Briggs's model assumes that there is a fixed period of time before individuals mature from one juvenile developmental stage to the next. In nature, there can be a great deal of variation in the time spent in any one particular stage. By incorporating a *maturation weighting function* into the model, Briggs *et al.* (1993) found that coexistence of the two parasitoid species is possible if there is sufficient variation in the time that different host individuals remain in each stage. In this way, coexistence is promoted by the developmental heterogeneity of the host population.

Although Briggs (1993) notes the difficulty of making comparisons with the results of May and Hassell (1981), where only one developmental stage of the host was explicitly modelled, she does consider the implications of the different approaches for multiple release strategies in biological control. In non-stage-structured models, such as those of May and Hassell (1981), and Hogarth and Diamond (1984), the winning parasitoid is the one that leads to the lowest abundance of the host. From this it is possible to conclude that there is no danger in releasing all possible parasitoid species. However, when a stage-structured model is used it becomes obvious that release of all possible natural enemy species may not lead to the lowest abundance of the pest stage of the host. Stage-structured population models can reveal other

---

Legend to facing page

Fig. 7.2 (a) Simulation of version 1 of the model of Briggs (1993) using $a_P = 1.0$, $a_Q = 2.0$. In this region of the parameter space, parasitoid P wins in competition, but parasitoid Q leads to a lower host density. The simulation starts with only parasitoid Q present. After an initial perturbation, the system settles at the equilibrium possible with parasitoid Q. After 40 time steps, P is introduced into the system. Species Q is displaced and the system settles at a new equilibrium, with a higher adult host density. (b) Diagram of the parameter space for version 1 of the model showing the outcome of competition between parasitoid species P and Q. The variable a_P is the attack rate of parasitoid P on the egg stage of the host; a_Q is the attack rate of the parasitoid Q on the larval stage of the host. In area *I*, parasitoid Q wins in competition and excludes parasitoid P from the system. In areas III and IV, parasitoid P wins and excludes Q. In area II, either P or Q will win depending on the starting conditions. In the shaded area IV, parasitoid P will lead to a lower host density than Q. (From Briggs 1993.)

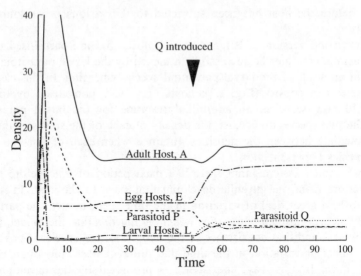

Fig. 7.3 Simulation of version 2 of the model of Briggs (1993) with parameters $a_P = 1.3$, $a_{Q1} = 1.0$, $a_{Q2} = 0.5$. All other parameters are as in Fig. 7.2. With these parameters, the two parasitoids coexist, but parasitoid P alone would lead to the lowest density of all of the stages of the host. The simulation starts with only parasitoid P and the host present. After an initial perturbation, the system settles at the equilibrium set by parasitoid P. After 50 time units, Q is introduced. The system settles at a new equilibrium with both parasitoid species present and higher abundances of all host stages. (From Briggs 1993.)

interesting properties that are not apparent in simpler models. For instance, Godfray and Waage (1991), in modelling the interaction between the mango mealy bug *Rastrococcus invadens* Williams (Hemiptera: Pseudococcidae) and two of its encyrtid parasitoids in West Africa, found that host equilibrium densities decreased as the severity of the density-dependence in parasitism increased. This is the opposite trend to that expected from simple discrete-generation, host–parasitoid models. Murdoch and Stewart-Oaten (1989) discuss the reasons why discrete and continuous models may vary in this respect. The difference seems to result from the absence, in Nicholson–Bailey models, of parasitoid re-aggregation in response to changing host distribution within a generation.

A common basis for specialist coexistence

Turnbull and Chant (1961) suggested that the result of multiparasitoid–single host interactions depended, to a very large extent, on the degree of niche overlap. Implicit in May and Hassell's (1981) model is the assumption of partial niche segregation, since the clumping distribution of attacks by the two parasitoid species are independent of one another. In Briggs's (1993) study the only form of potential niche partitioning that was considered was the specialisation on different host developmental stages. Kakehashi *et al.*

(1984) explored a range of possible patterns of niche overlap, represented by the covariance in the probability distribution of hosts being attacked by the different parasitoid species, culminating in a situation where both parasitoid species respond in exactly the same way to host cues. Kakehashi *et al.* concluded that although the effect of niche overlap on stability of the models is negligible, when there is considerable niche overlap, with niche being defined in terms of the parasitoid's searching behaviour, a single parasitoid has a higher potential for lowering the host equilibrium level. However, for interactions where niches are segregated it becomes generally the case that additional parasitoid species in the guild complement each other in reducing host abundance. Further, Shimada and Fujii (1985) showed experimentally that the resource utilisation curves, in terms of host development stage attacked, of two parasitoid species tended to diverge under conditions of interspecific competition. Such niche shifting promotes stability within a single host–multiparasitoid system.

Generalist and specialist parasitoids

Many, if not most, insect populations are attacked by both specialist and generalist parasitoid species. Populations of specialist natural enemies may show pronounced fluctuations coupled with the host's dynamics; those of generalists are more likely to be independent of the dynamics of any one of their hosts. This arises from the ability of the generalist to switch and attack either elsewhere or on other host species (Murdoch 1969; Royama 1971*a*; Latto and Hassell 1988). When combined with a Type II functional response, a generalist's numerical response results in density-dependent rates of attack over a range of host densities. At host densities above this range the percentage of parasitism declines. Such patterns are generally recognised as possessing the potential to regulate a population, although not beyond some critical host density (Murdoch and Oaten 1975; Southwood and Comins 1976).

One specialist–one generalist

Hassell and May (1986) explored theoretically the dynamics of a parasitoid–host interaction where the host is attacked by both a generalist and a specialist parasitoid species. For the generalist they assumed a Type II functional response and, instead of random encounters between host and natural enemy (Holling 1959*a*, 1959*b*; Royama 1971*b*; Rogers 1972), a negative binomial distribution of encounters was assumed (May 1978). The generalist's numerical response was assumed to approach an upper asymptote as described by Southwood and Comins (1976): $G_t = h(1 - \exp(-N_t/b))$, where h is the saturation number of generalist parasitoids and b determines the typical host density at which this maximum is approached. The number of hosts surviving attack by G_t generalists is given by $N_t g(N_t)$ where $g(N_t) = (1 + a_G G_t/k)^{-k}$.

To this system can then be added a specialist parasitoid, P. With more than one mortality acting, the relative timing of generalist and specialist in the host's life cycle are of the utmost importance (May and Hassell 1981). With the specialist acting before the generalist, we have:

$$N_{t+1} = \lambda N_t f(P_t) g(N_t f(P_t))$$

$$P_{t+1} = c N_t (1 - f(P_t))$$

where the function $f(P_t)$ is given by $f(P_t) = (1 + a_P P_t / k)^{-k}$. With the generalist acting first, the model becomes:

$$N_{t+1} = \lambda N_t f(P_t) g(N_t)$$

$$P_{t+1} = c N_t g(N_t)(1 - f(P_t)).$$

From this study, Hassell and May (1986) concluded that

(1) a specialist can invade an existing host–parasitoid interaction only if the generalist is not too efficient; a specialist can also invade and co-exist more easily if it acts before the generalist in the host's life cycle;

(2) in some cases the establishment of a specialist leads to *higher* host populations than existed previously with only the generalist acting;

(3) a stable three-species system can readily exist where the host–generalist interaction alone would be unstable or have no equilibrium at all, and

(4) a range of complicated dynamics are possible, either alternating between two-species or three-species states, or between different three-species states.

Thus, in contrast to two-species interactions which tend to have relatively straightforward dynamics, the combined three-species system presents a much wider range of dynamical behaviour.

A case study   One recent case study in which long-term data have been used to produce models comparable with those of Hassell and May (1986) explores the effects of generalist and specialist parasitoids on the dynamics of the cabbage root fly, *Delia radicum* (Jones *et al.* 1993).

The cabbage root fly lays its eggs in soil crevices around the roots of the host plant (Hughes and Salter 1959). At the end of a 3–4 week larval feeding period, third instar larvae move away from the roots to pupate in the soil. There are generally two generations of cabbage root fly per year. The first generation of adults emerge in late April and early May and their pupae occur in late June and early July; the second generation adults emerge from the puparia within 2 weeks and their pupae enter diapause and overwinter in the soil. *Delia radicum* pupae of the second generation, along with their natural enemies, have been censused from the soil around 40 brassica plants at Silwood Park each year between 1981 and 1989, and the mortality from two natural enemies quantified (Jones 1986; Jones and Hassell 1988; Jones *et al.* 1993): *Trybliographa rapae* (Westw.) (Hymenoptera: Cynipoidea), a specialist parasitoid of first and second instar cabbage root fly larvae, and *Aleochara bilineata* (Gyll.), a staphylinid beetle that is a generalist parasitoid attacking the pupae of *D. radicum*. Multiparasitism of the same host individual has only very occasionally been observed (Reader and Jones 1990). A simple model to describe this generalist–specialist interaction is given by:

$$N_{t+1} = N_t \lambda F_P F_G$$

$$P_{t+1} = sN_t(1 - F_P).$$

The *D. radicum* population in the next generation is obtained from the number in the previous generation (N_t) which are assumed to grow geometrically in the absence of parasitism at rate λ. The fraction of hosts that escapes attack by *T. rapae* is given by F_P, and F_G is the fraction that also escapes parasitism by *A. bilineata*. The population density of *T. rapae* in the next generation (P_{t+1}) is given by the number of hosts parasitised by *T. rapae* ($N_t(1 - F_P)$) times the survivorship term, s.

The forms of functions F_P and F_Q are determined from census data. For the specialist parasitoid, *T. rapae*, there is evidence from field data of the existence of within-generation spatial patterns, where parasitism is directly correlated with host density per plant (Jones 1986, Jones and Hassell 1988). The severity of this trend, measured by μ, the aggregation index, tends to decline as *T. rapae* increases in abundance (Fig. 7.4a). The local density (p) of

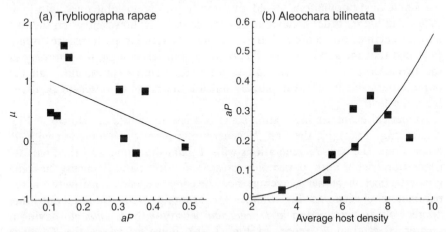

Fig. 7.4 (a) Estimates of μ and the corresponding values of aP_t (where a is the searching efficiency and P_t is the mean parasitoid abundance) for *Trybliographa rapae* ($y = 1.279-2.630x$; $r = -0.552$; $P = 0.123$). (b) Estimates of 'overall efficiency', aP_t, for *Aleochara bilineata* and the mean cabbage root fly density per year (ln $y = -6.511 + 2.601$ ln x; $r = 0.862$, $P = 0.003$). Parameter values are estimated for each year. (From Jones *et al.* 1993.)

searching *T. rapae* on a brassica plant with *n* hosts can be represented by the simple aggregation model:

$$p = cP_t \left(\frac{n}{N_t}\right)^{\mu}$$

where *c* is a normalisation constant and μ is given by $CP_t + D$ from the regression in Fig. 7.4*a*. The estimated abundance of *A. bilineata* in a given year increases with mean density (Fig. 7.4(b)). Assuming that the distribution of *D. radicum* follows a negative binomial $Q(n/N_t)$, the functions F_P and F_G may be written as:

$$F_P = \sum_{n=0}^{\infty} Q(n/N_t) \frac{n}{N_t} \exp(-P_t(n/N_t)^{CP_t+D})$$

$$F_G = \exp(-AN_t^B)$$

where the fraction of hosts surviving *T. rapae*, F_P, is simply the average over all hosts and *n* is the local host density. The fraction of hosts surviving *A. bilineata* is obtained from the zero term of a Poisson distribution with mean AN_t^B as defined in Fig. 7.4(b).

Exploration of the model shows that were *T. rapae* to act alone, this spatial heterogeneity in parasitism would promote stability, but only within a very narrow range of net host rate of increase, λ. In contrast, *A. bilineata*, the generalist natural enemy, acts as a simple, between-generation density-dependent factor. Were it present alone, as suggested by the functional form F_G, stable host populations would give way to locally unstable ones with limit cycles and higher order behaviour as λ increases. It is, however, when the two natural enemies are combined that the most interesting patterns are found. Provided that the survivorship of *T. rapae* is sufficiently high, the interplay of the two natural enemies can lead to alternative stable states, although the population levels involved lie largely outside the range of observed densities from the field.

Numerical examples illustrating some of these dynamics are shown in Fig. 7.5. In Fig. 7.5(a),(b), the first 75 generations show the interaction with *T. rapae* alone, the next 75 generations with *A. bilineata* alone, and the final 100 generations with both parasitoids present. Which natural enemy has the greatest effect in reducing average host densities depends sensitively on the parameters chosen. In Fig. 7.5(a), for example, both species contribute appreciably to a reduction in *D. radicum* density with *A. bilineata* having a greater effect than *T. rapae*. In Fig. 7.5(b), however, where the *T. rapae* survival is much higher, almost all the effect on *D. radicum* populations comes from the cynipid rather than *A. bilineata*. The simulation in Fig 7.5(c). illustrates the existence of alternative, three-species stable states.

These patterns are very similar to those found by Hassell and May (1986) except that, in their particular case, the parameter combinations required for

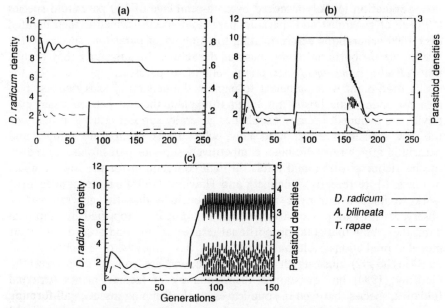

Fig. 7.5 Numerical examples illustrating the range of dynamics between cabbage root fly and its parasitoids. (a) First 75 generations with *T. rapae* alone, followed by 75 generations with *A. bilineata* alone, and the final 100 generations with both *T. rapae* and *A. bilineata* present (λ = 1.3, *s* = 0.08). The equilibrium cabbage root fly density is reduced by replacing *T. rapae* with *A. bilineata* and further reduced with both natural enemies present. (b) As for (a), but now illustrating the greater effectiveness of *T. rapae* at reducing cabbage root fly densities when its survival, *s*, is higher *(λ* = 1.7, *s* = 0.4). (c) An example of alternative, three species stable states. Following a perturbation after the first 75 generations, the locally stable three-species equilibrium gives way to a higher, locally unstable one showing limit cycles (λ = 1.8, *s* = 0.5). (From Jones *et al.* 1993.)

alternative three-species states required rather extreme values of λ. In both cases, instead of the relatively straightforward dynamics exhibited by the paired interactions (e.g. Hassell 1978; Hassell and May 1988; Hassell *et al.* 1991*a*), the combined three-species systems present a wider range of properties. Whether such alternative states do actually occur in nature remains unanswered and has recently been questioned by Grover and Lawton (1994) who comment that 'the existence of, and mechanisms generating these phenomena in the field remain poorly researched.'

More than two parasitoid species

Developing on the theme of coexistence of generalists and specialists, Hochberg and Hawkins (1992, 1994) have shown that refuges from parasitoid attack (arising from spatial heterogeneity and the invulnerability to parasitism of a proportion of the host population) are important in determining parasitoid species richness. In Hochberg and Hawkins (1992) they simulate a host population initially attacked by an assemblage of 100 parasitoid species (50 each of generalist and specialist parasitoid species) which is then iterated over 1000 generations before relating the number of parasitoid species to the size of the proportional component of the refuge. In this study they argued that '*all else being equal*', structural refuges to parasitism provided by the host's food-plant were sufficient to explain the general species richness pattern for endophytic hosts, but failed to explain the full pattern when exophytes are included, because exophytics generally support either too many or too few species. They showed how, when hosts occupy no refuge, and parasitoid attack rates are high, competitively superior parasitoids can reduce species richness to minimal levels through competition for the low numbers of hosts. More recently, Hochberg and Hawkins (1994) have expanded their work to include exophytic insects and conclude that the general pattern observed with endophytes also holds for exophytes. Supported by empirical data, they show that total proportional refuges are in good accord with their model's predictions. Although the sensitivity analysis of Hochberg and Hawkins (1992) suggests that their predictions are robust, more recently, Godfray (1994) has questioned their equating of feeding niches (external folivores, species that feed in spun leaves etc, leaf mining insects, gall-forming insects, internal feeders and root feeders) with increasing proportional refuges.

'Diffusive' dispersal

The majority of studies, such as those described above, assume that the variation in attack rate from patch to patch is derived ultimately from 'environmental inhomogeneities' (May 1995). Recent work (Hassell *et al.* 1991*b*; Comins *et al.* 1992) has suggested that a range of spatial patterns may, in fact, arise from simple, deterministic rules about the local, inter-generation movement of hosts and parasitoids within an environmentally homogeneous world (see also Allen 1975; Solé *et al.* 1992). The studies (Hassell *et al.* 1991*b*; Comins *et al.* 1992) illustrate that local movement in patchy environments, in contrast to global dispersal throughout the area, can be an important factor promoting persistence in otherwise unstable interactions. The dynamic spatial behaviour exhibited by these models may range from complex patterns of spiral waves or spatially chaotic variation to static crystal lattice patterns (or, of course, the population may become extinct). More recently, Hassell *et al.* (1994) have extended this work to multispecies interactions. In particular, they have attempted to determine how the simple process of 'diffusive' dispersal in a patchy environment may promote the persistence of complex, multispecies systems, and then attempt to unravel the underlying mechanisms of any coexistence that occurs.

In their model, Hassell *et al.* (1994) assume that the habitat is divided into a large number of discrete patches (e.g. food plants for a herbivorous insect) within which adult hosts oviposit and the adult parasitoids search for feeding, immature hosts to parasitise. Both hosts and parasitoids are assumed to have discrete generations and the parasitoid life cycle is assumed to be synchronous with that of the host. As in previous studies of this type, the basic Nicholson and Bailey (1935) framework describes the host–parasitoid interaction within each patch. In the dispersal phase, a certain fraction of adults of each species leaves the patch from which they emerged, while the remainder stay behind to reproduce in their original patch.

Coexistence of all three species in the system is hardest to obtain if the two competing parasitoid species have very similar attack and dispersal rates. The picture is very different if the dispersal rates of the two parasitoid species differ more markedly; coexistence occurs and this tends to be associated with a marked degree of self-organising niche separation between the competing species (Fig 7.6). In particular, the less dispersive species tends now to be confined to the central foci of the spirals where it may be the most abundant species (Fig. 7.6(c)), and the more dispersive species occupies the remainder of the 'trailing arm' of the spirals (Fig. 7.6(b)). The less dispersive species therefore appears to occur only in isolated, small islands within the habitat and persists with total densities very much lower than that of the more dispersive species. In other words, the competitively inferior species occupies a spatial niche created by the basic two-species dynamics. Hassell *et al.* (1994) go on further to explain that 'as the relative attack rate of the less dispersive parasitoid is increased, it becomes to dominate larger and larger areas around the foci, in other words, the niche of the less dispersive species spreads further into the arms of the spirals.' Eventually, competitive exclusion becomes the norm, the winning species being determined by the balance of the advantages conferred by relatively fast dispersal and by relatively high searching efficiency.

(a)   (b)   (c)

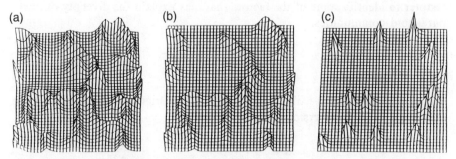

Fig. 7.6 Spatial density distribution (with linear scales) of hosts (a), highly dispersive parasitoids (b) and sedentary parasitoids (c), in a snapshot from the dynamics of a persistent single host-two parasitoid system with $\lambda = 2$, $\mu_N = 0.5$, $\mu_{P1} = 0.5$, $\mu_{P2} = 0.05$, $\alpha = 1.3$. The grids must be mentally superimposed in order to perceive the relationships between the densities of the various species. Spiral foci exist at the ends of the 'mountain ranges' in the left-hand figure (excluding ends at the edge of the grid). In the time evolution of the system the 'mountain ranges' are the peaks of population density waves, and are thus in continuous motion. The foci, by contrast, remain in almost exactly the same place for indefinitely long times.

Conclusion

Using these three general examples we have attempted to discuss some of the ways in which competing parasitoid species may coexist within a community. In many cases, instead of the relatively straightforward dynamics exhibited by simple single host–single parasitoid interactions, the addition of another parasitoid to the system gives rise to a far wider range of properties. Developmental heterogeneity (variation in the time that different host individuals remain in each developmental stage) (Briggs *et al.* 1993) and the ability of the second parasitoid to attack a host previously parasitised, and to succeed (Briggs 1993), have both been found to promote coexistence. The community of parasitoids attacking the gall-forming midge *Rhapalomyia californica* (Felt) (Diptera: Cecidomyiidae) exemplifies the latter case. The egg parasitoid *Platygaster californica* (Ashmead) (Hymenoptera: Platygasteridae) is an internal parasitoid that may not be detected by any of the larval parasitoid species. As the larval parasitoid develops, it kills the host and any internal parasitoid present within it (Force 1970). In another study, Hassell and May (1986) found that the interaction between generalist and specialist parasitoids may lead to a three-species stable system that can exist where the host–specialist or host–generalist interactions on their own would be unstable. Finally, it has been shown (Hassell *et al.* 1994) how it is possible for a third, relatively sessile species to coexist stably within the spatial dynamics generated by a two-species host–parasitoid interaction.

We believe that developing a general, all-encompassing theory to explain species richness within host–parasitoid communities remains a distant prospect. However, by exploring and understanding the dynamics of specific community components using simple models, we have attempted in this chapter to identify some of the factors that may explain the diversity of host–parasitoid communities.

Acknowledgements

We are very grateful to Mike Bonsall and Cheryl Briggs for their constructive comments on earlier versions of this manuscript.

References

Allen, J. C. (1975). Mathematical models of species interactions in time and space. *American Naturalist*, **109**, 319–342.

Askew, R. R. and Shaw, M. R. (1986). Parasitoid communities: their size, structure and development. In: *Insect parasitoids* (ed. J. Waage and D. Greathead), pp. 225–264. Academic Press, London.

Briggs, C. J. (1993). Competition among parasitoid species on a stage-structured host and its effect on host suppression. *American Naturalist*, **141**, 372–397.

Briggs, C. J., Nisbet, R. M. and Murdoch, W. W. (1993). Coexistence of competing parasitoid species on a host with a variable life cycle. *Theoretical Population Biology*, **44**, 341–373.

Comins, H. N., Hassell, M. P. and May, R. M. (1992). The spatial dynamics of host–parasitoid systems. *Journal of Animal Ecology*, **61**, 735–748.

Force, D. C. (1970). Competition among four hymenopterous parasites of an endemic insect host. *Annals of the Entomological Society of America*, **63**, 1675–1688.

Godfray, H. C. J. (1994). *Parasitoids. Behavioural and Evolutionary Ecology*. Princeton University Press.

Godfray, H. C. J. and Waage, J. K. (1991). Predictive modelling in biological control: the mango mealy bug (*Rastrococcus invadens*) and its parasitoids. *Journal of Applied Ecology*, **28**, 434–453.

Grover, J. P. and Lawton, J. H. (1994). Experimental studies on community convergence and alternative stable states: comments on a paper by Drake *et al. Journal of Animal Ecology*, **63**, 484–487.

Hassell, M. P. (1978). *The dynamics of arthropod predator–prey systems*. Princeton University Press.

Hassell, M. P. and May, R. M. (1973). Stability in insect host–parasite models. *Journal of Animal Ecology*, **42**, 693–726.

Hassell, M. P. and May, R. M. (1986). Generalist and specialist natural enemies in insect predator–prey interactions. *Journal of Animal Ecology*, **55**, 923–940.

Hassell, M. P. and May, R. M. (1988). Spatial heterogeneity and the dynamics of parasitoid–host systems. *Annales Zoologici Fennici*, **25**, 55–61.

Hassell, M. P., Comins, H. N. and May, R. M. (1991*b*). Spatial structure and chaos in insect population dynamics. *Nature (London)*, **353**, 255–258.

Hassell, M. P., May, R. M., Pacala, S. W. and Chesson, P. (1991*a*). The persistence of host–parasitoid associations in patchy environments. I. A general criterion. *American Naturalist*, **138**, 568–583.

Hassell, M. P., Comins, H. N. and May, R. M. (1994). Species coexistence via self-organizing spatial dynamics. *Nature*, **370**, 290–292.

Hawkins, B. A. and Lawton, J. H. (1987). Species richness for the parasitoids of British phytophagous insects. *Nature*, **326**, 788–790.

Hawkins, B. A. and Lawton, J. H. (1988). Species richness patterns: Why do some insects have more parasitoids than others? In: *Parasitoid insects* (ed. M. Bouletreau and G. Bonot), pp. 131–136. Les Colloques de l'INRA No. 48, Paris.

Hochberg, M. E. and Hawkins, B. A. (1992). Refuges as a predictor of parasitoid diversity. *Science*, **255**, 973–976.

Hochberg, M. E. and Hawkins, B. A. (1994). The implications of population dynamics to parasitoid diversity and biological control. In: *Parasitoid community ecology* (ed. B. A. Hawkins and W. Sheehan), pp. 451–472. Oxford University Press.

Hogarth, W. L. and Diamond, P. (1984). Interspecific competition in larvae between entomophagous parasitoids. *American Naturalist*, **124**, 552–560.

Holling, C. S. (1959*a*). The components of predation as revealed by a study of small mammal predation of the European pine sawfly. *Canadian Entomologist*, **91**, 293–320.

Holling, C. S. (1959*b*). Some characteristics of simple types of predation and parasitism. *Canadian Entomologist*, **91**, 385–398.

Hughes, R. D. and Salter, D. D. (1959). Natural mortality of *Erioischia brassicae* (Bouché) (Diptera: Anthomyiidae); life tables and their interpretation. *Journal of Animal Ecology*, **29**, 359–374.

Jones, T. H. (1986). Patterns of parasitism by *Trybliographa rapae* Westw., a cynipid parasitoid of the cabbage root fly. Unpublished PhD thesis, University of London.

Jones, T. H. and Hassell, M. P. (1988). Patterns of parasitism by *Trybliographa rapae*, a cynipid parasitoid of the cabbage root fly, under laboratory and field conditions. *Ecological Entomology*, **13**, 309–317.

Jones, T. H., Hassell, M. P. and Pacala, S. W. (1993). Spatial heterogeneity and the population dynamics of a host–parasitoid system. *Journal of Animal Ecology*, **62**, 251–262.

Jones, T. H., Hassell, M. P. and May, R. M. (1994). Population dynamics of host–parasitoid interactions. In: *Parasitoid community ecology* (ed. B. A. Hawkins and W. Sheehan), pp. 371–396. Oxford University Press.

Kakehashi, N., Suzuki, Y. and Iwasa, Y. (1984). Niche overlap of parasitoids in host–parasitoid systems: its consequences to single versus multiple introduction controversy in biological control. *Journal of Applied Ecology*, **21**, 115–131.

Kareiva, P. K. (1989). Renewing the dialogue between theory and experiments in population ecology. In: *Perspectives in ecological theory* (ed. J. Roughgarden, R. M. May and S. A. Levin), pp. 68–88. Princeton University Press.

Latto, J. and Hassell, M. P. (1988). Generalist predators and the importance of spatial density dependence. *Oecologia*, **77**, 375–377.

Mackauer, M. (1990). Host discrimination and larval competition in solitary endoparasites. In: *Critical issues in biological control* (ed. M. Mackauer, L. E. Ehler and J. Roland). Intercept, Andover. pp. 41–62.

May, R. M. (1973). *Stability and Complexity in Model Ecosystems*. Princeton University Press.

May, R. M. (1978). Host-parasitoid systems in patchy environments: a phenomenological model. *Journal of Animal Ecology*, **47**, 833–843.

May, R. M. (1981). Models for two interacting populations. In: *Theoretical ecology*. 2nd edition. (ed. R. M. May), pp. 78–104. Blackwell Scientific, Oxford.

May, R. M. (1995). Spatial chaos ... ecology and evolution. In: *Frontiers in mathematical biology* (Lecture Notes in Biomathematics Vol. 100), (ed. S. A. Levin), pp. 326–344. Springer-Verlag, New York.

May, R. M. and Hassell, M. P. (1981). The dynamics of multiparasitoid-host interactions. *American Naturalist*, **117**, 234–261.

Memmott, J. and Godfray, H. C. J. (1994). The use and construction of parasitoid webs. In: *Parasitoid community ecology* (ed. B. A. Hawkins and W. Sheehan), pp. 300–318. Oxford University Press.

Murdoch, W. W. (1969). Switching in general predators: Experiments on predator and stability of prey populations. *Ecological Monographs*, **39**, 335–354.

Murdoch, W. W. and Oaten, A. (1975). Predation and population stability. *Advances in Ecological Research*, **9**, 2–131.

Murdoch, W. W. and Stewart-Oaten, A. (1989). Aggregation by parasitoids and predators: effects on equilibrium and stability. *American Naturalist*, **134**, 288–310.

Nicholson, A. J. and Bailey, V. A. (1935). The balance of animal populations. Part 1. *Proceedings of the Zoological Society of London 1935*, 551–598.

Reader, P. M. and Jones, T. H. (1990). Interactions between an eucoilid (Hymenoptera) and a staphylinid (Coleoptera) parasitoid of the cabbage root fly. *Entomophaga*, **35**, 241–246.

Rogers, D. J. (1972). Random search and insect population models. *Journal of Animal Ecology*, **41**, 369–383.

Royama, T. (1971a). Evolutionary significance of predator's response to local differences in prey density: a theoretical study. Dynamics of Populations. *Proceedings of the Advanced Study Institute on Dynamics of Numbers in Populations (Oosterbeek, 1970)*, 344–357.

Royama, T. (1971*b*). A comparative study of models of predation and parasitism. *Researches in Population Ecology*, **1** (suppl), 1–91.

Shimada, M. and Fujii, K. (1985). Niche modification and stability of competitive systems. I. Niche modification process. *Researches in Population Ecology*, **27**, 185–201.

Solé, R. V., Valls, J. and Bascompte, J. (1992). Spiral waves, chaos and multiple attractors in lattice models of interacting populations. *Physics Letters A.*, **166**, 123–128.

Southwood, T. R. E. and Comins, H. N. (1976). A synoptic population model. *Journal of Animal Ecology*, **45**, 949–965.

Turnbull, A. L. and Chant, D. A. (1961). The practice and theory of biological control of insects in Canada. *Canadian Journal of Zoology*, **39**, 697–753.

van Alphen, J. J. M. and Visser, M. E. (1990). Superparasitism as an adaptive strategy for insect parasitoids. *Annual Review of Entomology*, **35**, 59–79.

van Strien-van Liempt, W. T. F. H. (1983). The competition between *Asobara tabida* (Nees von Esenbeck, 1834) and *Leptopilina heterotoma* (Thompson, 1862) in multiparasitised hosts. I. The course of competition. *Netherlands Journal of Zoology*, **33**, 125–163.

8

Structural constraints on food web assembly

Philip H. Warren

Introduction

Diversity can be quantified in many ways (Gaston 1994). Food webs are one representation of diversity which explicitly incorporates both taxonomic and functional aspects of community structure. Webs may be represented as binary structures, with links simply present or absent, or as 'functional webs' including information about energy flow or other measures of interaction strength. Most comparative analyses of food webs have utilised the former (e.g. Cohen 1978, Briand 1983, Pimm 1982, Pimm *et al.* 1991, Schoenly *et al.* 1991, Martinez 1992, Bengtsson, 1994; though see also Hairston and Hairston 1993).

Such data define the trophic 'scaffolding' of a community, representing interactions that are possible, or simply those that are most frequent (depending on the resolution of the data). Although just looking at the scaffolding provides limited information about the functional significance of each link, existence is a prerequisite for importance, and a connectance web defines the bounds of structure within which functionally important links must lie. Additionally, many of the factors determining the occurrence of a link will also be important in determining its strength (e.g. relative body size of predator and prey may both set limits to, and influence the frequency of, prey capture (e.g. Werner and Hall 1974, Kislalioglu and Gibson 1976)).

Perhaps more importantly, consistent structural differences in equally well resolved food webs, for example between webs from different habitat types, do convey meaningful information in much the same way as do species' presence/absence data. If such data can be found, they provide quite valid material for comparative analysis (Bengtsson 1994). The restrictions on what can be learned from these data are both in the quality and comparative resolution of the webs (Paine 1988, Lawton 1989, Martinez 1991, Polis 1991, Cohen *et al.* 1993) and in the type of questions which such data, even perfectly resolved, are appropriate to answer. Both are significant limitations, but these perhaps need to be balanced against the realistic thought that descriptive food webs, along with guild or functional group composition, may be the only level at which we will ever be able to quantify the trophic structure of entire complex systems, at something close to species-level resolution. While better data are clearly a necessity (Cohen *et al.* 1993), it is worth pursuing the question of what can and cannot be learned from such descriptive food web data, both for the light it may shed on trophic structure, and for the guidance it may provide toward the collection of the improved data.

Since feeding may, most obviously, have consequences for either energy flow, population dynamics, or both, it is common to assume that patterns evident in the structure of food webs must therefore be proximate consequences of such dynamic effects. Recent discussions of food webs review these hypotheses in detail (Pimm 1982, Lawton 1989, Pimm *et al.* 1991, Hall and Raffaelli 1993). Here I consider some less dynamically based ideas about how very general constraints on feeding adaptations of individuals may influence the way food webs are assembled, focusing in particular on the interactions between three major food web patterns: the density of linkage in a web (connectance), the lengths of food-chains and the ratios of predator to non-predator species.

Connectance and the cascade model

Cohen and Newman (1985) and Cohen *et al.* (1985) proposed that a number of the observed patterns in food webs could be adequately accounted for by a simple structural model (i.e. not explicitly modelling population dynamics or energy flow) they termed the 'cascade model'. This requires that links in a web are assigned independently and at random subject to two constraints: that species can be arranged in a hierarchy such that each can potentially feed only on those below it, and that the ratio of number of links to number of species is set at the value typically observed in natural food webs. The hierarchical arrangement may be justified on the basis of (among other things) body size (larger organisms generally eating smaller) (Warren and Lawton 1987, Cohen 1989*a*, Cohen *et al.* 1993), and the model's predictions seem fairly robust to moderate violations of this assumption (Warren 1988). A more fundamental problem is: what determines the number of links per species?

The density of linkage in a food web may be expressed simply as L/S, the number of links (L) divided by the number of species (S), or alternatively as connectance (C), the total number of actual links in a food web divided by the total number of possible links ($C = L/S^2$) (Martinez 1991). In various collections of food web data the number of links in a web is roughly proportional to the number of species with L/S almost equal to 2 (Cohen and Briand 1984, Cohen 1989*a*), or, more or less equivalently, C declines roughly hyperbolically with S (Rejmanek and Stary 1979, Yodzis 1980, Pimm 1982, Auerbach 1984). Subsequent evidence, however, suggests that L/S may be underestimated in many of the webs in the original collection and that it may increase with web size (Schoener 1989, Warren 1990, Martinez 1992). High values of connectance, especially in larger webs (i.e. if connectance doesn't decrease with S), quite markedly affect the results of the cascade model, in particular predicting food-chains that are longer than observed (Warren 1988, 1989, Cohen 1990). This dependence on connectance highlights the need for an independent explanation for linkage density, without which the assembly rules defined by the cascade model cannot be viewed as complete.

Explanations for connectance

Ideas proposed to explain the observed patterns in connectance have generally sought to account for the observation of constant L/S (decreasing connectance in larger webs), though more recent hypotheses have been suggested which predict that L/S should increase with web size. These ideas are reviewed elsewhere (Pimm 1982, Lawton 1989, Warren 1994), but the main ones can be viewed as variants on four basic ideas.

(1) Stability: some mathematical models of food webs indicate that more highly connected webs are less likely to be stable (May 1973, Pimm 1982, Cohen and Newman 1988).

(2) Morphology: there are biological restrictions on the range of resources on which a species can be adapted to feed, and on the range of consumers a species can be adapted to avoid (Pimm 1982, Schoener 1989, Warren 1990).

(3) Sampling: because the number of potential links scales as S^2, it is harder to record all the links in a 100 species web than in a 10 species web, and hence links are likely to be disproportionately unrecorded in larger webs (Paine 1988, Cohen and Newman 1988, Kenny and Loehle 1991).

(4) Combinatorics: if links are assigned between species at random, there are more possible ways of getting some food web configurations than others, and these configurations should be observed most frequently in the real world (Kenny and Loehle 1991; see also, Auerbach 1984).

A number of recent studies of natural food webs have emphasised aspects of the second of these, the biological constraints on feeding, and avoiding being fed upon (Schoener 1989, Warren 1990, Martinez 1991, Havens 1992). These are related to the suggestion by Pimm (1982) that any consumer species can only feed on a fixed maximum number of other species, however big the community, because of limits to the range of resource types on which a species can be adapted to feed (it is difficult to be a 'jack-of-all-trades' feeding effectively on very different resources). However, while Pimm's scheme predicts that linkage density will remain constant with increasing species number (and hence connectance decline), the other ideas generally predict an increase in linkage density (Warren 1994). These ideas, or developments of them, form the starting point for some simple biological assembly rules for food webs.

Biological constraints on feeding links

A particular species, or life-stage of a species, can in principle be defined in terms of its various morphological, chemical, behavioural, physiological, and other, characteristics. If the possible range of each characteristic is represented as an axis in a multidimensional niche space, then a species may be defined as a point (or more usually a region) in this space, i.e. the point defined by the particular combination of morphology, behaviour, environmental tolerance etc. exhibited by that organism. In addition to an organism occupying part of such space, its feeding adaptations (mouthpart morphology, reaction characteristics, speed of movement, microhabitat use, etc.) delimit another region which defines the types of resources it is capable of capturing and eating (its fundamental feeding niche). For a link to occur between two species, one of them must have characteristics which fall in the region which the other is capable of exploiting. Several species may occur in the region exploitable by another, and hence all will contribute links to that consumer. In essence, this argument simply extends the conventional depiction of the niche—defining what a species requires to survive (what might be termed the 'demand' niche) to multiple trophic levels, by incorporating the characteristics of each species as a potential resource for others (the 'supply' niche) into the same space.

Assume that the total niche space is finite and imagine a community assembling from a single basal resource (see Fig. 8.1). For a species to enter the system its demand niche must to some extent overlap with the supply niche of another organism in the system (it must have food). As more species are added, the supply niche of an invader may also fall within the demand niche of a species already present. Thus, links accrue both upwards and downwards as species arrive (Fig. 8.1).

Several things are evident from this process. If species are arranged at random in niche space, then the expected number of links for a single species is pS, where p is the proportion of niche space exploited by the species. The total number of links in a web is therefore pS^2, where p is now the average of the proportions of space exploited by all species (Warren 1990). Directed connectance, $C = L/S^2$, with certain biases, is an estimate of p. As communities assemble (S increases) species will become more closely packed in niche space. All else being equal, this will result in an increase in the average number of species falling within the demand niches of others and, consequently, an increase in the number of links per species. The actual number of links per species will be affected by the size of the feeding niches of each species. Webs having high proportions of generalist species (large p) will tend to have higher linkage density than those with a high proportion of specialists. The assembly of a single community will follow a particular links:species relationship of the form $L = pS^2$, but p will almost certainly vary between systems, resulting in a range of $L:S$ relationships. A single food web is a single point off the appropriate relationship for that particular type of system

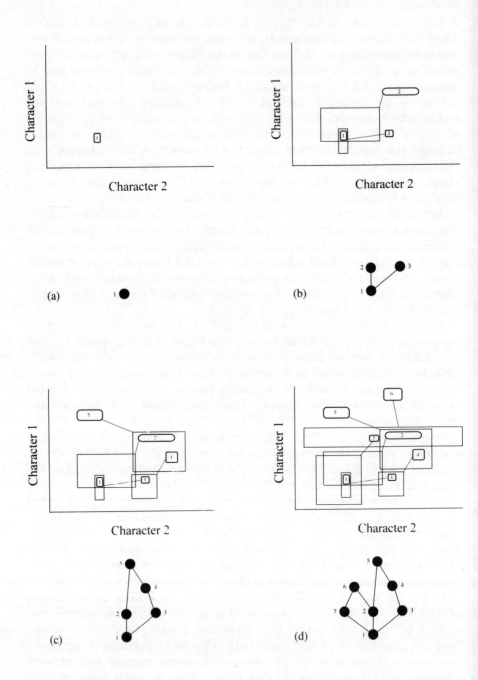

(e.g. Warren 1990), resulting, with data from many different communities, in a spreading scatter of data points, following a generally upward curvilinear trend (Winemiller 1989, Schoener 1989, Martinez 1991, Havens 1992).

Although the scheme above suggests the general form of the relationship between L and S, it does not predict specific values. Investigation of another aspect of web structure, however, does suggest a way to develop such predictions.

Predator:non-predator ratios

One pattern that has received considerable attention in trophic ecology is the observation of a correlation between the numbers of predatory and non-predatory (often termed 'prey') species in a wide range of communities (e.g. Jeffries and Lawton 1985, Warren and Gaston 1992). The correlation of predator:non-predator species richness has been construed as evidence for the structuring effects of biological interactions in communities ('assembly rules' (Lawton 1987)), though the idea that the ratio (along with other web statistics) is 'constant' has been justly criticised (Hall and Raffaelli 1993). Five explanations for the correlation have been advanced (Warren and Gaston 1992):

(1) more non-predators provide more predator food niches;

(2) apparent competition limits the number of prey species that can coexist with each predator (Jeffries and Lawton 1985);

(3) the ratios of energy availability to predator and non-predator species in a system constrain the ratios (Warren and Gaston 1992);

Legend to facing page

Fig. 8.1 A simple graphical illustration of the assembly of a food web in terms of occupancy and exploitation of niche space, and the corresponding binary food web structure. The *upper part* of each diagram indicates a two-dimensional niche space in which species are defined by their range of characteristics on the two niche axes (e.g. size) (denoted by the round-cornered, numbered boxes), and exploit certain regions of the same space—their feeding niche (denoted by the square-cornered, unnumbered boxes linked to the species by a thin line). Each species falling within the feeding niche of another is assumed to be a potential prey of that species. The corresponding food web for each stage in the assembly is shown in the *lower part* of each diagram. The assembly starts with some basal resource (species 1) which has no feeding niche on these axes. The assembly to a seven species web includes some generalist species (2,5,6,7) and more specialist species (3,4). Species can only invade if there are resources available in the region of niche space they are able to exploit. The representation here is an illustrative, simple case of the process; variations and other assumptions are discussed in the text.

(4) non-trophic determinants of diversity act similarly, and independently on predator and non-predator sections of a community (Warren and Gaston 1992); and

(5) the observed ratios correspond to the expectation of a random draw from the pool of available species (Cole 1980).

These explanations are not easily separable, since all make similar predictions, nor are they mutually exclusive. Individually, or in combination, these factors seem likely to reduce the likelihood of extreme dissociation of predator and prey species richness and it is hard to imagine how some correlation between the two could be avoided.

Although most attention has focused on predator–prey ratios, there is also some evidence from food webs (Cohen 1989a, Schoenly et al. 1991) and from other studies (Gaston 1992, and refs therein) that the number of basal species in a community forms a roughly scale-invariant proportion of the total number of species in a community, at least where the number of species is not small (though see Martinez 1994 for more extensive discussion of this point).

How do these observations relate to connectance? Grouping species by general feeding type defines broad areas of niche space they can exploit. This, in combination with the body size restrictions discussed above, constrains the number and arrangement of potential trophic links in a web. Basal species (B) feed on no others in the web. Primary consumers (C_1) can feed on basal species (max. links = BC_1). Secondary consumers (C_2) can feed on primary consumers, but there are no reciprocal links. Of these possible links only those from larger to smaller species occur (for predators). Assuming size to be independent of whether a species is a primary or secondary consumer, approximately half the links will satisfy this condition (expected max. links = $(C_1C_2)/2$). Links among secondary consumers occur subject to size restrictions as above (expected max. links = $C_2^2/2$ (allowing cannibalism)). The expected maximum number of potential links in a web is the sum of these parts: $L_{max} = BC_1 + (C_1C_2)/2 + C_2^2/2$. For simplicity, independence of size and feeding type and a lack of omnivory (species feeding on both animal and plant material) and parasites are assumed. The estimate is obviously for the maximum expected number of links; other factors will determine which of these potential links actually occur.

Predictions of web structure

The basic prediction from the constraints above is that connectance should be essentially constant with a value of about 0.2 (taking the proportions of basal species from existing food web data (≈ 0.2) and the proportions of non-basal species that are predators from data in Warren and Gaston (1992) (≈ 0.3)). To fully explore the effects of these constraints on web assembly, food webs were assembled on a computer, following the general form of rules indicated in Fig 8.1.

Based on existing food web data (Cohen 1989*b*) the proportion of basal species was selected at random from an approximately normal distribution, with the lower tail truncated at zero (mean = 0.196, s.d. = 0.11). The proportion of non-basal species that were non-predators was selected at random from the distribution of this statistic, taken from the data in Warren and Gaston (1992), which again was approximately normal, but truncated at zero and one (mean = 0.707, s.d. = 0.12). Each non-basal species was allocated a body size drawn at random from a lognormal distribution. In the same way each species was allocated a value, drawn from a uniform distribution, on a second niche axis (with no specific meaning) termed 'morphology'. These two values define the species position in niche space. Species were then also allocated values for the region of niche space they exploit. For size, this was a random fraction of the size axis smaller than the consumer, and was assigned for predators only; for morphology, it was simply a random fraction of the morphology axis, independent of the morphological value of the consumer, and was assigned for predators and non-predators alike. Each species was added to the food web (in order from basal, through non-predator to predator), and links were allocated between species where the feeding niche of a consumer overlapped the position in niche space of a potential resource. If a species turned out to have no feeding links, it was rejected and another (of the same feeding type) was tried. Thus the final webs have basal:non-predator:predator ratios which vary but are not systematically affected by the assembly process (necessary because the values are set using data from real assembled systems and not the true species pool). Niche positions and feeding niches may however be affected, since values which increase the probability of a species having others to exploit (e.g. a large feeding niche) will be favoured when S is small.

Connectance The least restrictive application of these rules, in which primary consumers are able to feed on all basal resources and secondary consumers on any other animal smaller than themselves (with random size ordering of consumers) gives essentially the result derived from the calculation above (Fig. 8.2). Maximum expected connectance is roughly constant and rather higher than that for the data from published food webs. Reducing the size range and the range of morphological types of resources on which a consumer can feed, will obviously reduce the number of links in a web. Figure 8.2 shows the expected connectance of webs, assembled subject to the constraints described above, in which the proportion of resource sizes (for those smaller than a consumer) and the proportion of morphological types which a consumer can exploit are selected at random, and independently, in the range 0–1 for each consumer. Connectance declines with S to a roughly constant value of about 0.1 (the value Martinez (1992) finds as the 'mean directed connectance' in well-resolved food web data). Further specialization (where the proportion of each axis being exploited is a random value from a uniform distribution, in the interval 0–0.4 times consumer size, for the size

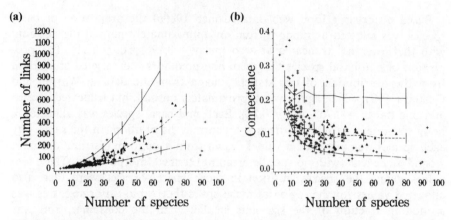

Fig. 8.2 The relationships of (a) trophic links and (b) connectance, to species numbers for simulated food webs (lines) and real data (points). The lines join the mean values of links or connectance for each species number (each mean is from 60 simulated webs). The error bars indicate 1 standard deviation about the mean. Three simulations are shown, in all cases the predator:non-predator:basal species proportions were selected at random from the same empirically derived distribution (see text), (i) solid line—expected maximum linkage—see text; (ii) short-dashed line—size and morphology feeding niches (i.e., proportion of the possible size range and range of resource morphologies over which a consumer actually feeds) chosen independently and at random from a uniform distribution (interval 0–1); (iii) long-dashed line—size and morphology feeding niches chosen at random on the restricted interval 0–0.4. The data are from the ECOWEB collection (Cohen 1989*b*) with the addition of webs from Warren (1989), Martinez (1991), Hall and Raffaelli (1991), Goldwasser and Roughgarden (1993), and Havens (1992)—points in the latter data-set are indicated by the solid triangles.

axis, and 0–0.4 on the morphological axis) results in connectance around the lowest values in real webs (Fig. 8.2). Thus it seems that the range of feeding niche sizes, with respect to prey size and morphology, used here predicts patterns and values of connectance which roughly bound the real (although imperfect) data. If this is so, then are these values sensible? A thorough investigation of this question is beyond the scope of this discussion but, at least with respect to predator–prey body size ratios, examination of data from various taxa (e.g. Vezina 1985, Warren and Lawton 1987, Cohen *et al.* 1993) suggests the values used here are reasonable. However, despite various general discussions of predator–prey size effects (e.g. Sabelis 1992), good comparative data for a wide range of species, establishing the precise nature of the prey size distribution (with a range of prey types and with predator and prey measured in comparable units) is surprisingly scarce. Scarcer still are data on what proportion of the prey types falling within the prey size-range distribution of a predator are actually taken (values for the morphological axis, above). I shall return to an attempt to estimate such parameters later.

Food chain length   The model outlined above can also be used to examine other aspects of web structure. One pattern of particular interest is the observation that food-chains are typically short ($\approx$ 4–5 links) in natural food webs (Hutchinson 1959, Pimm 1982, Lawton 1989, Pimm *et al.* 1991). The structural constraints provided by the cascade model predict the generally observed pattern of chain lengths, but only if limited values of connectance, which decrease with S, are used in the model. Using the assembly model outlined here, some interesting observations concerning chain length emerge. For the version in which both size and morphology are allowed to vary at random (Fig. 8.2), the predicted mean chain lengths for webs from six to eighty species show a slight increase with S with the expected values for the largest webs of between 3 and 4 links (Fig. 8.3). These values are well within the range of values observed in the ECOWEB data (indeed many real chains are longer), and the slight increase in mean chain length with S has been noted by others in well-resolved data sets (e.g. Hall and Raffaelli 1993). The unrestricted model predicts food-chains that are rather longer (Fig. 8.3), though, interestingly, over the range of web sizes for which the bulk of data are available the predictions are remarkably good, suggesting that for much of the existing web data chain, lengths fall within the range expected even at maximum connectance. More restricted (specialist) webs produce shorter food-chains (Fig. 8.3). Thus it seems possible to obtain structural constraints on webs which give acceptable predictions of food-chain length, along with predictions of approximately constant connectance.

The assembly of a 'real' food web

One way to examine the role of the feeding adaptations of different species in

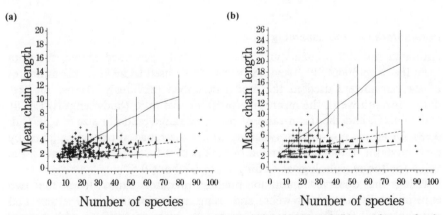

Fig. 8.3   The relationships of (a) mean food-chain length and (b) maximum food-chain length, to species numbers for simulated food webs (lines) and real data (points). All symbols and data as for Figure 8.2.

structuring food web patterns is to examine the way in which food web structure changes as it assembles in the absence of any energetic or population dynamic constraints. I have looked at these effects using organisms from freshwater systems (Warren 1995) to investigate the potential structure of a food web assembled as far as possible at random from the organisms available in a regional species pool. Food webs were assembled by collecting individual species (selected at random from field-compiled species lists for each site) from fifteen freshwater habitats in the area around Sheffield (northern England). The potential feeding links between each pair of species were examined in laboratory feeding trials and on the basis of this information two hypothetical food webs constructed.

Patterns

The experimentally assembled webs make some interesting points. Firstly, the relationships between the number of links and number of species as each web is assembled follow the predicted $L = pS^2$ pattern. Secondly, the values of connectance ($C = L/[S(S - 1)]$) fall well within the range of those observed in webs from the existing collections. This suggests that the expected value for connectance in this type of system lies somewhere around 0.14–0.16, and that many real webs are no more connected than would be expected on the basis of the morphological feeding limitations of their potential constituent species. Additionally, the experimentally assembled webs also have mean and maximum food-chain lengths and proportions of intermediate and top species (allowing for the constraint of a single basal species), which are similar to those observed, for similarly sized webs, in the ECOWEB collection (Cohen 1989b); and have predator/non-predator richness ratios that are well within the range observed for other British freshwater systems (Warren and Gaston 1992).

Parameters for web assembly

Although not the original purpose of the study described above, the data from the experimentally assembled webs can be used to make crude estimates of the parameters used in the model described previously. Across all predators in both webs, the mean prey/predator size ratio (body lengths) is 0.61 (s.d. = 0.25), and the mean ratio of prey size range/predator size is 0.56 (s.d. 0.38). Of all the non-predators falling within the range of prey sizes taken by each predator, a mean proportion of 0.52 (s.d. = 0.26) are eaten, that is, the value for the morphological feeding niche is about 0.5.

Using the observed basal:non-predator:predator ratios from the two experimentally assembled webs, and using approximations to the size and morphological feeding niche sizes above, the range of numbers of links and food-chain lengths expected under the simple assembly rules defined earlier can be predicted. One thousand model webs were assembled for each of the two experimental webs (differing only in the proportions of predators and non-predators). Body sizes were drawn at random from the same lognormal distribution (size distributions of predators and non-predators were slightly, but not significantly, different in the actual webs). The width of the size feeding niches of species (as a proportion of consumer size) were drawn at random (from a uniform distribution) on the interval 0.3 to 1, and positioned randomly on the size axis; the morphological feeding niche was selected as an absolute value from the interval 0.2 to 1 (where the morphological axis has a range from 0 to 1) and again positioned at random. These limits and means expected from them approximate to those observed in the experimentally assembled webs, though the distribution of values within the limits may not do so exactly. The average number of links from all the simulated webs is 35.8 (s.d. = 7.0). The experimentally assembled webs have 36 and 40 links. The average mean food-chain length from all the simulated webs is 2.8 (s.d. = 0.46), and the average maximum length is 4.4 (s.d. = 0.9). The experimental webs have values of 2.76 and 3.13 (mean lengths) and 4 and 5 (or 6) (maximum lengths). The predictions are altered, as might be expected, by variation in the values of the feeding niches and the distributions of relative predator and prey size, though for none of the plausible parameter values from the experimentally assembled webs examined do the actual webs fall outside two standard deviations from the mean prediction, and the predicted values are generally lower than the observed. If we assume that these parameters have some generality for freshwater invertebrates then it is worth examining the predictions for larger webs. Using the same parameters to simulate webs of 89 species (300 webs generated) predicts a mean number of links of 857 (s.d. = 154) and a mean chain length of 4.8 (s.d. = 1.4). Few comparably sized webs are documented, and there is only one freshwater web with a substantial benthic invertebrate component, namely that of Little Rock Lake (Martinez 1991). This has 93 (trophic) species, slightly over 1000 links, and a mean chain length of about 7. The predictions are rather lower than the Martinez' web, but this web's values fall within two standard deviations of the mean prediction.

What can be read into the correspondence, or disparity, of the experimental and theoretical assembly? The obvious suggestion is that the linkage and food-chain lengths of many real food webs fall within the ranges expected from some simple assembly rules, parameterized by some rather crude data. Are more complex explanations unnecessary? It is perhaps premature to draw such a general conclusion at this stage, but the results outlined here do suggest that examination of the consequences of individual or species-level feeding constraints on the distribution of links in a food web may repay further investigation.

Implications

The ideas outlined here indicate that there is potentially considerable non-independence among at least three food web patterns: connectance, ratios of trophic types and food-chain length. These observations, and the mechanisms they imply, have a number of possible consequences for analysing and interpreting natural food webs; in particular, for the temporal and spatial scales at which explanations for web structure should be sought, the potential for webs to respond to environmental effects, and the effect of biological interactions on web structure. The following are a few speculative suggestions on these points.

Much of the above discussion places the emphasis firmly on an evolutionary route to understanding and predicting many differences in food webs. Evolutionary explanations for aspects of web structure have been advanced before (e.g. Hutchinson 1959, Hastings and Conrad 1979, Yodzis 1984, Stenseth 1985). These studies deal with evolutionary effects such as design constraints, the trade-offs which restrict niche breadth, especially the likelihood of feeding on both animal and plant material, and the consequences of general evolutionary stable strategies of feeding for food-chain length. The question of what determines the potential feeding niche breadth of a species is an evolutionary one and, if web structure is affected by niche breath, then understanding the factors which influence niche breadth in different habitat types or taxa is one important component of predicting web structure. For example, such data as there are suggest that freshwater benthic invertebrates are typically generalist, which leads to the prediction that webs composed of such species should be comparatively highly connected and, perhaps, have some rather long food-chains. By the same token, webs dominated by more specialist species (e.g. plants, insect herbivores and parasitoids) should be less connected (e.g. Rejmanek and Stary 1979). Such links are obvious, but potentially tautologous. The selective pressures on an organism do not exist entirely outside the context of the community in which it normally occurs, so aspects of web structure (e.g. stability) could potentially feed back to species' niche breadths. However, many other aspects of a species environment will affect its feeding adaptations, and community turnover will typically occur on a faster time-scale than fundamental changes in feeding biology, so proceeding from feeding adaptations to web structure seems justified. This approach also makes explicit a route from individual natural selection to emergent properties of entire food webs.

The question clearly posed by such a conclusion is: what is the basis for the evolutionary change of species' feeding ranges? Various ideas have been advanced, including the influences of environmental stability and resource availability, resource disparity, and competition with other organisms (e.g. MacArthur and Levins 1967, Glasser 1982, Crawley 1983, Stenseth 1985, Jaenike 1990, Zwolfer and Arnold-Rinehart 1993, Pianka 1994). Optimal foraging theory makes some well-established predictions about the shifts in diet breadth with changes in patterns of resource availability, in particular suggesting that an increase in resource abundance should result in a reduction of niche breadth as less valuable resource types are dropped from the diet (Pyke et al. 1977, Crawley and Krebs 1992). Generally, however, the evolution of dietary breadth seems to be an involved problem. Most work has focused the basis of specialism in herbivores, especially herbivorous insects (Crawley 1983, Jaenike 1990), and comparatively little on, for example, predatory arthropods (Sabelis 1992). Jaenike (1990) reviews a range of factors with the potential to produce selection for narrow, or broad, host ranges among phytophagous insects, many of which are potentially applicable to other types of organism. In particular, the role of resource predictability (and 'apparency') may have wide applicability to the evolution of diet breadth (Glasser 1982), with the suggestion from plant–insect studies that predictable and abundant resources favour specialism (Jaenike 1990, Zwolfer and Arnold-Rinehart 1993). Other discussions concentrate on the disparity of resources (and hence the adaptations required by a consumer to exploit them) in relation to their abundance, with specialists dominating where resources are abundant, and either specialists or generalists, depending resource disparity, where resources are scarce (Stenseth 1985). Pressure on species to efficiently exploit limited resources in competition with other species is also hypothesised to play a part in selecting for specialism, under the assumption that specialists are of higher efficiency than generalists (Fox and Morrow 1981, Crawley 1983, Jaenike 1990, Pianka 1994).

If correspondence could be established between the selection for different feeding strategies and environmental conditions, there would be a basis for predictions about the relationship between web structure and environment. For example, species in temporally unstable habitats, where chance plays a strong role in determining species composition (and hence resource types) might experience selection for generalist feeding habits, and consequently food webs from such environments might be more connected than those from more predictable systems (the opposite prediction to that from the population stability hypothesis, e.g. Briand 1983). Similarly, if enrichment of a habitat promotes more specialist species, or foraging behaviour, then increasing energy input might reduce connectance, with a consequent tendency to produce shorter food-chains, countering any effect of increased energy availability at higher trophic levels.

In addition to factors influencing potential diet breadth, web structure will also be affected by the dispersion of species' morphologies in niche space. Where several similar species occur, potential consumers of one are also likely to feed on the others, resulting in higher numbers of links than if resources were less aggregated in niche space. Patterns of species occurrence in niche space could be affected by phylogeny; individual communities may tend to have larger numbers of species from the more speciose taxa, and these species will not be morphologically independent. Patterns may also be affected by environmental conditions. In harsh or extreme habitats, only a limited range of taxa, or ecological types, may be able to adapt to that environment; this could also produce clustering in niche space, possibly increasing connectance in food webs from such habitats. Biological interactions could be important. If strong competition for resources (or apparent competition, for 'enemy free space' (Jeffries and Lawton, 1984)) occurs, we would expect species to be more dissimilar than expected by chance, and the resulting overdispersion of species in niche space to reduce connectance.

A final point concerns the widely cited problem that binary food webs give little information about the strength of interactions (Paine 1980, 1988). While this is clearly the case, it is also true that the feeding niche, simplified here to a simple range where resources falling within the range are always taken, is actually a set of probability distributions whose joint effect reflects the likelihood of a resource of particular characteristics being taken at each instance when the potential opportunity for that link occurs. These probabilities will therefore be a reflection of the relative frequency of each interaction, which will affect the energy flow along that link and possibly its strength (in a functional sense, e.g. Paine 1980). This has something in common with the suggestion of Hairston and Hairston (1993) that aspects of trophic structure (resulting from morphological factors such as predator–prey size ratios) may determine some of the observed patterns of energy flow, rather than energy flow determining trophic structure, as is often assumed. In this context, biologically generated structural constraints may be useful in setting the framework for modelling web dynamics (e.g. Cohen et al. 1990).

Conclusion

The focus of this paper has been on the extent to which aspects of the structure of natural food webs may result from constraints on the numbers and arrangements of links provided by assembling communities with particular proportions of basal:non-predator:predator species, and restricted feeding niches. Do these ideas, connectance food webs, or the area of food web ecology of which these are a part, have significant things to say about the determinants and conservation of biodiversity (in the wide sense of this volume)? Despite optimistic predictions about future possibilities in recent reviews (e.g. Cohen 1989a, Pimm et al. 1991) the direct contribution of the sort of food web analyses discussed here to applied issues has, with a few exceptions (e.g. Rasmussen et al. 1990, Cabana et al. 1994), been fairly limited.

It has been suggested or implied that food webs may be used empirically as a basis for identifying systems of unusual structure, perhaps reflecting damage or stress to a system, or as a template for the construction, or reconstruction of communities in artificial, or natural, habitats (Cohen 1989a, Jenkins and Kitching 1990, Pimm 1991, Havens 1991, Locke 1992). Use of webs as indicators of change depends substantially on the sensitivity and accuracy of the derived patterns and also on whether the same problem could be addressed by other means. For example, many food web patterns are not independent of numbers of species (Martinez 1994, Bengtsson 1994). Species number might provide an equally good and easier to collect measure of community change, and may make interpretation of change in other food

web statistics more difficult. Given the difficulty of collecting good comparable food web data (Cohen *et al.* 1993) it is likely to be a very labour-
intensive method for use as a working index of ecosystem state. Deciding
whether to use an aspect of web structure as a measure of community change
or disruption depends substantially on understanding the causes and correlates of that pattern. This applies equally to scale-dependent and scale-
invariant patterns. Basing the assembly of a new community, or management
of existing ones, on scale-invariant properties (Cohen 1989*a*) may be a sensible precaution, but if those properties are unavoidable consequences of
simple structural assembly rules such as those embodied in the cascade
model, then effort should perhaps be concentrated on addressing more
functionally critical aspects of a food web in the selection of suitable components. Despite these caveats, accurate data on pattern in natural webs
could provide one component of a template for assembling and maintaining
communities, even without a complete understanding of the functional basis
for the pattern. This especially applies to patterns based on the distributions
of species functional types (e.g. producers, herbivores, predators etc.) rather
than designations that can change as other components of the web change
(e.g. intermediate/top species). Such distributions have already been discussed
in the context of estimating biodiversity (May 1990, Gaston 1992, and
references therein), though these approaches do not generally utilise food web
information as such, and the related approach of using higher taxa may be
more practical (Gaston 1993). Overall, the empirical use of food webs as
indicators of community function or predictors of aspects of diversity is likely
be limited by the problems of data collection and the availability of alternative routes to the same goal. This is not to say that analysis of changes in
food webs, for example in response to environmental perturbations, is not
useful, but its value must lie principally in helping to develop an understanding of how communities function rather than as a system for monitoring.

As stated at the outset, food webs are a tool and not an end in themselves.
The contributions of food web analyses to the wider goal of understanding
how natural communities work, and how their composition and function may
be maintained, are therefore largely indirect. Food webs, especially connectance food webs, provide a way of defining the trophic structure of a
community in such a way that it can be quantified, and hence provide data
for comparative analyses and for testing of theoretical predictions. The
ultimate goal is to understand the significance of these processes. This
understanding may in turn have implications for predictive conservation
management, and for sorting out the competing hypotheses to account for
patterns in relative biodiversity. Pimm (1991) presents perhaps the most
substantial example of this process with respect to food webs, in which,
among other things, conclusions drawn from the population dynamics
approach to food web theory (e.g. Pimm 1982) are used to explore problems
of conservation biology from extinction in small populations to the impact of

introduced species on communities. In doing this much depends on whether or not the original conclusions are correct, and many of these conclusions are based on the (often qualitative) correspondence between modelling studies and observed community, or binary food web, structure. If aspects of web structure are consistent with predictions from several, functionally very different processes (e.g. the debate over the length of food-chains; Lawton 1989) then it is critical to distinguish which process, or combinations of processes, are actually responsible for the patterns observed, before those processes are used as the basis for developing theories about biodiversity. Food webs, along with other quantitative descriptions of communities (e.g. species abundance distributions), are one tool in this testing procedure. Understanding what makes food webs the way they are may well generate the information to understand many other aspects of community and ecosystem function.

Acknowledgements

I should like to thank Kevin Gaston, John Lawton and Dave Raffaelli for helpful and critical discussions; Gerard Lacroix, Kevin Gaston and Neo Martinez for comments on the manuscript; L'Ecole Normale Superieure for hospitality and the British Council for financial support.

References

Auerbach, M. J. (1984) Stability, probability, and the topology of food webs. In Strong, D. R., Simberloff, D., Abele, L. G. and Thistle, A. B. (eds.) *Ecological communities: conceptual issues and the evidence*. Princeton University Press, pp. 413–436.

Bengtsson, J. (1994) On comparative analyses in community ecology, confounding variables and independent observations in food web studies. *Ecology* **75**, 1282–1288.

Briand, F. (1983) Environmental control of food web structure. *Ecology* **64**, 253–263.

Cabana, G., Tremblay, A., Kalff, J. and Rasmussen, J. B. (1994) Pelagic food-chain structure in Ontario lakes, a determinant of mercury levels in lake trout (*Salvelinus namaycush*). *Can. J. Fish. Aquat. Sci.* **51**, 381–389.

Cohen, J. E. (1978) *Food webs and niche space*. Princeton University Press.

Cohen, J. E. (compiler) (1989a) *ECOWeB. Machine readable database of food webs*. Version 1.00. Rockefeller University, New York.

Cohen, J. E. (1989b) Food webs and community structure. In Roughgarden, J., May, R. M. and Levin, S. A. (eds.) *Perspectives in ecological theory*. Princeton University Press, pp. 181–202.

Cohen, J. E. (1990) A stochastic theory of community food webs. VI. Heterogeneous alternatives to the cascade model. *Theor. Pop. Biol.* **37**, 55–90.

Cohen, J. E. and Briand, F. (1984) Trophic links of community food webs. *Ecology* **81**, 4105–4109.

Cohen, J. E. and Newman, C. M. (1985) A stochastic theory of community food webs I. Models and aggregated data. *Proc. Roy. Soc. Lond. B* **224**, 421–448.

Cohen, J. E., Newman, C. M. and Briand F (1985) A stochastic theory of community food webs II. Individual webs. *Proc. Roy. Soc. Lond. B* **224**, 449–461.

Cohen, J. E. and Newman, C. M. (1988) Dynamic basis of food web organization. *Ecology* **69**, 1655–1664.

Cohen, J. E., Luczak, T., Newman, C. M. and Zhou, Z.-M. (1990) Stochastic structure and nonlinear dynamics of food webs, qualitative stability in a Lotka-Volterra cascade model. *Proc. Roy. Soc. Lond. B* **240**, 607–627.

Cohen, J. E., Beaver, R. A., Cousins, S. H., DeAngelis, D. L., Goldwasser, L., Heong, K. L., *et al.* (1993) Improving food webs. *Ecology.* **74**, 252–258.

Cohen, J. E., Pimm, S. L., Yodzis, P. and Saldana, J. (1993) Body sizes of animal predators and animal prey in food webs. *J. Anim. Ecol.* **62**, 67–78.

Cole, B. J. (1980) Trophic structure of a grassland insect community. *Nature* **288**, 76–77.

Crawley, M. J. (1983) *Herbivory, the dynamics of animal-plant interactions*. Blackwell Scientific Publications, Oxford.

Crawley, M. J. and Krebs, J. R. (1992) Foraging theory. In Crawley, M. J. (ed.) *Natural enemies, the population biology of predators, parasites, and diseases*. Blackwell Scientific Publications, Oxford, pp. 90–116.

Fox, L. R. and Morrow, P. A. (1981) Specialization, species property or local phenomenon? *Science* **211**, 887–893.

Gaston, K. J. (1992) Regional numbers of insect and plant species. *Funct. Ecol.* **6**, 243–247.

Gaston, K. J. (1993) Mapping the world's species—the higher taxon approach. *Biodiversity Letters* **1**, 2–8.

Gaston, K. J. (1994) Biodiversity—measurement. *Progress in Physical Geography* **18**, 565–574.

Glasser, J. W. (1982) A theory of trophic strategies, the evolution of facultative specialists. *Am. Nat.* **119**, 250–262.

Goldwasser, L. and Roughgarden, J. (1993) Construction and analysis of a large Caribbean food web. *Ecology* **74**, 1216–1233.

Hairston, N. G. and Hairston, N. G. (1993) Cause–effect relationships in energy flow, trophic structure, and interspecific interactions. *Am. Nat.* **142**, 379–411.

Hall, S. J. and Raffaelli, D. G. (1993) Food webs, theory and reality. *Adv. Ecol. Res.* **24**, 187–239.

Hastings, H. M. and Conrad, M. (1979) Length and evolutionary stability of food-chains. *Nature* **282**, 838–839.

Havens, K. E. (1991) Crustacean zooplankton food web structure in lakes of varying acidity. *Can. J. Fish. Aquat. Sci.* **48**, 1846–1852.

Havens, K. E. (1992) Scale and structure in natural food webs. *Science* **257**, 1107–1109.

Hutchinson, G. E. (1959) Homage to Santa Rosalia or why are there so many kinds of animals? *Am. Nat.* **93**, 145–159.

Jaenike, J. (1990) Host specialization in phytophagous insects. *Ann. Rev. Ecol. Syst.* **21**, 243–273.

Jeffries, M. J. and Lawton, J. H. (1984) Enemy free space and the structure of ecological communities. *Biol. J. Linn. Soc.* **23**, 269–286.

Jeffries, M. J. and Lawton, J. H. (1985) Predator–prey ratios in communities of freshwater invertebrates, the role of enemy free space. *Freshwat. Biol.* **15**, 105–112.

Jenkins, B. and Kitching, R. L. (1990) The ecology of water-filled treeholes in Australian rainforests, Food web reassembly as a measure of community recovery after disturbance. *Aust. J. Ecol.* **15**, 199–205.

Kenny, D. and Loehle, C. (1991) Are food webs randomly connected? *Ecology* **72**, 1794–1799.

Kislalioglu, M. and Gibson, R. N. (1976) Prey 'handling time' and its importance in food selection by the 15-spined stickleback, *Spinachia spinachia* (L). *J. Exp. Mar. Biol. Ecol.* **25**, 151–158.

Lawton, J. H. (1987) Are there assembly rules for successional communities? Gray, A. J., Crawley, M. J., Edwards, P. J. (eds.) *Colonization, succession and stability.* Blackwell Scientific Publications, Oxford, pp. 225–244.

Lawton, J. H. (1989) Food webs. In, Cherrett, J. M. (ed.) *Ecological concepts.* Blackwell Scientific Publications, Oxford, pp. 43–78.

Locke, A. (1992) Factors influencing community structure along stress gradients, zooplankton responses to acidification. *Ecology* **73**, 903–909.

MacArthur, R. and Levins, R. (1967) The limiting similarity, convergence, and divergence of coexisting species. *Am. Nat.* **101**, 377–385.

Martinez, N. D. (1991) Artifacts or attributes? Effects of resolution on the little Rock Lake food web. *Ecol. Monogr.* **61**, 367–392.

Martinez, N. D. (1992) Constant connectance in community food webs. *Am. Nat.* **139**, 1208–1218.

Martinez, N. D. (1994) Scale-dependent constraints on food-web stucture. *Am. Nat.* **144**, 935–953.

May, R. M. (1973) *Stability and complexity in model ecosystems.* Princeton University Press, Princeton, NJ.

May, R. M. (1990) How many species? *Phil. Trans. Roy. Soc. Lond. B* **330**, 293–304.

Paine, R. T. (1980) Food webs, linkage, interaction strength and community infrastructure. *J. Anim. Ecol.* **49**, 667–685.

Paine, R. T. (1988) Food webs, road maps of interactions or grist for theoretical development? *Ecology* **69**, 1648–1654.

Pianka, E. R. (1994) *Evolutionary ecology.* (5th Ed.) Harper Collins College Publishers, New York.

Pimm. S. L. (1982) *Food webs.* Chapman and Hall, London.

Pimm, S.L (1991) *The balance of nature? Ecological issues in the conservation of species and communities.* University of Chicago Press.

Pimm, S. L., Lawton, J. H. and Cohen, J. E. (1991) Food web patterns and their consequences. *Nature* **350**, 669–674.

Polis, G. A. (1991) Complex trophic interactions in deserts, an empirical critique of food web theory. *Am. Nat.* **138**, 123–155.

Pyke, G. H., Pulliam, H. R. and Charnov, E. L. (1977) Optimal foraging, a selective review of theory and tests. *Q. Rev. Biol.* **52**, 137–54.

Rasmussen, J. B., Rowan, D. J., Lean, D. R. S., and Carey, J. H. (1990) Food chain structure in Ontario lakes deternines PCB levels in lake trout (*Salvelinus namycush*). *Can. J. Fish. Aquat. Sci.* **47**, 2030–2038.

Rejmanek, M. and Stary, P. (1979) Connectance in real biotic communities and critical values for stability in model ecosystems. *Nature* **280**, 311–313.

Sabelis, M. W. (1992) Predatory arthropods. In, Crawley, M. J. (ed.) *Natural enemies, the population biology of predators, parasites and diseases.* Blackwell Scientific Publications, Oxford, pp. 225–264.

Schoener, T. W. (1989) Food webs, from the small to the large. *Ecology* **70**, 1559–1589.

Schoenly, K., Beaver, R. A. and Heumier, T. A. (1991) On the trophic relations of insects, a food web approach. *Am. Nat.* **137**, 597–638.

Stenseth, N. C. (1985) The structure of food webs predicted from optimal food selection models, and alternative to Pimm's stability hypothesis. *Oikos* **44**, 361–364.

Vezina, A. F. (1985) Empirical relationships between predator and prey size among terrestrial vertebrate predators. *Oecologia* **67**, 555–565.

Warren, P. H. (1988) The structure and dynamics of a freshwater benthic food web. Unpublished DPhil. thesis. University of York, England.

Warren, P. H. (1989) Spatial and temporal variation in the structure of a freshwater food web. *Oikos* **55**, 299–311.

Warren, P. H. (1990) Variation in food-web structure, the determinants of connectance. *Am. Nat.* **136**, 689–700.

Warren, P. H. (1994) Making connections in food webs. *Trends Ecol. Evol.* **9**, 136–141.

Warren, P. H. (1995) Estimating morphologically determined connectance and structure for food webs of freshwater invertebrates. *Freshwater Biology* **33**, 213–221.

Warren, P. H. and Lawton J H (1987) Invertebrate predator–prey body size relationships, an explanation for upper triangular food webs and patterns in food web structure? *Oecologia* **74**, 231–235.

Warren, P. H. and Gaston, K. J. (1992) Predator–prey ratios, a special case of a general pattern? *Phil. Trans. Roy. Soc. Lond. B* **338**, 113–130.

Werner, E. E. and Hall, D. J. (1974) Optimal foraging and the size selection of prey by the bluegill sunfish. *Ecology* **55**, 1042–1052.

Winemiller, K. O. (1989) Must connectance decrease with species richness? *Am. Nat.* **134**, 960–968.

Yodzis, P. (1980) The connectance of real ecosystems. *Nature* **284**, 544–545.

Yodzis, P. (1984) How rare is omnivory? *Ecology* **65**, 321–323.

Zwolfer, H. and Arnold-Rinehart, J. (1993) The evolution of interactions and diversity in plant–insect systems, the *Urophora-Eurytoma* food web in galls on Palearctic Cardueae. *Ecol. Studies* **99**, 211–229.

9

Trophic interactions, nutrient supply, and the structure of freshwater pelagic food webs

Gérard Lacroix, Françoise Lescher-Moutoué and Roger Pourriot

Introduction

Limnological studies in the first half of the century largely considered the effects of abiotic factors on aquatic communities, with those few studies on biotic interactions tending to focus on the lowest trophic levels (Persson *et al.* 1988). In the early 1960s, limnological studies were conducted for a large part in the context of 'bottom-up' tropho-dynamic concepts (Liebig's law of limiting factors, efficiency of energy transfers between trophic levels...), in particular within the scope of the International Biological Program. This bottom-up view was reinforced in the late 1960s with the understanding of the major role of nutrient loading in lake eutrophication.

During this same period, some key papers (Hutchinson 1961, Hrbacek *et al.* 1961, Brooks and Dodson 1965) were published, stimulating for three decades an impressive number of descriptive, experimental, and theoretical studies on the respective roles of predation and competition in plankton communities. These studies led to substantial progress in the comprehension of organisms' life histories, population dynamics, species-specific interactions, and coexistence within aquatic food webs (Kerfoot 1980, Tilman 1982, Lampert 1985, Sommer 1989).

By the early 1980s, it had become clear that both bottom-up (resource-mediated) and top-down (predator-mediated) forces may simultaneously influence the general organization of food webs and the characteristics of aquatic ecosystems. However, most studies still focused on a few species with clear trophic links, single trophic levels, or particular taxonomic groups (Kerfoot and DeAngelis 1989). Efforts towards a more synthetic view on food webs led to the appearance of several related concepts, such as the 'trophic cascade' theory and the 'biomanipulation' method. The trophic cascade hypothesis (Carpenter *et al.* 1985) is based on the observation that much of the variability in primary production of temperate lakes remains unexplained by nutrient loading, turnover time of water, and vertical mixing. Carpenter *et al.* (1985) hypothesized that this unexplained variability could be due to cascading trophic interactions at the community and eco-system levels. They proposed that fluctuations in piscivory propagate down through the food web, causing changes in planktivory, herbivory, and primary production. The biomanipulation method involves a series of top-down manipulations, in particular modifications of the structure of fish populations, as possible alternate methods to nutrient diversion for reducing eutrophication and improving water quality (see Shapiro and Wright 1984, Gophen 1990).

An integration of bottom-up and top-down effects was proposed by Persson et al. (1988). They suggested that the theoretical model of Fretwell (1977) and Oksanen et al. (1981), a generalization of the HSS hypothesis (Hairston, Smith and Slobodkin 1960), could be applied to aquatic food webs. It predicts that the equilibrium biomass of each trophic level is dependent upon the number of trophic levels, the top-predators always being limited by resources and controlling, in turn, their prey ('top–down' view). As such, trophic chains with an odd number of levels should produce 'green' ecosystems (primary producers controlled by resources), while trophic chains with an even number of levels should produce 'barren' ecosystems (primary producers controlled by grazing). For example, an increase in nutrient load should induce an increase in the equilibrium phytoplankton biomass in a 3-level aquatic food web (phytoplankton–zooplankton–planktivorous fish), whereas no increase in algal biomass should be observed in systems with two (phytoplankton–zooplankton) or four (phytoplankton–zooplankton–planktivorous fish–piscivorous fish) trophic levels (Persson et al. 1988). Arditi and Ginzburg (1989), however, have shown that when taking into account a more realistic functional response of predators (ratio-dependent model), an increase in the primary productivity of the system induces a proportional increase of the equilibrium densities of all trophic levels, whatever the length of the food chain. According to Arditi et al. (1991), the prey-dependent (top-down) model should apply for homogeneous systems with random encounters between prey and predators, while the ratio-dependent model should apply under more heterogeneous conditions.

Although different in their aims and their structures, the trophic cascade hypothesis, the prey-dependent model of Oksanen et al. (1981), and the biomanipulation technique all assume that algal biomass is strongly regulated by top-down forces (Carpenter and Kitchell 1992, 1993). DeMelo et al. (1992) strongly questioned the validity of these models by reviewing 50 papers on aquatic food web experiments. Ginzburg and Akçakaya (1992) analysed the relationships between the nutrient input and the biomasses of phytoplankton, zooplankton, and fish across lakes. In accordance with the ratio-dependent model, an increase in nutrient input resulted in a parallel increase of all trophic levels. These results contradict the proposition of Persson et al. (1988) on the regulation of trophic levels in lakes. However, several studies have also provided valuable evidence of strong top-down mediated effects on lower trophic levels (see Carpenter et al. 1987, Gulati et al. 1990, Carpenter and Kitchell 1992, 1993).

Thus, as emphasized by Hunter and Price (1992), Menge (1992) and Power (1992), a strong dichotomous view between bottom-up and top-down controlled systems could lead to artificial debates on the importance of these forces in natural communities. On the other hand, there is need for a synthesis on the factors that mediate resource limitation and predator control within ecosystems. In the past decade, it has become clear that the multiple pathways which link organisms and their resources in natural food webs result in complex indirect effects (cascading effects, indirect facilitations, time-lags, bottlenecks, behavioural and morphological responses...) (Kerfoot and Sih 1987, Carpenter 1988). These effects may interfere with direct predator–prey interactions in the regulation of food webs. Our aim is to review and analyse recent advances clarifying how the trophic characteristics of organisms at each trophic level, and induced indirect effects, should magnify or buffer cascading trophic interactions in freshwater pelagic food webs.

Direct and indirect effects of predators

Fish and piscivory

Small fish are more vulnerable to predation by piscivorous fish than large ones (Tonn and Paszkowski 1986, Post and Evans 1989), suggesting that an increase in piscivory in fish assemblages will preferentially eliminate small-bodied individuals and species. In European lakes, increased piscivory has been shown to be associated with a strong reduction in recruitment of young cyprinids, a decrease in their density, and a shift to large-bodied individuals (Gerdeaux 1985, Tonn et al. 1990, 1992). This increase in prey body size is generally coupled with a reduction of planktivory.

Piscivorous predators often drive juvenile zooplanktivorous prey into refuges in the inshore vegetation (Werner and Hall 1988, Persson 1991, Tonn et al. 1992), leading in some cases to reduced exploitation of the offshore zooplankton. As such, habitat shifts could respond to a trade-off between foraging benefits and predation risk (Werner and Hall 1988, Persson 1993). Other behavioural and morphological responses to predation have been analysed in recent years (see Walls et al. 1990), some resulting in complex indirect effects, such as the modification of competitive interactions between prey species (Persson 1991, 1993).

Zooplankton and fish planktivory

Zooplanktivorous fish tend to select the largest zooplanktonic individuals, although young fish often prefer smaller-sized categories of prey (for a synthesis, see Lazzaro 1987). As a result, average prey body size decreases as the biomass of planktivorous fish increases (Hrbacek 1962, Brooks and Dodson 1965, Vanni 1987). This reduction in mean body size corresponds to species replacement and to demographic shifts in some prey species towards smaller size at first reproduction. These processes and compensatory increase in zooplankton birth rates mean that the total biomass of zooplankton responds in a complex and unpredictable way to modifications in predation pressure (Soranno *et al.* 1993).

Specific impacts of predators depend upon their mode of feeding. While visual feeders typically select large (e.g. the cladoceran *Daphnia*) or active (e.g. calanoid copepods) zooplankton (Crowder and Cooper 1982, O'Brien *et al.* 1984), less discriminant filter-feeding fish tend to suppress slow-moving zooplanktonic forms, and favour species with higher escape ability (Drenner *et al.* 1986, personal observations, Fig. 9.1). Several species are able to switch from a particle-feeding mode to a filter-feeding one, according to fish size, density and size distribution of prey, and light conditions (Hoogenboezem *et al.* 1992).

The size-efficiency hypothesis (Brooks and Dodson 1965, Hall *et al.* 1976) explains shifts in the size structure of zooplankton communities as being due to a greater competitive ability of the largest species. For example, under low predation pressure by planktivorous fish, the smallest zooplanktonic species (e.g. rotifers) appear to be strongly limited by either exploitative and/or interference competition with the largest cladocerans, in particular the genus *Daphnia* (Gilbert and Stemberger 1984). It seems also that the trade-off between competitive ability and vulnerability to vertebrate predators exists frequently within the cladocerans, although shifts in competitive advantage with environmental changes are common (Bengtsson 1987). An alternative explanation to the observed shift in zooplankton size with variation in predation pressure by planktivorous fish proposes a balance between the effects of invertebrate predators, which select small prey, and the effects of vertebrate predators, which select large prey (Dodson 1974).

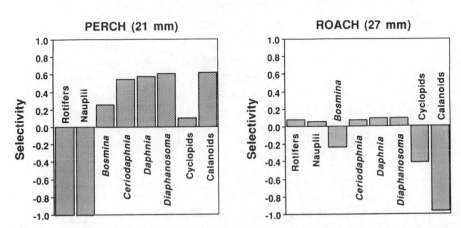

Fig. 9.1 Food selection of a facultative filter-feeding fish (*Rutilus rutilus*, Cyprinidae) and a visual-feeder (*Perca fluviatilis*, Percidae), estimated with the Jacobs' electivity index (Jacobs 1974) in laboratory-replicated experiments. Starved fish were provided with zooplankton of Créteil Lake (France) at natural abundances in 16-litre aquariums (24 roach individuals in 4 trials, and 21 perch individuals in 3 trials).

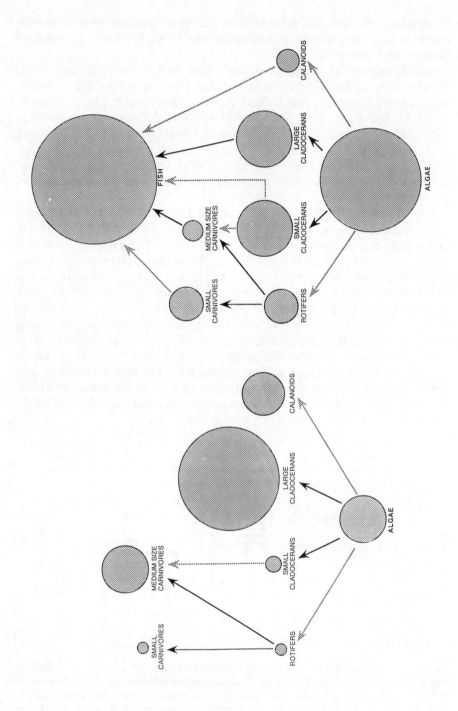

Besides body size, zooplanktonic organisms have developed an impressive number of anti-predator adaptations which involve behavioural responses (migration, swarming, escape reactions), morphological alterations, and modifications of life-history traits (for a review, see Walls *et al.* 1990).

Indirect 'mutualism' between vertebrate and invertebrate predators

As indicated previously, there is now a large body of empirical evidence indicating that certain large-bodied crustacean grazers (especially *Daphnia* spp.) have the ability to reduce inferior competitors and monopolize resources (for more details, see Kerfoot and DeMott 1985). A reduction in the numbers of these large cladocerans by planktivorous fish favours small grazers (rotifers, small crustaceans) and allows the development of subordinate food chains with secondary invertebrate predators, such as carnivorous copepods (Dodson 1970, Kerfoot and DeMott 1985). Dodson's paper inspired the theoretical study of Levine (1976), who showed that two consumers that are specialized on two competing resources may lead to a net reciprocal benefit ('indirect mutualism'). However, as invertebrate predators are also potential prey for zooplanktivorous fish, the effect of such a scenario should depend upon the balance between bottom-up positive effects and top-down negative ones. Such a balance is clearly dependent on the body size of the invertebrate predators (unpublished experimental results, Fig. 9.2).

As emphasized by Kerfoot and DeMott (1985), planktivorous fish typically act as a guild of keystone predators (Paine 1966), by limiting the dominant competitors and increasing zooplankton diversity. Thus, along an increasing gradient of vertebrate planktivory, zooplankton communities classically range from communities dominated by large herbivorous crustaceans (in particular *Daphnia* spp.), to more diversified communities, comprising rotifers, small crustaceans, and cyclopoids (Fig. 9.2).

Herbivorous zooplankton and algal biomass

According to a large number of studies (see Shapiro and Wright 1984), food webs dominated by large zooplanktonic crustaceans (mainly *Daphnia*) are generally characterized by lower algal biomass than food webs dominated by

---

Legend to facing page

Fig. 9.2 Structure of the planktonic community in the presence and absence of juvenile planktivorous roach (20 individuals of *Rutilus rutilus* per 9 m^3 enclosure), in an experimental mesocosm study (Créteil Lake, July–August 1990). Each treatment was run in three replicates, with phosphorus and nitrogen loading rates of respectively 3.16 and 63.2 μg l^{-1} d^{-1} (for more details on the experimental design, see Lacroix and Lescher-Moutoué 1991). The circle areas are proportional to the average instantaneous biomasses of the different planktonic groups in three replicated enclosures on 23 July 1990. Grey and black arrows indicate respectively 'low' and 'high' abilities to control the prey density.

small zooplankton (rotifers, small cladocerans, and small copepods) (Fig. 9.2).

In contradiction with the general expectations of the trophic cascade and prey-dependent model hypotheses, biomass and grazing rate of herbivorous zooplankton are not reliable predictors of algal biomass. As emphasized previously, there is no clear positive relationship between average body size and total biomass of zooplankton. Moreover, allometric relationships for zooplankton (Peters and Downing 1984) or cladocerans (Mourelatos and Lacroix 1990) indicate that large organisms have lower filtering rates per unit of biomass than small ones; thus an increase in zooplankton biomass and average body size could possibly be associated with decreased grazing. Interestingly, Cyr and Pace (1992) could not detect a relationship between zooplankton size distribution and total grazing rate in 30 zooplanktonic communities.

On the other hand, large *Daphnia* can ingest autotrophic or heterotrophic organisms ranging from picoplankton (a few micrometres) to microplankton (several hundred micrometres in length, for elongated cells), and are mostly non-selective feeders within the size range of their food spectrum. In contrast, rotifers and small cladocerans eat preferentially algal particles of less than a few tens of a micrometres. Moreover, several rotifers, the small cladoceran *Bosmina*, and copepods are selective feeders (DeMott 1986).

Thus, because large cladocerans are able to exploit non-selectively a wider range of particles than small zooplankton (Burns 1968, Gliwicz 1977, Kerfoot and DeMott 1985), they are probably able to depress algal communities more uniformly and extensively than are other zooplanktonic species. As a consequence, size of zooplankton is generally a better predictor of algal biomass (or water transparency) than is the total biomass of zooplankton (Soranno *et al.* 1993; unpublished experimental results, Fig. 9.3).

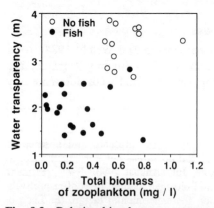 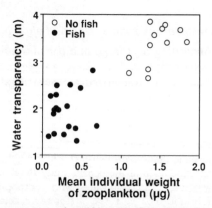

Fig. 9.3 Relationships between water transparency and total zooplankton biomass, and water transparency and mean individual weight of zooplankton, in 12 fishless and 18 fish enclosures with various nutrient supply levels, on 23 July 1990 (for more details on the experimental design, see Lacroix and Lescher-Moutoué, 1991).

Herbivory and trophic bottlenecks

In spite of their ability to ingest a larger range of algal particles than small zooplankton, large-bodied grazers do not completely control phytoplankton populations, since (1) some phytoplankton cells with thick walls or gelatinous sheaths pass relatively unharmed through the guts of *Daphnia* (Porter 1975, Kerfoot and DeMott 1985), and (2) some filamentous blue greens (cyanobacteria) are toxic for grazers; they can form large colonies, and reduce by mechanical interference the filtering rate of *Daphnia* (De Bernardi and Giussani 1990).

Contradictory results are found concerning *Daphnia*-phytoplankton interactions and the predominance of such grazing-resistant algae. Grazing pressure exerted by large crustaceans has been frequently cited as a factor contributing to the shift of algal assemblages towards more inedible phytoplankton species (Haney 1987, Kerfoot *et al.* 1988, Vanni and Temte 1990, Elser and Goldman 1991). On the other hand, several experimental studies manipulating planktivorous fish indicate that a high biomass of large crustaceans and the dominance of cladocerans prevent cyanobacteria from forming large colonies (Andersson *et al.* 1978, Shapiro and Wright 1984, Christoffersen *et al.* 1993). Direct *Daphnia* manipulations in large enclosures have shown that intense grazing promoted the transition from edible to other edible species and delayed further succession to grazing-resistant, filamentous, blue greens (Sarnelle 1993). Therefore, the effects of herbivory on algal succession were not predictable from the relative susceptibilities of phytoplankton species to grazers. Sarnelle (1993) suggested the importance of indirect effects, as the modification of the nutrient supply ratio by the grazer, in determining the outcome of herbivore–plant interactions.

Nutrient supply ratio, zooplankton stoichiometry, and structure of the algal community

The relative supply rates of different chemical resources have been shown to be important in explaining the coexistence of algal species and the structure of phytoplankton communities (Tilman 1982, Tilman *et al.* 1982). In particular, the nitrogen:phosphorus supply ratio is often negatively correlated with the development of cyanobacteria (Tilman *et al.* 1982, Smith 1983, Varis 1991). Thus, direct modifications of N:P supply ratio could alter the composition of algae and modify the capacity of zooplankton to regulate phytoplankton abundance (Neill 1988). However, in recent factorial experiments crossing various N and P loading supplies (with N:P molar ratio ranging from 4.42 to 442), we found that algae were always controlled by large grazers in the absence of fish (Fig. 9.4). In contrast, we observed a strong increase of algal biomass with phosphorus inputs in the presence of fish. The increase of phytoplankton was particularly strong for the lowest levels of nitrogen input (Fig. 9.4). This suggests that the N:P supply ratio is important in controlling the biomass of the algal community in the presence of fish.

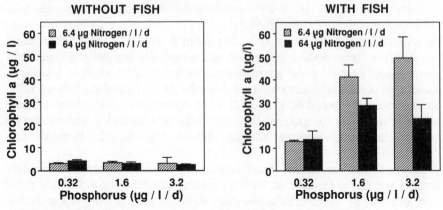

Fig. 9.4. Relationships between chlorophyll *a* biomass, nutrient supply, and presence of juvenile roach in a 2 (nitrogen) × 3 (phosphorus) × 2 (absence or presence of roach) factorial experiment run with four replicates in Créteil Lake (July–September 1992). Each chlorophyll *a* biomass represents the average value of 6 successive sampling dates (± standard error of the mean of the four enclosures).

During the last few years, several studies have demonstrated that the elemental composition of zooplankton shows rather low intraspecific variability, but clear interspecific variability (Andersen and Hessen 1991, Hessen and Lyche 1991, Sterner *et al.* 1992). This stable composition of zooplankton species should have several consequences:

(1) the growth and reproduction of zooplankton could become limited by a minor element instead of energy under some food conditions (Sommer 1992);

(2) zooplankton should assimilate a limiting nutrient with higher efficiency, and thus decrease its excretion (Sterner *et al.* 1992);

(3) the N:P recycling ratio for algae should depend of the N:P ratio of herbivorous zooplankton (Sterner *et al.* 1992);

(4) the limiting nutrient for primary producers might be determined from the ratio of N:P recycled by grazers (Sterner 1990, Sterner *et al.* 1992).

According to Hessen and Lyche (1991), and Sterner *et al.* (1992), copepods have higher N:P ratios (39–52) than cladocerans (12–29). Moreover, *Daphnia* spp. seem to be characterized by very low N:P ratios (12–14) when compared to other cladoceran taxa (19–29). Urabe and Watanabe (1992) also found larger N:P ratios for the small cladoceran *Bosmina* than for larger *Daphnia*. Although the chemical composition of rotifers is unknown, it seems that their N:P excretion ratio is lower than that for *Daphnia* (Ejsmont-Karabin 1984, in Sarnelle 1992).

Phytoplankton N:P generally varies between 15 and 22. Thus, herbivorous assemblages dominated by rotifers, copepods and small cladocerans could drive N-limitation in algal populations. In contrast, zooplanktonic communities dominated by large cladocerans, especially *Daphnia*, could lead to P-limitation (Sterner *et al.* 1992).

Support for the influence of zooplankton composition on the N:P supply ratio has been demonstrated in two lakes where zooplanktivorous and piscivorous fish were manipulated (Elser *et al.* 1988). The authors found that an increase in zooplankton body size induced a shift of the algal community towards P-limitation, whereas a decrease in zooplankton body size induced N-limitation. Moreover, in mesocosms, shifts between N- and P-limitation occurred only a few days after the manipulation of zooplankton. Sarnelle (1992) obtained congruent results through the manipulation of *Daphnia* in large enclosures, and whole-lake observations before and after fishkill.

Classical aquatic food-chains and microbial loops

Until recently, the major trophic flows in pelagic ecosystems were generally considered to occur along a linear food chain, from primary producers to piscivorous fish, with recycled nutrients at different points along this pathway. During the past decade, this traditional view on aquatic food webs has been profoundly altered by the discovery that major pathways of carbon flow to the upper levels are linked to the activity of bacteria and their main consumers: heterotrophic flagellates and ciliate protozoans (see Stockner and Porter 1988, Güde 1989, Stone and Weisburd 1992). Limnologists are now beginning to incorporate this microbial loop in analysing top-down and bottom-up process in aquatic food webs (Riemann 1985, Scavia and Fahnenstiel 1988, Christoffersen *et al.* 1993, Pace 1993). Two contrasting situations can be drawn. First, in communities typical of high fish predation, the microbial loop seems to be linked to the classic food web by rotifers, which are, in turn, ingested by copepods. Copepods do not eat bacteria and ingest the smallest micrograzers with low efficiency. Second, in zooplanktonic communities typical of low fish predation, the dominant *Daphnia* are able to ingest autotrophic or heterotrophic organisms ranging from picoplankton to microplankton. Bacteria could be partially controlled by cladoceran species such as *Diaphanosoma* and *Daphnia* (Pace *et al.* 1983, Porter *et al.* 1983, Vaqué and Pace 1992). Heterotrophic flagellates are probably strongly limited by *Daphnia* (Güde 1989, Christoffersen *et al.* 1993, Pace 1993). Thus, the presence of large cladocerans probably shortens the microbial food web and increases the efficiency of energy transfer to higher trophic levels (Stockner and Porter 1988, Pace *et al.* 1990).

From simple food chains to complex food webs

The relative importance of top-down vs. bottom-up forces will partly depend on the strength of predator–prey coupling. Strong top-down control is more probable in simple food webs, with homogeneous, poorly differentiated and highly edible trophic levels, while food web complexity and differentiation of interactions at the same trophic level should favour damping of the system (Strong 1992).

Aquatic food webs, as analysed by ecologists, are generally rough caricatures of actual webs (Crowder *et al.* 1988, Paine 1988); some components are still largely ignored (epibionts of crustacean zooplankton, parasites, coupling of pelagic and benthic species...). Moreover, quantitative approaches often involve the aggregation of species into large functional groups or trophic levels. As indicated by Crowder *et al.* (1988), these aggregations may be reasonable simplifications as long as the components included in a given compartment are substitutable with regard to the question addressed. According to the main direct and indirect effects of predation described previously, contrasting results emerge in freshwater systems.

Food webs characterized by two main trophic levels

Food webs with only two main trophic levels (phytoplankton, zooplankton) might be frequently simplified to linear food chains, as long as large *Daphnia* are the dominant zooplankton and do not favour the development of inedible algae. Invertebrate predators, such as copepods, should not be able to control zooplankton grazers and should function as a donor-controlled trophic level, owing to a large reduction in their feeding efficiency with increases in prey size. The effect of larger and more efficient invertebrate predators, such as *Chaoborus* larvae or the predaceous cladoceran *Leptodora kindtii*, could be more important. Experimental studies with *Chaoborus* larvae showed however that there is intense resource limitation for juveniles, which feed on small rotifers, which are scarce in presence of *Daphnia* (Neill 1988). Under conditions of high nutrient loading, where rotifers reproduce in sufficient numbers to eliminate the bottleneck in juvenile recruitment, this large predator could play a more important role on herbivorous zooplankton (Neill 1988).

Food webs characterized by three main trophic levels

In contrast, hypotheses of bottom-up or top-down control cannot be easily formulated in terms of linear, vertical food chains in aquatic food webs characterized by three main trophic levels (phytoplankton, zooplankton, and planktivorous fish). Compensatory growth of small zooplankton and non-selected algae, enhancement of some invertebrate predators, adaptive responses to predation pressure, longer and less direct microbial loops, and induced modifications in nutrient supply are more in accordance with the predictions of the general ratio-dependent model proposed by Arditi and Ginzburg (1989).

The theoretical consequences of trophic level diversity have been analysed by Kretzschmar *et al.* (1993) in a 3-species predator–prey model, with 'edible' and 'inedible' algal categories and *Daphnia* as a grazer. When the different species were allowed to interact directly (by reduction of the attack rate of both algal categories in the presence of inedible algae) and indirectly (interspecific competition between the prey categories), the presence of inedible algae was shown to stabilize the predator–prey system. The authors also studied the behaviour of the model under nutrient enrichment and observed an increase in the equilibrium densities of both phytoplankton and zooplankton, with a decrease in the edible fraction in the system. Consistent with these results, McCauley *et al.* (1988) observed that levels of edible and inedible algae increase with increasing nutrient status in a large number of lakes, while the fraction of edible algae decreases within the total algal population.

Food webs characterized by four main trophic levels

In a 4-level system (phytoplankton, zooplankton, planktivorous fish, and piscivorous fish), strong top-down cascading effects should depend upon the capacity of piscivorous fish to regulate the zooplanktivores and drive the zooplanktonic community towards the dominance of large *Daphnia*, despite size refuges and behavioural responses of planktivores. Several studies indicate that very large modifications in fish populations should be necessary in order to cascade towards lower trophic levels (Benndorf 1990, McQueen 1990). In contrast to manipulations in which planktivorous fish were totally removed, Persson *et al.* (1993) did not observe differences in seasonal average phytoplankton biomass despite strong reductions of the biomass of cyprinids in two consecutive whole-lake experiments.

Conclusion

The relative importance of top-down and bottom-up forces in aquatic ecosystems appears to be strongly linked to food web complexity and prey differentiation within a given trophic level. Major shifts in the structure of pelagic ecosystems seem to be dependent firstly upon the elimination of the keystone guild of fish planktivores, which allows the dominance of large generalist zooplanktonic grazers, with high ability to control the algal community. Moreover, it appears that such apparent trophic cascades could be strongly favoured by an induced modification of the nutrient supply for algae. Thus, not only direct predator–prey relationships, but also ecosystemic processes, should be taken simultaneously into account in the understanding of food web organization and dynamics.

When analysing trophic interactions within food webs, ecologists often have a population-oriented point of view, focusing on direct predator–prey relationships, functional responses, or induced defences. However, the different species of a food web also participate in nutrient cycling, both as sources through excretion, egestion and mortality, and as sinks through ingestion and growth. Several effects which have been mainly interpreted in terms of trophic cascades are confounded with induced complex changes in nutrient cycles (Threlkeld 1987, DeMelo *et al.* 1992).

The facility of constructing a synthetic theory of food web structure may be promoted by the strong hierarchical effects of body size in aquatic food webs: most predators are larger than their prey, and predation is most often size-selective; in a general way, there is an inverse relationship between organism body size and metabolic rate; the stoichiometric composition of zooplankton seems to be size-related. As emphasized by Cyr and Pace (1993), allometric theory and patterns of size distribution could provide a solid ground for comparing the implications of food web structure on energy flow, productivity, and nutrient cycling at the level of populations, communities and aquatic ecosystems.

Acknowledgements

We gratefully acknowledge Drazen Borcic, Inma Martin-Castano, and Claude Rougier for their insights and their contribution to the acquisition of data.

References

Andersen, T. and Hessen, D. O. (1991). Carbon, nitrogen and phosphorus content of freshwater zooplankton. *Limnology and Oceanography*, **36**, 807–14.

Andersson, G., Berggren, H., Cronberg, G. and Gelin, C. (1978). Effects of planktivorous and benthivorous fish on organisms and water chemistry in eutrophic lakes. *Hydrobiologia*, **59**, 9–15.

Arditi, R. and Ginzburg, L. R. (1989). Coupling in predator–prey dynamics: ratio-dependence. *Journal of Theoretical Biology*, **139**, 311–26.

Arditi, R., Perrin, N. and Saïah, H. (1991). Functional responses and heterogeneities: an experimental test with cladocerans. *Oikos*, **60**, 69–75.

Bengtsson, J. (1987). Competitive dominance among Cladocera: are single-factor explanations enough? *Hydrobiologia*, **145**, 245–57.

Benndorf, J. (1990). Conditions for effective biomanipulation; conclusions derived from whole-lake experiments in Europe. In *Biomanipulation—Tool for water management* (eds R. D. Gulati, E. H. R. R. Lammens, M.-L. Meijer and E. Van Donk), pp. 187–203. *Hydrobiologia*, 200/201.

Brooks, J. L., and Dodson, S. I. (1965). Predation, body size and composition of plankton. *Science*, **150**, 28–35.

Burns, C. W. (1968). The relationship between body size of filter-feeding Cladocera and the maximum size of particle ingested. *Limnology and Oceanography*, **13**, 675–78.

Carpenter, S. R. (ed.) (1988). *Complex interactions in lake communities*. Springer-Verlag, New York.

Carpenter, S. R., Cottingham, K. L. and Schindler, D. E. (1992). Biotic feedbacks in lake phosphorus cycles. *Trends in Ecology and Evolution*, **7**, 332–36.

Carpenter, S. R. and Kitchell, J. F. (1992). Trophic cascade and biomanipulation: Interface of research and management—A reply to the comment by DeMelo *et al. Limnology and Oceanography*, **37**, 208–13.

Carpenter, S. R. and Kitchell, J. F. (eds) (1993). *The trophic cascade in lakes*. Cambridge University Press.

Carpenter, S. R., Kitchell, J. F. and Hodgson, J. (1985). Cascading trophic interactions and lake productivity. *Bioscience*, **35**, 634–39.

Carpenter, S. R., Kitchell, J. F., Hodgson, J. R., Cochran, P. A., Elser, J. J., Elser, M. M., Lodge, D. M., Kretchmer, D., He, X. and Von Hende, C. N. (1987). Regulation of lake primary productivity by food web structure. *Ecology*, **68**, 1863–76.

Christoffersen, K., Rieman, B., Klysner, A. and Sondergaard, M. (1993). Potential role of fish predation and natural populations of zooplankton in structuring a plankton community in eutrophic lake water. *Limnology and Oceanography*, **38**, 561–73.

Crowder, L. B. and Cooper, W. E. (1982). Habitat structural complexity and the interaction between bluegills and their prey. *Ecology*, **63**, 1802–13.

Crowder, L. B., Drenner, R. W., Kerfoot, W. C., McQueen, D. J., Mills, E. L., Sommer, U., Spencer, C. N. and Vanni, M. J. (1988). Food web interactions in lakes. In *Complex interactions in lake communities* (ed. S. R. Carpenter), pp. 141–60. Springer-Verlag, New York.

Cyr, H. and Pace, M. L. (1992). Grazing by zooplankton and its relationship to community structure. *Canadian Journal of Fisheries and Aquatic Sciences*, **49**, 1455–65.

Cyr, H. and Pace, M. L. (1993). Allometric theory: extrapolations from individuals to communities. *Ecology*, **74**, 1234–45.

De Bernardi, R. and Giussani, G. (1990). Are blue-green algae a suitable food for zooplankton? An overview. In *Biomanipulation—Tool for water management* (eds R. D. Gulati, E. H. R. R. Lammens, M.-L. Meijer and E. Van Donk), pp. 29–41. *Hydrobiologia*, 200/201.

DeMelo, R., France, R. and McQueen, D. J. (1992). Biomanipulation: Hit or myth? *Limnology and Oceanography*, **37**, 192–207.

DeMott, W. R. (1986). The role of taste in food selection by freshwater zooplankton. *Oecologia*, **69**, 334–40.

Dodson, S. I. (1970). Complementary feeding niches sustained by size-selective predation. *Limnology and Oceanography*, **15**, 131–37.

Dodson, S. I. (1974). Zooplankton competition and predation: an experimental test of the size-efficiency hypothesis. *Ecology*, **55**, 605–13.

Drenner, R. W., Threlkeld, S. T. and McCracken, M. D. (1986). Experimental analysis of direct and indirect effects of an omnivorous filter-feeding clupeid on plankton community structure. *Canadian Journal of Fisheries and Aquatic Sciences*, **43**, 1935–45.

Elser, J. J., Elser, M. M., MacKay, N. A. and Carpenter, S. R. (1988). Zooplankton-mediated transitions between N- and P-limited algal growth. *Limnology and Oceanography*, **33**, 1–14.

Elser, J. J. and Goldman, C. R. (1991). Zooplankton effects on phytoplankton in lakes of contrasting trophic status. *Limnology and Oceanography*, **36**, 64–90.

Fretwell, S. D. (1977). The regulation of plant communities by the food chains exploiting them. *Perspectives in Biology and Medicine*, **20**, 169–85.

Gerdeaux, D. (1985). Evolution des populations de Gardon et de Sandre dans le lac de Créteil. *Verhandlungen der internationale Vereinigung für theoretische und angewandte Limnologie*, **22**, 2605–10.

Gilbert, J. and Stemberger, R. (1984). Control of *Keratella* populations by interference competition by *Daphnia*. *Limnology and Oceanography*, **30**, 180–88.

Ginzburg, L. R. and Akçakaya, H. R. (1992). Consequences of ratio-dependent predation for steady-state properties of ecosystems. *Ecology*, **73**, 1536–43.

Gliwicz, Z. M. (1977). Food size selection and seasonal succession of filter feeding zooplankton in an eutrophic lake. *Ekologia Polska*, **25**, 179–225.

Gophen, M. (1990). Biomanipulation: retrospective and future development. In *Biomanipulation—Tool for water management* (eds R. D. Gulati, E. H. R. R. Lammens, M.-L. Meijer and E. Van Donk), pp. 1–11. *Hydrobiologia*, **200/201**.

Güde, H. (1989). The role of grazing on bacteria in plankton succession. In *Plankton ecology: Succession in plankton communities* (ed. U. Sommer), pp. 337–64. Springer-Verlag, Berlin.

Gulati, R. D., Lammens, E. H. R. R., Meijer, M.-L. and Van Donk, E. (eds) (1990). *Biomanipulation—Tool for water management*. Proceeding of an International Conference, Amsterdam, 8–11 August 1989. *Hydrobiologia*, **200/201**, 628.

Hairston, N. G., Smith, F. E. and Slobodkin, L. B. (1960). Community structure, population control, and competition. *The American Naturalist*, **94**, 421–25.

Hall, D. J., Threlkeld, S. T., Burns, C. W. and Crowley, P. H. (1976). The size-efficiency hypothesis and the size structure of zooplankton communities. *Annual Review of Ecology and Systematics*, **7**, 177–208.

Haney, J. F. (1987). Field studies on zooplankton–cyanobacteria interactions. *New Zealand Journal of Marine and Freshwater Research*, **21**, 467–75.

Hessen, D. O. (1992). Nutrient element limitation of zooplankton production. *The American Naturalist*, **140**, 799–814.

Hessen, D. O. and Lyche, A. (1991). Inter- and intraspecific variations in zooplankton element composition. *Archiv für Hydrobiologie*, **121**, 343–53.

Hoogenboezem, W., Lammens, E. H. R. R., van Vugt, Y. and Osse, J. W. M. (1992). A model of switching between particulate-feeding and filter-feeding in the common bream, *Abramis brama*. *Environmental Biology of Fishes*, **33**, 13–21.

Hrbacek, J. (1962). Species composition and the amount of zooplankton in relation to fish stock. *Rozpravy Ceskoslovenske Akademie Ved, Rada mathematicko-prirodovedecka*, **72**, 1–116.

Hrbacek, J., Dvorakova, M., Korinek, V. and Prochazkova, L. (1961). Demonstration of the effect of the fish stock on the species composition and the intensity of the metabolism of the whole plankton association. *Verhandlungen der internationale Vereinigung für theoretische und angewandte Limnologie*, **14**, 192–95.

Hunter, M. D. and Price, P. W. (1992). Playing chutes and ladders: heterogeneity and the relative roles of bottom-up and top-down forces in natural communities. *Ecology*, **73**, 724–32.

Hutchinson, G. E. (1961). The paradox of the plankton. The *American Naturalist*, **95**, 137–46.

Jacobs, J. (1974). Quantitative measurement of food selection–A modification of the forage ratio and Ivlev's electivity index. *Oecologia*, **14**, 413–17.

Kerfoot, W. C. (ed.) (1980). *Evolution and ecology of zooplankton communities*. University Press of New England, Hanover.

Kerfoot, W. C. and DeAngelis, D. L. (1989). Scale-dependent dynamics: zooplankton and the stability of freshwater food webs. *Trends in Ecology and Evolution*, **4**, 167–71.

Kerfoot, W. C. and DeMott, W. R. (1984). Food web dynamics: dependent chains and vaulting. In *Trophic interactions within aquatic ecosystems* (eds D. G. Meyers and J. R. Strickler), pp. 347–82. AAAS Selected Symposium #85, Washington D.C. Westview Press, Boulder.

Kerfoot, W. C., Levitan, C. and DeMott W. R. (1988). *Daphnia*–phytoplankton interactions: density-dependent shifts in resource quality. *Ecology*, **69**, 1806–25.

Kerfoot, W. C. and Sih, A. (eds) (1987). *Predation: direct and indirect impacts on aquatic communities*. Univ. Press of New England, Hanover and London.

Kretzschmar, M., Nisbet, R. M. and McCauley, E. (1993). A predator–prey model for zooplankton grazing on competing algal populations. *Theoretical Population Biology*, **44**, 32–66.

Lacroix, G. and Lescher-Moutoué, F. (1991). Interaction effects of nutrient loading and density of young-of-the-year cyprinids on eutrophication in a shallow lake: an experimental mesocosm study. In *Ecosystem research in freshwater environment recovery* (eds G. Giussani, L. Van Liere and B. Moss). *Memorie dell'Istituto Italiano di Idrobiologia*, **48**, 53–74.

Lampert, W. (ed.) (1985). Food limitation and the structure of zooplankton communities. *Archiv für Hydrobiologie Beiheft*, **21**, 497.

Lazzaro, X. (1987). A review of planktivorous fish: their evolution, feeding behaviours, selectivities, and impacts. *Hydrobiologia*, **146**, 97-167.

Levine, S. H. (1976). Competitive interactions in ecosystems. *The American Naturalist*, **110**, 903–10.

McCauley, E., Murdoch, W. W. and Watson S. (1988). Simple models and variation in plankton densities among lakes. *The American Naturalist*, **132**, 383–403.

McQueen, D. J. (1990). Manipulating lake community structure: where do we go from here? *Freshwater Biology*, **23**, 613–20.

Menge, B. A. (1992). Community regulation: under what conditions are bottom-up factors important on rocky shores? *Ecology*, **73**, 755–65.

Mourelatos, S. and Lacroix, G. (1990). In situ filtering rates of Cladocera: effect of body length, temperature and food concentration. *Limnology and Oceanography*, **35**, 1101–11.

Neill, W. E. (1988). Complex interactions in oligotrophic lake food webs: responses to nutrient enrichment. In *Complex interactions in lake communities* (ed. S. R. Carpenter), pp. 31–44. Springer-Verlag, New York.

O'Brien, W. J., Buchanan, C. and Luecke, C. (1984). Apparent size choice of zooplankton by planktivorous sunfish: exceptions to the rule. *Environmental Biology of Fishes*, **13**, 225–33.

Oksanen, L., Fretwell, S. D., Arruda, J. and Niemelä, P. (1981). Exploitation ecosystems on gradients of primary productivity. *The American Naturalist*, **118**, 240–61.

Pace, M. L. (1984). Zooplankton community structure, but not biomass, influence the phosphorus-chlorophyll *a* relationship. *Canadian Journal of Fisheries and Aquatic Sciences*, **41**, 1089–96.

Pace, M. L. (1993). Heterotrophic microbial processes. In *The trophic cascade in lakes* (eds S. R. Carpenter and J. F. Kitchell), pp. 252–77. Cambridge University Press.

Pace, R. T., McManus, G. B. and Findlay, S. E. G. (1990). Planktonic community structure determines the fate of bacterial production in a temperate lake. *Limnology and Oceanography*, **35**, 795–808.

Pace, M. L., Porter, K. G. and Feig, N. Y. S. (1983). Species and age specific differences in bacterial resource utilisation by two co-occurring cladocerans. *Ecology*, **64**, 1145–56.

Paine, R. T. (1966). Food web complexity and species diversity. *The American Naturalist*, **100**, 65–75.

Paine, R. T. (1988). Food webs: road maps of interactions or grist for the theoretical mill? *Ecology*, **69**, 1648–54.

Persson, L. (1991). Behavioral response to predators reverses the outcome of competition between prey species. *Behavioral Ecology and Sociobiology*, **28**, 101–5.

Persson, L. (1993). Predator-mediated competition in prey refuges: the importance of habitat dependent prey resources. *Oikos*, **68**, 12–22.

Persson, L., Andersson, G., Hamrin, S. F. and Johansson, L. (1988). Predator regulation and primary production along the productivity gradient of temperate lake ecosystems. In *Complex interactions in lake communities* (ed. S. R. Carpenter), pp. 45–65. Springer-Verlag, New York.

Persson, L., Johansson, L., Andersson, G., Diehl, S. and Hamrin, S. F. (1993). Density dependent interactions in lake ecosystems: whole lake perturbations experiments. *Oikos*, **66**, 193–208.

Peters, R. H. and Downing, J. A. (1984). Emperical analysis of zooplankton filtering and feeding rates. *Limnology and Oceanography*, **29**, 763–84.

Porter, K. G. (1975). Viable gut passage of gelatinous green algae ingested by *Daphnia. Verhandlungen der internationale Vereinigung für theoretische und angewandte Limnologie*, **19**, 2840–50.

Porter, K. G., Feig, Y. and Vetter, E. (1983). Morphology, flow regimes, and filtering rates of *Daphnia, Ceriodaphnia* and *Bosmina* fed natural bacterioplankton. *Oecologia*, **58**, 156–63.

Post, J. R. and Evans, D. O. (1989). Experimental evidence of size-dependent predation mortality in juvenile yellow perch. *Canadian Journal of Zoology*, **67**, 521–23.

Power, M. E. (1992). Top-down and bottom-up forces in food webs: do plants have primacy? *Ecology*, **73**, 733–46.

Riemann, B. (1985). Potential influence of fish predation and zooplankton grazing on natural populations of freshwater bacteria. *Applied Environmental Microbiology*, **50**, 187–93.

Sarnelle, O. (1992). Contrasting effects of *Daphnia* on ratios of nitrogen to phosphorus in a eutrophic, hard-water lake. *Limnology and Oceanography*, **37**, 1527–42.

Sarnelle, O. (1993). Herbivore effects on phytoplankton succession in a eutrophic lake. *Ecological Monographs*, **63**, 129–49.

Scavia, D. and Fahnenstiel, G. L. (1988). From picoplankton to fish: complex interactions in the great lakes. In *Complex interactions in lake communities* (ed. S. R. Carpenter), pp. 85–97. Springer-Verlag, New York.

Shapiro, J. and Wright, D. I. (1984). Lake restoration by manipulation. Round Lake, Minnesota—the first two years. *Freshwater Biology*, **14**, 371–83.

Smith, V. H. (1983). Low nitrogen to phosphorus ratio favor dominance by blue-green algae. *Science*, **221**, 669–71.

Sommer, U. (ed.) (1989). *Plankton Ecology: Succession in plankton communities*. Springer-Verlag, Berlin.

Sommer, U. (1992). Phosphorus-limited *Daphnia:* Intraspecific facilitation instead of competition. *Limnology and Oceanography*, **37**, 966–73.

Soranno, P. A., Carpenter, S. R. and He, X. (1993). Zooplankton biomass and body size. In *The trophic cascade in lakes* (eds S. R. Carpenter and J. F. Kitchell), pp. 172–188. Cambridge University Press.

Sterner, R. W. (1990). The ratio of nitrogen to phosphorus resupplied by herbivores: zooplankton and the algal competitive arena. *The American Naturalist*, **136**, 209–29.

Sterner, R. W., Elser, J. J. and Hessen, D. O. (1992). Stoichiometric relationships among producers, consumers and nutrient cycling in pelagic ecosystems. *Biochemistry*, **17**, 49–67.

Stockner, J. G. and Porter, K. G. (1988). Microbial food webs in freshwater planktonic ecosystems. In *Complex interactions in lake communities* (ed. S. R. Carpenter), pp. 69–83. Springer-Verlag, New York.

Stone, L. and Weisburd, S. J. (1992). Positive feedback in aquatic ecosystems. *Trends in Ecology and Evolution*, **7**, 263–67.

Strong, D. R. (1992). Are trophic cascades all wet? Differentiation and donor-control in speciose ecosystems. *Ecology*, **73**, 747–54.

Threlkeld, S. T. (1987). Experimental evaluation of trophic-cascade and nutrient-mediated effects of planktivorous fish on plankton community structure. In *Predation: direct and indirect impacts on aquatic communities* (eds W. C. Kerfoot and A. Sih), pp. 161–73. University Press of New England, Hanover and London.

Tilman, D. (1982). *Resource competition and community structure*. Princeton University Press.

Tilman, D., Kilham, S. S. and Kilham, P. (1982). Phytoplankton community ecology. The role of limiting nutrients. *Annual Review of Ecology and Systematics*, **13**, 349–72.

Tonn, W. M. and Magnuson, J. J. (1982). Patterns in the species composition and richness of fish assemblages in northern Wisconsin lakes. *Ecology*, **63**, 1149–66.

Tonn, W. M., Magnuson, J. J., Rask, M. and Toivonen, J. (1990). Intercontinental comparison of small-lake fish assemblages: the balance between local and regional processes. *The American Naturalist*, **136**, 345–75.

Tonn, W. M. and Paszkowski, C. A. (1986). Size-limited predation, winterkill, and the organization of *Umbra-Perca* fish assemblages. *Canadian Journal of Fisheries and Aquatic Sciences*, **43**, 194–202.

Tonn, W. M., Paszkowski, C. A. and Holopainen, I. J. (1992). Piscivory and recruitment: mechanisms structuring prey populations in small lakes. *Ecology*, **73**, 951–58.

Urabe, J. and Watanabe, Y. (1992). Possibility of N or P limitation for planktonic cladocerans: an experimental test. *Limnology and Oceanography*, **37**, 244–251.

Vanni, M. J. (1987). Effect of food availability and fish predation of a zooplankton community. *Ecological Monographs*, **57**, 61–88.

Vanni, M. J., and Findlay, D. L. (1990). Trophic cascades and phytoplankton community structure. *Ecology*, **71**, 921–37.

Vanni, M. J. and Temte, J. (1990). Seasonal patterns of grazing and nutrient limitation of phytoplankton in a eutrophic lake. *Limnology and Oceanography*, **35**, 697–709.

Vaqué, D. and Pace, M. L. (1992). Grazing on bacteria by flagellates and cladocerans in lakes of contrasting food web structure. *Journal of Plankton Research*, **14**, 307–21.

Varis, O. (1991). Associations between lake phytoplankton community and growth factors–a canonical correlation analysis. *Hydrobiologia*, **210**, 209–16.

Walls, M., Kortelainen, I. and Sarvala, J. (1990). Prey responses to fish predation in freshwater communities. *Annales Zoologici Fennici*, **27**, 183–99.

Werner, E. E. and Hall, D. J. (1988). Ontogenetic habitat shifts in the bluegill sunfish (*Lepomis macrochirus*): the foraging rate-predation risk trade-off. *Ecology*, **69**, 1352–66.

10

Linking communities and ecosystems: trophic interactions as nutrient cycling pathways

James P. Grover and Michel Loreau

Introduction

Almost the entire field of community ecology has been based on the implicit assumption that communities can be understood in terms of the populations that constitute them and their direct interactions (essentially consumption and competition). There is no doubt that this approach has been fruitful in the analysis of simple systems of few species, but one of its basic weaknesses in understanding large systems is that it ignores the constraints imposed by ecosystem functioning, such as matter and energy balances, decomposition processes and nutrient cycling. Although ecological systems are organized hierarchically (Allen and Starr 1982; O'Neill *et al.* 1986), so that community properties can be partly independent of general ecosystem-level constraints, the latter should affect the biological components of ecosystems and the communities they form. Significant organizing processes seem to occur from the bottom-up through vertical (trophic) interactions, which are tied to energy and nutrient flow, rather than through horizontal (competitive) interactions (Loreau 1994*a*). Therefore community organization has ultimately to be studied within the broader framework of ecosystem organization.

From an ecosystem viewpoint, trophic interactions correspond to matter and energy flows. Although the emphasis has often been on energy as a sort of universal currency for ecosystem processes, nutrients frequently appear to be the major limiting factors in many ecosystems (DeAngelis 1992). If nutrient cycling is a key process in ecosystem functioning, then it is important to consider trophic interactions as nutrient cycling pathways. This shift in perspective can have a profound influence on our views of both the dynamical properties of communities and food webs, and the functional role of species in their biotic environment.

Nutrient cycling as an organizing process

Influence on food web stability

Cycles of material flow buffer ecosystems against external fluctuations in material input and thus increase their homeostasis (Odum 1969; Loreau 1994b). Incorporating nutrient cycling in model food webs increases the probability that these are locally stable at a steady state (DeAngelis *et al.* 1989; DeAngelis 1992). On the other hand, as the degree of material cycling increases relative to material exchanges with the external world so that an ecosystem approaches an idealized, closed system, the resilience of the ecosystem decreases. That is, it returns more slowly to its locally stable steady state following a perturbation (DeAngelis 1980, 1992; DeAngelis *et al.* 1989), a result which seems to contradict the former prediction. This apparent paradox can be resolved by the fact that what decreases is only the resilience of the total quantity of matter within the ecosystem, but not the resilience of its internal structure (Loreau 1994b).

This is easily illustrated with the general model ecosystem portrayed in Fig. 10.1, consisting of consumers, producers and decomposers. The openness of this system with respect to material exchange is parameterized as a flow of nutrient at a constant turnover rate e, analogous to the dilution rate of a chemostat, for which the inflowing concentration of nutrient is S. In the model considered here, S is also the total quantity of nutrient in the system at steady state. S and the sizes of all compartments are measured in units of nutrient (e.g. mol per unit area or volume), and biomass in each compartment is assumed to be proportional to nutrient mass. The total quantity of

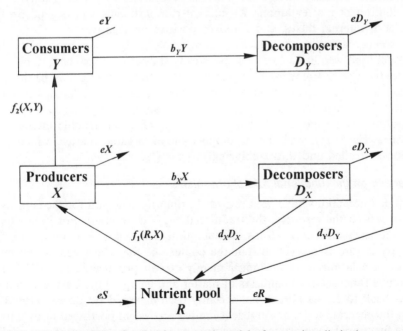

Fig. 10.1 Flow diagram of a simple, general model of a nutrient-limited ecosystem. Notation is defined in the text, and formal models specified as ordinary differential equations are easily constructed from the nutrient fluxes shown.

nutrient in the system, T, is the sum of available nutrient, R, nutrient bound in producers, X, nutrient bound in consumers, Y, and that bound in the two types of decomposers associated with these functional groups, D_X and D_Y.

Because we assume that each compartment suffers losses to the external world at a rate e, the dynamics of total nutrient are determined only by exchanges across the boundary between the system and the external world, which obey the linear differential equation:

$$\frac{dT}{dT} = e(S - T).$$ (10.1)

Solving this equation shows that the total quantity of nutrient reaches a stable steady state $T^* = S$; for the limiting case of a closed ecosystem ($e = 0$), S is simply defined as the total quantity of nutrient.

For open systems, total nutrient approaches the value S at a rate e, which thus measures resilience of the total nutrient stock of the ecosystem. This resilience decreases as e becomes smaller, that is, as the system closes, and internal recycling becomes rapid relative to boundary exchanges. On the other hand, the resilience of the system's internal structure, as expressed by the relative sizes of the compartments, is determined by the dynamics of nutrient flows among the various compartments. These internal fluxes become virtually independent of e, when it becomes small. Thus the influence of nutrient recycling on the resilience of ecosystems to perturbations in the distribution of matter among its internal compartments will depend on the details of internal fluxes, and could be positive or negative.

It seems reasonable that the same conclusion holds for the probability that a model food web has a locally stable steady state. Enhanced local stability in model food webs following incorporation of nutrient cycling may be due not so much to the degree of nutrient recycling *per se* as to related, otherwise neglected, constraints of ecosystem functioning, in particular nutrient limitation, which has a well-known stabilizing effect on trophic interactions (Rosenzweig 1971), and the decomposition of organic matter, which is a donor-controlled and thus stabilizing process (Hearon 1968).

Influence on the functional role of organisms

Trophic interactions have been regarded primarily as a gain for the consumer and a loss to the resource, the traditional $+,-$ interaction. For individuals, but not necessarily populations, consumption is indeed a $+,-$ interaction by its very nature. However, within the context of the whole ecosystem, consumer populations may no longer merely exploit producers, but might play an active function as maximizers of matter and energy flow. This idea, which dates back to Lotka (1925), has recently been explored by Loreau (1995).

In the general model of a nutrient-limited ecosystem portrayed in Fig. 10.1, three functional compartments constitute its minimum structure: the available nutrient pool (R), the producers (X), and their decomposers (D_X). Consumers (Y) and their decomposers (D_Y) are then added to this basic structure.

Nutrient recycling to the nutrient pool by the decomposers is assumed to be a linear donor-controlled process; b_X and b_Y are rates of transfer of nutrient from the producers and consumers to the decomposers as detritus; and d_X and d_Y are the rates at which nutrients are released to the pool by decomposition. Total fluxes of nutrient from producers and consumers to decomposers are thus $b_X X$ and $b_Y Y$, respectively, while total fluxes to the nutrient pool from the respective decomposers are $d_X D_X$ and $d_Y D_Y$. Decomposers are divided into two functional compartments to allow for different decomposition rates of the detritus from the producers and the consumers; this however, does not preclude part of the two decomposition processes from being affected by the same decomposer species.

There has been considerable argument about the nature of material flows between successive trophic levels, which are represented here by the functions f_1 and f_2. Control of these flows can vary from complete donor control to complete recipient control. The *per capita* uptake rate of nutrient by autotrophs is usually modelled by a Michaelis–Menten function (DeAngelis 1992). A Lotka–Volterra form

$$f_1(R,X) = aRX \qquad (10.2)$$

where a measures the affinity of producers for nutrient, leads to the same qualitative conclusions, but is simpler to analyse and is used here. Both forms imply that autotrophs actively control the size of the nutrient pool in the absence of consumers. At the ecosystem level, donor control of the trophic interaction between plants and consumers may be a good approximation near equilibrium because of the superposition of a large number of individual interactions (Patten 1975; Strong 1992). Accordingly, a linear donor-controlled form will be used here for f_2:

$$f_2(X,Y) = cX, \qquad (10.3)$$

where c is the consumption rate of producers by herbivores. Loreau (1995) analysed several other forms for the functions f_1 and f_2, with qualitatively similar results.

Since all the energy flowing in the ecosystem enters through the producer compartment and drives the material flow from the nutrient pool to the producers, energy flow in the ecosystem (Φ) is measured simply by the material flow from the nutrient pool to the producers, f_1 (eqn 10.2), assuming an appropriate transformation of units. Solving for the steady-state sizes of the compartments, energy flow at steady-state is

$$\Phi = \frac{(b_X + c + e)(S - R_{(XY)})}{b_X \omega_X + cy}, \qquad (10.4)$$

where

$$y = \frac{b_Y + d_Y + e}{(b_Y + e)(d_Y + e)}.$$

When the ecosystem is closed ($e = 0$), eqn 10.4 reduces to the simpler and more interpretable expression:

$$\Phi = \frac{S - R^*_{(XY)}}{\bar{\omega}}. \tag{10.5}$$

In eqn 10.4 and eqn 10.5, $R^*_{(XY)}$ is the steady-state size of the available nutrient pool when both X and Y are present:

$$R^*_{(XY)} = \frac{b_X + c + e}{a}.$$

Clearly, S must be greater than $R^*_{(XY)}$ for this ecosystem to be viable, which requires:

$$aS > b_X + c + e. \tag{10.6}$$

Although energy limitation is not taken into account explicitly in this model, it is reasonable to assume that primary productivity is positively related to a, and increases with the amount of solar radiation. Equation 10.6 thus shows that persistence of the system requires sufficient amounts of both available energy (incorporated in a) and nutrient (S).

The ω parameters in eqn 10.4 and eqn 10.5 represent the rates of nutrient recycling. First,

$$\omega_i = \frac{1}{b_i} + \frac{1}{d_i + e}, \qquad i = X, Y$$

represents the mean transit time of a unit quantity of nutrient through the paths $X \rightarrow D_X \rightarrow R$ or $Y \rightarrow D_Y \rightarrow R$. Second,

$$\bar{\omega} = \frac{b_X \omega_X + c \omega_Y}{b_X + c}$$

represents its average transit time through these two paths, that is, through the whole biomass. Thus eqn 10.5 has a straightforward interpretation in the case of a closed ecosystem: the flow of nutrient and energy is the product of the quantity of nutrient in the biota ($S - R^*_{(XY)}$), and its rate of circulation through the biota (which is the reciprocal of $\bar{\omega}$).

What is the effect of consumers on energy flow? As the consumption rate c increases, $R^*_{(XY)}$ increases until it equals S; at this point energy flow becomes zero and the ecosystem collapses. But at small or moderate consumption rates, energy flow can increase with c. The condition for energy flow to be maximized at intermediate values of c is that the partial derivative of energy flow Φ with respect to c be positive at $c = 0$. This condition is found to be, in the case of a closed ecosystem:

$$\frac{\omega_Y}{\omega_X} < \frac{S - 2R^*_{(X)}}{S - R^*_{(X)}}, \tag{10.7}$$

where $R^*_{(X)} = (b_X + e)/a$ is the threshold quantity of nutrient necessary for persistence of producers in the absence of herbivores. The equivalent condition is similar, but less transparent, for an open ecosystem. Equation 10.7 requires that: (a) ω_Y be sufficiently smaller than ω_X, that is, the presence of consumers must sufficiently accelerate nutrient cycling; (b) $S > 2R^*_{(X)}$, that is, the quantity of matter must be at least twice the threshold necessary for the persistence of the producers.

The curve of energy flow as a function of the consumption rate is illustrated in Fig. 10.2 for two values of the ratio ω_Y/ω_X in an open system. As expected from eqn 10.7, when this ratio is equal to 1 (Fig. 10.2(a)), energy flow decreases monotonically with increasing c, but when this ratio is equal to 0.1 (Fig. 10.2(b)), energy flow is maximized at an intermediate value of c. On the other hand, the steady-state sizes of the various compartments always follow the same pattern as a function of c: the size of the nutrient pool increases, the biomass of producers decreases, and the biomass of consumers peaks at an intermediate value of c. Similar results are obtained with other trophic functions, and when a predator trophic level is added (Loreau 1995).

This analysis shows that consumers can indeed play an important functional role as maximizers of matter and energy flow in nutrient-limited ecosystems, and that the qualitative conditions under which they play such a role are quite general: (a) The first condition is that consumers act to increase the circulation of matter within the ecosystem through accelerated nutrient cycling; (b) the second condition is that consumption be moderate; (c) the third condition is that the total quantity of nutrient in the ecosystem be

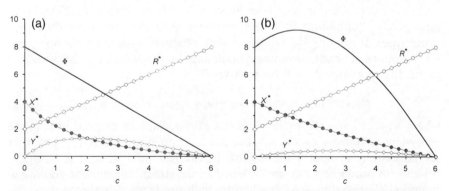

Fig. 10.2 Energy flow (Φ) and steady state compartment sizes in the model ecosystem of Fig. 10.1, as a function of the consumption rate c, with trophic interactions described by equations (10.2) and (10.3). (a) $\omega_Y = \omega_X$; consumption always decreases energy flow. (b) $\omega_Y = 0.1 \, \omega_X$; energy flow is maximized at intermediate consumption rates. Parameter values: $S = 8$ and $a = b_X = d_X = 1$ in both cases; $b_Y = d_Y = 1$ in (a); $b_Y = 10$ and $d_Y = 19$ in (b).

higher than some threshold value. This final condition, however, is dependent on the functional form of nutrient uptake by producers; it disappears in the case of complete donor control (Loreau 1995).

These conclusions provide a theoretical foundation for empirical studies of the effects of consumers on nutrient cycling and ecosystem functioning. For instance, the present model can be used to address the so-called 'grazing optimization' hypothesis (e.g. Owen and Wiegert 1976; McNaughton 1979; Hilbert *et al.* 1981; Belsky 1986; Belsky *et al.* 1993) which proposes that consumption by herbivores often maximizes energy flow in ecosystems. As shown in Fig. 10.2(b) the model outlined here exactly reproduces the grazing optimization curve of Hilbert *et al.* (1981), and emphasizes the ecosystem properties that produce such optimization. These properties may be common: DeAngelis (1992) compiled examples from a variety of ecosystems where patterns consistent with grazing optimization have been reported. For some of these, there is good evidence that herbivores accelerate nutrient recycling (e.g. savannas: Ruess and McNaughton 1987; plankton: Sterner 1986).

Selection of competing producers in nutrient-limited ecosystems

R*-rules under donor-dependence in herbivory

To explore the issue of grazing optimization further, we must disaggregate the producer trophic level in our model ecosystem (Fig. 10.1) into separate compartments representing different producer types. This leads to models describing selection between homogeneous populations of producers with fixed properties, which might approximate the dynamics of different species or clones. To begin, imagine that there are two producer types, X_1 and X_2, which have growth functions $a_j R X_j$ ($j = 1, 2$), and suffer abiotic losses $e X_j$, and donor-dependent losses to decomposers and herbivores of $b_{Xj} X_j$ and $c_j X_j$, respectively. We assume that herbivores are generalists ingesting both producer types. If our original ecosystem with producer X_1 is at equilibrium, and we introduce a small, invading population of producer X_2, the invader's population growth rate will be positive if

$$R^*_{(1Y)} = (b_{X1} + c_1 + e)/a_1 > (b_{X2} + c_2 + e)/a_2 = R^*_{(1Y)} \qquad (10.8)$$

where $R^*_{(jY)}$ denotes the equilibrium size of the available nutrient pool in the ecosystem containing producer j, but not its competitor. Therefore, producer X_2 successfully invades if the size of the nutrient pool when it is at equilibrium with producer X_1 absent, is less than the size of the nutrient pool when the original ecosystem is at equilibrium with producer X_1 only.

Turning the situation around, it is easily seen that if producer X_2 is at equilibrium in our model ecosystem with producer X_1 absent, the latter successfully invades if $R^*_{(1Y)} < R^*_{(2Y)}$. Thus mutual invasibility and mutual non-invasibility are impossible. This suggests that when two producers compete for a single nutrient while suffering donor-dependent herbivory, one producer unconditionally excludes the other, and there is neither herbivore-mediated coexistence nor priority effects. This result is, of course, qualified by our assumption of a spatially homogeneous environment at steady-state.

Tentatively then, we conclude that selection between producer types might follow a universal rule. This is an 'R^*-rule', analogous to the familiar rule applying to resource competition among producers in the absence of herbivory (Tilman 1982). The producer type that persists at equilibrium with the smallest available nutrient pool, in the face of all its losses including herbivory, competitively excludes other producer types. One way that a producer type can increase its probability of persisting in such a system is to reduce c, its loss rate to herbivory. Successive selection of such fitter producer types will, *ceteris paribus*, move the system down the abscissae of Figs. 10.2(a) and (b). Depending on other system parameters, and the initial and final positions of the selective trajectory, there could be either increases or decreases in both energy flow and herbivore biomass. If there is counter-vailing selection pressure on herbivores to increase the value of c, it is conceivable that the net result would tend to maintain the moderate values of c that are required for grazing optimization. Selection for producer properties that reduce losses to herbivory does not necessarily contradict the ecosystem-level potential for grazing optimization. Exploring fully the evolutionary consequences of this line of argument will eventually require explicit consideration of the genetic bases of critical parameters, such as c. Some progress has been made in using population-genetic models to study resource-dependent evolution of producers alone (Yi and Songling 1992), but we are not aware of studies of such evolution in an ecosystem context.

R^*-rules for other forms of herbivory

The relatively straightforward result that competition between producer types follows an R^* rule depends critically on the donor-dependence of herbivory. Results are potentially more complicated if other forms for the function f_2 are used. For example, Holt *et al.* (1994) explored the consequences of herbivory according to a Lotka–Volterra function:

$$f_{2j}(X_jY) = \tilde{c}_jX_jY, \qquad j = 1,2, \tag{10.9}$$

where $\tilde{c}_j$ is the attack rate of herbivores on producer type j. They also simplified the description of decomposition and nutrient recycling to achieve analytical tractability. In their system, no decomposer compartments are explicitly represented, and nutrient recycling from all populations goes directly and instantaneously to the available pool, R. Only three biological populations are represented, two producers and one herbivore, feeding on both producers. Their system follows a mass-balance constraint, so that variations in the nutrient pool are not independent of other compartment sizes, effectively reducing the system to three dimensions, and allowing the application of permanence theory (Hutson and Law 1985) to judge competitive outcomes between producers.

Holt *et al.* (1994) found that unilateral dominance by one producer type occurs when that producer both exploits the nutrient and withstands attack by herbivores better than its competitor. This dominance is sometimes predicted by an R^*-rule, such that the winner reduces the available nutrient pool to a smaller size than the loser, when growing at equilibrium with the herbivore, but without its competitor. Unlike the donor-dependent case, however, this R^*-rule sometimes fails and when it does, an alternative rule applies, based on Y^*, the size of the herbivore population supported by a given producer at equilibrium, without its competitor. When this rule applies, the producer supporting the larger Y^* excludes its competitor. This latter case is an example of 'apparent competition' (Holt 1977), an indirect interaction between prey consumed by a common predator. When the functions f_{2j} describing herbivory are not donor-dependent, apparent competition and associated Y^*-rules may play a role in selection among producer types.

In the model of Holt *et al.* (1994) herbivore-mediated coexistence and priority effects are also possible. Coexistence requires three conditions: (a) the superior exploiter in the absence of herbivores (i.e. the producer with the lowest value of $R^*_{(x)}$) is more vulnerable to herbivory; (b) the superior exploiter has a smaller compartment size (i.e. ties up less nutrient in its own population) than the inferior exploiter when growing at equilibrium with the herbivore but without its competitor (this allows the inferior exploiter to enjoy a richer resource base if it is invading the system of superior exploiter and herbivore); and (c) nutrient supply is moderate. At sufficiently low nutrient supplies, the superior exploiter wins, and at sufficiently high nutrient supplies the inferior exploiter wins.

These shifts in competitive outcomes with nutrient supply are accompanied by shifts in the applicability of R^*- and Y^*-rules and by a monotonic increase in herbivore density (Fig. 10.3). At low nutrient supply, the dominance of the superior exploiter is predicted by a Y^*-rule. Here, the producer that exploits nutrient best in the absence of herbivores is also highly vulnerable to herbivores, and thus it is a better 'transducer' of the abiotic resource base from their standpoint. This producer supports high herbivore densities while itself having a low density, making it a formidable 'apparent competitor'. At high nutrient supply, dominance of the inferior exploiter in the absence of herbivores is predicted by an R^*-rule. This is because its low vulnerability to herbivores reduces its nutrient demand below that of its competitor, under the conditions of very high herbivore density associated with high nutrient supply. For the intermediate nutrient supplies permitting coexistence, there is a trade-off in the two types of rules, in that one producer has a lower R^*, and one a higher Y^*, reflecting the balance between exploitative and apparent competition required for coexistence.

In the scenario sketched above, it may seem paradoxical that a Y^*-rule based on apparent competition applies at low nutrient supplies, while an R^*-rule based on resource competition applies at high nutrient supply. This outcome is a consequence of the superior exploiter's high susceptibility to

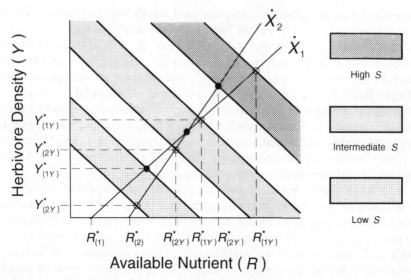

Fig. 10.3 Under the assumptions of Holt *et al.* (1994), zero-net-growth isoclines for producers (labelled $\dot{X}_1 = 0$) and $\dot{X}_2 = 0$) are linear on the plane of available nutrient (R) and herbivore density (Y). For a given nutrient supply, the feasible space for equilibria is delimited by the shaded region. Circles indicate equilibria: filled = stable; open = unstable. Equilibria on the boundaries of the feasible space correspond to food chains in which one producer is missing. If stable, boundary equilibria are not invasible by the missing producer, and competitive exclusion occurs. If unstable, boundary equilibria are invasible by the missing producer. Mutual invasibility is possible only for intermediate nutrient supplies. The food chain equilibria $R^*_{(jY)}$ and $Y^*_{(jY)}$ correspond to the rules for predicting competitive outcomes explained in the text.

herbivory, coupled with the strong control of producer density by herbivores which results from Lotka–Volterra grazing. The resulting strong impact of herbivory on the producer that dominates in the absence of herbivory can be interpreted as countering that producer's exploitative advantage. Thus, as a habitat is enriched, and herbivore density and impact increase, the direction of exploitative advantage is eventually reversed, and this is the basic tradeoff driving coexistence of producers in this model.

The model of Holt *et al.* (1994) predicts priority effects when the second condition for coexistence is reversed, so that the superior exploiter has a larger compartment size than the inferior exploiter, when growing at equilibrium with the herbivore but without its competitor. Again, the competitive outcomes depend on nutrient supply, with the superior exploiter dominating at low nutrient supply, the inferior exploiter dominating at high nutrient supply, and priority effects occurring only at intermediate nutrient supply. Here, an R^*-rule predicts the winner at low nutrient supply, and a Y^*-rule predicts the winner at high nutrient supply. In this case, the superior exploiter when herbivores are absent does not support a high enough herbivore density at low nutrient supply to benefit from apparent competition.

From examining the model system of Holt *et al.* (1994) we learn that, unlike the case of our original ecosystem with donor-dependent herbivory, selection between producer types is not governed by a universally applicable R^*-rule. Instead, this rule has a limited domain of applicability, as does the Y^*-rule that emerges in relation to apparent competition. Selection between producer types, and the associated R^*- and Y^*-rules also depend on nutrient supply. Nevertheless, explicitly accounting for both population and ecosystem processes leads to predictions that are biologically transparent and testable in principle. Furthermore, the system of Holt *et al.* (1994) is formally equivalent to a Lotka–Volterra model of two competitors and one predator, which in general can support cyclic or chaotic dynamics (Vance 1978). However, these possibilities are prevented by the ecosystem structure imposed by Holt *et al.* (1994), because only parameters consistent with the hypothesized nutrient fluxes and mass-balance are allowed. Therefore, imposition of ecosystem structure on population models sometimes simplifies the universe of possible outcomes.

Holt *et al.* (1994) provide a paradigm of how competing producers fare under generalist herbivory in nutrient-limited ecosystems. Specialist herbivores were considered by Grover (1994), who again assumed direct and instantaneous nutrient recycling. With the important restriction that all viable subcommunities have stable equilibria, he found that any number of producers, each grazed by its own specialist herbivore, can coexist in a community consisting of parallel food chains—all based on a single, limiting nutrient. However, real communities do not arise as a whole, but are assembled more or less sequentially, as evolution and biogeography provide new colonists to a local habitat. Even if a large number of species can potentially coexist in the model system of Grover (1994), the maximally diverse system can only be assembled by one sequence of species invasions. Basically, the producer that is the best exploiter must enter the community first, followed by its herbivore, which must impose sufficiently high losses to raise the nutrient demand of the first producer. Then, a second producer, less effective at exploiting the nutrient, can invade, followed by its herbivore, and so on. If the necessary order of invasion of various species is not followed, then the sequence gets 'stuck' in a community configuration which is not maximally diverse. If invasion sequences are random and equally likely, Grover (1994) suggests that such communities would most likely be dominated by a few highly exploitative plants and their specialist herbivores.

It is not surprising that the forms of the functions f_1 and f_2 describing nutrient uptake and herbivory strongly influence population-level phenomena in ecosystems. The Lotka–Volterra functions used by Holt *et al.* (1994) probably apply best in well-mixed, spatially uniform systems, or on small spatial scales. Explaining coexistence and diversity in heterogeneous (especially terrestrial) environments or on a larger scale may require donor- or ratio-controlled coupling (Arditi and Ginzburg 1989). Alternatively, explicit descriptions of herbivore spatial behavior, plant patterning, and spatial transport of nutrients may be required (e.g. Huston and DeAngelis 1994).

It is also possible that the functional description of linkages with decomposers will prove important, but this has received comparatively little theoretical attention. The simplified representation of nutrient cycling used by Holt *et al.* (1994) and Grover (1994) is equivalent to assuming (a) that standing stocks of decomposers are negligible compared to other compartments, (b) that kinetics of decomposition are rapid relative to other system dynamics, and (c) that decomposers are always net mineralizers of nutrient, rather than consumers. Only the third assumption is shared with our original nutrient-limited ecosystem model (Fig. 10.1). The first two assumptions are probably not critical to the qualitative conclusions of Holt *et al.* (1994) and Grover (1994), but the third, which may be contradicted in many aquatic ecosystems (Bratbak and Thingstad 1985), requires more exploration. Under conditions in which organic carbon supply to decomposers is high, they may become net consumers of inorganic nutrients, and hence competitors with producers for the same resources. Thus there is a possibility, as yet poorly understood, for decomposers as well as consumers to alter the selective environment experienced by producers.

Conclusion

It is time the two main streams of ecology, population and community ecology on the one hand, and ecosystem ecology on the other hand, joined in a new synthesis. Populations, interspecific interactions, communities and food webs describe biotic components of ecosystems, and thus have to be incorporated conceptually within the broader framework of ecosystems. This shift in perspective can indeed drastically alter our view of the structure, organization and dynamical properties of communities and can highlight certain indirect interactions, such as grazing optimization. Reciprocally, this shift in perspective can change our view of ecosystems and their functioning, which may well be more dependent upon the biological diversity they contain than has been traditionally thought (Schulze and Mooney 1993; Naeem *et al.* 1994). Given the rate at which this diversity is currently decreasing, we must urgently explore the implications for continued provision of 'ecosystem services'.

References

Allen, T. F. H. and Starr, T. B. (1982). *Hierarchy: perspectives for ecological complexity*. The University of Chicago Press.

Arditi, R. and Ginzburg, L. R. (1989). Coupling in predator–prey dynamics, ratio dependence. *Journal of Theoretical Biology* **139**, 311–326.

Belsky, A. J. (1986). Does herbivory benefit plants? A review of the evidence. *American Naturalist* **127**, 870–892.

Belsky, A. J., Carson, W. P., Jensen, C. L. and Fox, G. A. (1993). Overcompensation by plants, herbivore optimization or red herring? *Evolutionary Ecology* **7**, 109–121.

Bratbak, G. and Thingstad, T. F. (1985). Phytoplankton–bacteria interactions, an apparent paradox? Analysis of a model system with both competition and commensalism. *Marine Ecology—Progress Series* **25**, 23-30.

DeAngelis, D. L. (1980). Energy flow, nutrient cycling, and ecosystem resilience. *Ecology* **61**, 764–771.

DeAngelis, D. L. (1992). *Dynamics of nutrient cycling and food webs.* Chapman and Hall, London.

DeAngelis, D. L., Mulholland, P. J., Palumbo, A. V., Steinman, A. D., Huston, M. J. and Elwood, J. W. (1989). Nutrient dynamics and food-web stability. *Annual Review of Ecology and Systematics* **20**, 71–95.

Grover, J. P. (1994). Assembly rules for communities of nutrient-limited plants and specialist herbivores. *American Naturalist* **143**, 258–282

Hearon, J. Z. (1968). Theorems on linear systems. *Annals of the New York Academy of Sciences* **108**, 38–68.

Hilbert, D. W., Swift, D. M., Detling, J. K. and Dyer, M. I. (1981). Relative growth rates and the grazing optimization hypothesis. *Oecologia (Berlin)* **51**, 14–18.

Holt, R. D. (1977). Predation, apparent competition, and the structure of prey communities. *Theoretical Population Biology* **12**, 197–229.

Holt, R. D., Grover, J. P. and Tilman, G. D. (1994). Simple rules for interspecific dominance in systems with exploitative and apparent competition. *American Naturalist* **144**, 741–771.

Huston, M. A. and DeAngelis, D. L. (1994). Competition and coexistence, the effect of resource transport and supply rates. *American Naturalist* **144**, 954–977.

Hutson, V. and Law, R. (1985). Permanent coexistence in general models of three interacting species. *Journal of Mathematical Biology* **21**, 285–298.

Loreau, M. (1994*a*). Ground beetles in a changing environment, determinants of species diversity and community assembly. *Biodiversity, temperate ecosystems, and global change*, T. J. B. Boyle and C. E. B. Boyle, eds., Springer-Verlag, Berlin (in press).

Loreau, M. (1994*b*). Material cycling and the stability of ecosystems. *American Naturalist* **143**, 508–513.

Loreau, M. (1995). Consumers as maximizers of matter and energy flow in ecosystems. *American Naturalist* **145**, 22–42.

Lotka, A. J. (1925). *Elements of physical biology.* Reprinted as *Elements of mathematical biology.* Dover Publications, New York (1956).

Naeem, S., Thompson, L. J., Lawler, S. P., Lawton, J. H. and Woodfin, R. M. (1994). Declining biodiversity can alter the performance of ecosystems. *Nature* **368**, 734–737.

McNaughton, S. J. (1979). Grazing as an optimization process, grass-ungulate relationships in the Serengeti. *American Naturalist* **113**, 691–703.

Odum, E. P. (1969). The strategy of ecosystem development. *Science* **164**, 262–270.

O'Neill, R. V., DeAngelis, D. L., Waide, J. B. and Allen, T. F. H. (1986). *A hierarchical concept of ecosystems.* Princeton University Press.

Owen, D. F. and Weigert, R. G. (1976). Do consumers maximize plant fitness? *Oikos* **27**, 488–492.

Patten, B. C. (1975). Ecosystem linearization, an evolutionary design problem. *American Naturalist* **109**, 529–539.

Rosenzweig, M. L. (1971). Paradox of enrichment, destabilization of exploitation ecosystems in ecological time. *Science* **171**, 385–387.

Ruess, R. W. and McNaughton, S. J. (1987). Grazing and the dynamics of nutrient and energy regulated microbial processes in the Serengeti grasslands. *Oikos* **49**, 101–110.

Schulze, E.-D. and Mooney, H. A. (1993). *Biodiversity and ecosystem function.* Springer-Verlag, Berlin.

Sterner, R. W. (1986). Herbivores' direct and indirect effects on algal populations. *Science* **231**, 605–607.

Strong, D. R. (1992). Are trophic cascades all wet? Differentiation and donor-control in speciose ecosystems. *Ecology* **73**, 747–754.

Tilman, D. (1982). *Resource competition and community structure.* Princeton University Press.

Vance, R. R. (1978). Predation and resource partitioning in one predator–two prey model communities. *American Naturalist* **112**, 797–813.

Yi, T. and Songling, Z. (1992). Resource-dependent selection. *Journal of Theoretical Biology* **159**, 387–395.

Part 3 Large scale diversity patterns and conservation

Overview

John H. Lawton, Robert Barbault, Claude Combes and Richard D. Gregory

Imagine an internationally renowned museum, housing priceless objects, that does not know what they are, or in which rooms most of them are stored, has no acquisition strategy, and only a ramshackled and makeshift restoration and conservation programme. It is, of course, unthinkable! Unthinkable at least in the world of art, but not in the natural world. It is a disgrace that we still have only the haziest notion of how many species there are on earth, how they are distributed, and how to ensure the survival of most of them into the twenty-second century and beyond, in the face of relentless and rapidly expanding human pressures.

The chapters in this final part of the volume encapsulate the dilemmas and problems faced by biologists in their role as curators of biodiversity. It is impossible in six short chapters to do more than illustrate the nature of these problems, but the examples are telling. There are three key issues. What have we got and where is it? How was the diversity of life on earth generated, and how is it maintained? And how do we save as much of it as possible from human destruction?

Answering these questions is far from simple. One problem that immediately confronts us is that 'biodiversity' is not just a function of numbers of species. Although species are important, and form the principal focus of the first three chapters (Turner, Lennon and Greenwood; Gaston; Morand), species have phylogenetic histories, and some are more unusual (with fewer close relatives) than others. Put bluntly, all species are not 'born equal' if our aim is to conserve the greatest variety of life on earth (Williams). Nor are all individuals of a species the same, which forces biological curators to consider how many different populations need to be studied and protected. Genetic questions aside, the final chapter of this section (by Thomas) makes very plain that entrusting species survival to small, isolated populations on nature reserves is extremely unlikely to work. We will need appropriately managed, interconnected networks of reserves and the sympathetic, wise management by farmers and landowners of the countryside surrounding these reserves.

These are the main themes. Associated with them are a series of substantial sub-themes, each of which is important to a mature understanding of patterns of diversity and conservation.

Turner, Lennon and Greenwood (Chapter 11) review theories of how present latitudinal patterns of species diversity evolved (processes which they group together under the banner of 'Wallace theories'), and (simplifying their arguments) how diversity is maintained (referred to as 'Hutchinson theories'). Turner *et al.* pay particular attention to the energy hypothesis, in which the species richness of a region is controlled by the total or average amount of thermal energy entering the system. The chapter highlights some of the problems that biologists interested in large scale geographical processes have to confront. In particular, direct experimental tests of models and hypotheses are impossible—one cannot manipulate energy inputs and species distributions over entire countries or continents! Nevertheless, they show how carefully formulated hypotheses can be tested statistically. As all astronomers know, it is not necessary to do manipulative experiments to do science. Definitive answers to the questions posed by Turner *et al.* are not forthcoming, and probably cannot be achieved without further tests on alternative data sets. But the boldness of the question, and their approach to solving it are important.

Similar large scale problems are tackled by Gaston (Chapter 12), who asks an extremely simple question: at levels that range from landscapes to continents, how are patterns of species richness in one taxon (for instance, birds) correlated with patterns of species richness in a different taxon (for instance, higher plants)? Perhaps, because the question seems so obvious, it has rarely been answered rigorously! Problems of data quality and scale aside (neither of which, as Gaston points out, is trivial), we generally expect positive correlations (more plants means more birds, or more insects), but it need not always be so, and the reasons for such positive correlations are poorly understood and under-researched. They are expected under the Turner *et al.* energy hypothesis, for example, but other explanations are equally plausible. There are also clear conservation implications. Because data on the distribution and diversity of many taxa are extremely poor, decisions on where to place reserves must often be based on limited data from a few taxa, in the hope that an area 'good' for birds will also be rich in other less-loved groups—beetles or slugs, for instance. It is not at all clear, as Gaston points out, that this assumption is sound.

Morand (Chapter 13) takes a different, but again large scale perspective. He tackles the problem of parasite diversity and, among other questions asks how parasite life-cycles influence taxonomic richness. For nematodes, at least, he finds that average numbers of genera per family and species per genus are higher in parasitic than in free-living species; and that in parasitic species there are more families with complex life-cycles (CLCs—requiring more than one species of host to complete development) than there are families with simple ones (SLCs—requiring one species of host). He attributes the larger number of CLC compared to SLC taxa to the fact that complex life histories reduce extinction risks by stabilising population dynamics. It is possible to argue with this hypothesis. For example, CLC parasites may conceivably have reduced risks of extinction, but presumably this form of life-cycle is harder to evolve in the first place than that of SLC parasites. Hence, it is uncertain where the net balance of advantages lie. Nevertheless, Morand's is a first and valuable attempt to seek major patterns in nature linking the diversity of parasites to the form of their life-cycles.

Williams (Chapter 14) links the study of patterns in the distribution and species richness of organisms with attempts to make rational and effective decision about which taxa most deserve the greatest attention from conservationists. We cannot hope to conserve every species, but are there rational ways of deciding which to let go, and which to try and save? He argues that we should attempt to conserve the maximum variety of life on earth (coelacanths and tuataras are 'worth' more in conservation terms than yet another species of salmonid or *Anolis* lizard); and he reviews some of the ways and means of achieving these ends. The aims are laudable, but while we argue about what to do and where to do it, the destruction gathers pace. Time is not on our side.

Similar concerns about lack of time surround the chapter by Ennos (Chapter 15). He provides a succinct and insightful account of how we can measure and what is known about genetic diversity within plant species. He then goes on to summarise data on the considerable genetic and phenotypic variation that exists in one species of British pine tree, *Pinus sylvestris*, albeit one now confined to about five percent of its former range. The implicit messages that emerge from this chapter are stark. We must have already lost considerable genetic variation within this one species, even though it survives and is not threatened by extinction; and we cannot hope to discover, let alone do anything about, the genetic variation that exists and will be lost from many thousands of tree species elsewhere in the world. Ennos himself cautions against pursuing the single goal of maintaining genetic diversity in circumstances where this runs counter to pragmatic conservation. The time and resources are simply not there and are unlikely to be made available to save anything but a fraction of the genotypes of most species.

Finally, Thomas (Chapter 16) takes a small British butterfly as a case study and shows the complexities involved in understanding the persistence and extinction of populations and what must be done to deliver effective conservation. Populations on small, and/or isolated patches of suitable habitat are unable to persist, a fate that will undoubtedly befall many organisms the world over. The only difference is that we cannot hope to study anything but a fraction of the world's biota with the elegance and depth of Thomas's work on *Plebejus argus*. But we can try to act on its general message.

Drawing these arguments together, we do not wish to denigrate the value of Thomas's detailed studies; or Ennos's ideas on the genetics of individual species; or Williams's call to think rationally about what to conserve; or

Gaston's endeavour to establish patterns of diversity in different taxa. What links these four chapters is the realisation that time is not on the side of those who would have perfect, or at least very good information, on most organisms, before making decisions about what to conserve and where to conserve it. If we wait that long, there will be nothing left to protect! We have to be bold, and act upon what we can learn from the few species that *have* been well studied.

The needs are urgent, even desperate and are starkly (albeit mostly implicitly) outlined by the six chapters of this final section. They are these. First, we need to investigate and document, as quickly as possible, the major patterns of diversity of life on earth as we currently find them. They will not be here to study in 100 years time. Second, we need to promote the sympathetic and sustainable management of the larger countryside outside reserves and protected areas. There needs to be greater interaction between conservationists and the policy makers and planners. Third, we need to grab as much land as possible for conservation; almost anywhere reasonably untouched or gently used by humans will do—and we *do* know where key areas of high diversity exist for some taxa. Go for these. Finally, we must conserve areas that are as large as possible, and establish chains of reserves, thereby encompassing some of the genetic variation present in species and permitting some metapopulation processes to persist. It may be messy, but it is the best that the curators of life on earth can do in the time available.

11

Does climate cause the global biodiversity gradient?

John R. G. Turner, Jack J. Lennon and Jeremy J. D. Greenwood

Introduction

The fundamental question of biodiversity 'why are there so many kinds of organisms?' (Hutchinson 1959, May 1990) is easily answered in broad outline: the number of species on the planet must be governed by a dynamic equilibrium between the rate of speciation and the rate of global extinction. But a further outstanding fact about biodiversity is that it is unevenly distributed: in most groups of organisms (there are major exceptions—Hawkins 1990) there are many more species in the tropics than at higher latitudes. Diversity is very low indeed in the polar regions. Answering the question 'why does this diversity gradient occur?' might be an adjunct to the answer to the fundamental question, or it might contain a crucial insight into the answer to the fundamental question. It may require several different answers (Begon *et al.* 1986, Currie 1991). The *species–energy theory*, or as Turner (1992) has called it, the *dynamic theory*, is comparatively neglected among the many theories that have been advanced: for instance, very few front line textbooks mention it. Only since Lawton's (1990), Currie's (1991) and Rohde's (1992) reviews has the theory received wide publicity, after a slow recovery from relative neglect (Brown and Gibson 1983, Wright 1983, Currie and Paquin 1987, Turner *et al.* 1987, Wright *et al.* 1994). It is still misunderstood by many, and has been widely regarded as ridiculous, trivial or obvious.

Major hypotheses for the latitudinal gradient in species diversity

Theories which set out to explain the biodiversity gradient may be classed as Wallace theories (from Wallace 1878) or Hutchinson theories (Hutchinson 1959).

Wallace theories

Wallace theories explain the diversity gradient as a direct consequence of the same processes that determine the total number of species: their global birth and death rates, with or without the inclusion of migration. Thus it may be that speciation rates in the tropics are higher, that extinction rates at high

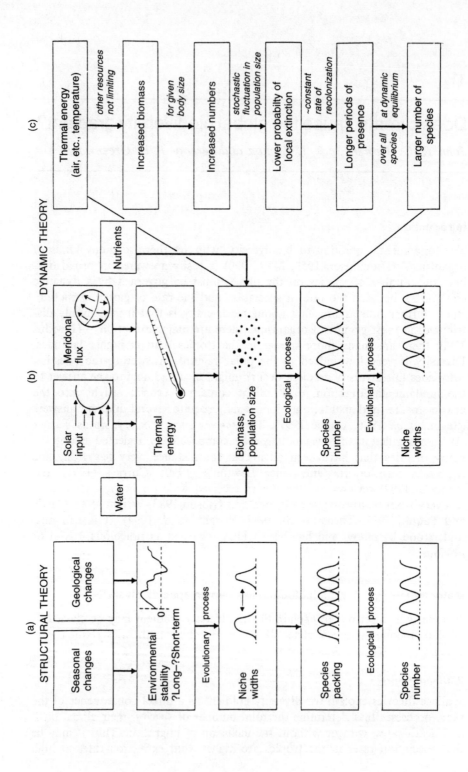

latitudes are higher, or both (Rohde 1992, Jablonski 1993). The theory may be one of permanent equilibrium between these processes or of diversity changing with geological time (Stehli *et al.* 1969), and it might assume that species remain in the latitude at which they originate. The model might, on the other hand, permit some or all species to spread from their point of origin and include progressive adaptation to unfamiliar latitudes, so that a further dynamic gradient is set up between net source areas for species birth and net sink areas of species death. Gray (1878) suggested that the present situation was displaced from its normal equilibrium, with the higher latitudes exceptionally impoverished by widespread extinction during the recent glaciations (also Adams and Woodward 1989). A particularly popular version of this theory in British natural history explains most of the modern distribution of the British biota purely as the result of steady recolonisation of the islands from further south (e.g. Ford 1945, Beirne 1952, Harrison Mathews 1952; Mathews 1955, Pennington 1969, Ragge 1988; or, for a more critical appraisal, see Dennis 1977). Silvertown (1985) has produced convincing evidence that this last view was faulty: the diversity gradient among British plants has persisted now for a long time, and seems not to be a still-advancing wave of colonisation.

Hutchinson theories

Hutchinson theories propose that the biodiversity gradient is produced by ecological or evolutionary processes other than the distribution of areas of global species birth and species death. In the extreme, the gradient could exist even if global birth and death were distributed evenly across the planet. Hutchinson's original formulation in his famous *Homage to Santa Rosalia* paper (Hutchinson 1959) suggested two forms for this theory:

1. The *structural theory* says that in tropical regions ecological niches are narrower, so that more species are packed in. The theory supposes that evolutionary processes determine the structure of ecological communities, and that this structure in turn controls their species richness (Fig. 11.1(a)). Thus once evolutionary forces have determined the ecological requirements or 'realised niche widths' of the species present, competition between species will eliminate those whose niches overlap too much with others. May (1973) developed this argument into a sophisticated mathematical theory.

There are many sub-variants of this theory (MacArthur and MacArthur 1961, Janzen 1970, MacArthur 1975; Rohde (1992) discusses many more, as

---

Legend to facing page

Fig. 11.1 Flow diagrams to explicate (a) the structural theory, (b) the dynamic or energy theory and (c) the stochastic process of population turnover which links the resource supply to species richness according to the dynamic theory. From Turner (1992); for a full mathematical treatment, see Wright *et al.* (1993).

well as various further theories that fall at the first fence); Fig. 11.1(a) shows one. It suggests that the relative stability of the environment in the tropics allows species to specialise more and to evolve narrower ecological niches. Some workers have seen the relative instability of the environment at high latitudes as wider short-term fluctuations in conditions (seasonality); some see the stability of the tropics as much smaller long-term fluctuations (stability in geological time), allowing evolutionary saturation of the niche space. Sometimes the stability of the tropics is explicitly described as a stability of the energy supply (Connell and Orias 1964).

Wallace theories and the structural theory, with their variants, are widely reviewed in textbooks (Pianka 1974, Begon *et al.* 1986, Colinvaux 1993).

2. The *dynamic* or *energy theory* proposes that ecological processes, such as population growth and extinction, being rapid, are initially in control of the dynamics of ecological communities (Turner *et al.* 1987, Turner 1992, Wright *et al.* 1994): it holds that the slow structuring of the community by natural selection, which normally requires a longer time, follows as a consequence of the ecological processes. Given a particular number of species in an eco-system, natural selection and adaptation will slowly adjust the resource requirements of the species; any ecological structuring is the consequence, not the cause, of the presence of large or small numbers of species. It is further proposed that the species richness of a region is fairly directly controlled by the available thermal energy, that is, by the total or average amount of energy and not by the stability of the energy supply (Brown and Gibson 1983, Wright 1983, Currie 1991, Turner 1992).

In short, the theory proposes that ecological niches are narrower at low latitudes because there are more species, not that there are more species because their niches are narrower.

The theory is summed up as a flow diagram in Fig. 11.1(b) (Turner 1992) to contrast with the flow diagram of the structural theory (Fig. 11.1(a)). Initially the availability of resources, such as water, nutrients and energy determines the total biomass and population size of the biota in a region. Through rather rapid ecological processes, population sizes determine the number of species, and then through slow evolutionary processes the number of species determines such structural properties of the community as niche widths and species packing. Clearly this theory reverses the central assumption of the structural theory, that evolutionary forces are the primary controllers of ecological communities, proposing instead that ecological processes like colonisation and local extinction have the upper hand.

Difficulties

The above outline of the dynamic or energy theory immediately suggests a number of problems. That the available resources will govern biomass and, for any given body size, the population size achieved by individual species is perhaps the part of this theory least likely to generate scepticism. Variation in any of the necessary resources, of energy, water or nutrients will affect these two properties of populations. We now come to the first of our difficulties: in what form does the organism experience the energy? Many workers interpret the species–energy theory as meaning that what matters is the flux of incident solar energy at the earth's surface. In fact, although some organisms may absorb energy from direct sunlight by basking or by the occupation of a suitable microhabitat, most of the energy experienced is in the form of the heat (in degrees above absolute zero) of the surrounding medium, be it water, soil or air. This, in turn, depends only indirectly on the flux of solar energy because of the transport of energy away from the tropics towards the poles by the circulation of the oceans and atmosphere.

Clearly the organism does not 'use' the energy of the ambient environment in the same way as the chemical energy in its food or the incident solar energy during photosynthesis. For reasons of fundamental physics, as well as evolutionary feasibility, the organism cannot operate a heat engine to exploit the ambient energy, whose influence is merely on the rate of chemical reaction in the organism's metabolism. Put simply: the warmer it is, the faster it will grow, other things being equal. This is obvious for a plant or an ectotherm. For an endotherm, at normal temperatures experienced on this planet, the organism maintains itself at a higher temperature than ambient, and the lower the ambient temperature the more of its energy budget is spent maintaining its body temperature. This entails less of its resources (energy, nutrients or time) being devoted to growth and reproduction; a higher environmental temperature reduces the organism's 'homeostatic load'. Very simply, the effect of extra atmospheric warmth for a plant is to promote growth, for an ectotherm is to permit more rapid intake and profitable use of nutrients and metabolic energy, and for an endotherm is to cause it to expend less of these in maintaining body temperature. The outcome for all organisms is thus the same: higher temperatures will, in general, lead to faster or greater growth, first of the individual and then of the population. What we are considering then is a relationship between the ecological properties of populations and the thermal energy of the immersing medium, measured by the appropriate environmental temperature.

Other things are not necessarily equal. Solar energy is transmitted to organisms by two *indirect* routes—the temperature of atmosphere and water and the rate of photosynthesis. Faster growth may be defeated by a limiting food supply. Thus the availability of minerals and water for plants and of water and food for animals may confound the effects of higher temperature. Higher up the food chain at least, we should be able to detect an effect of productivity (if it can be adequately measured) on diversity, additional to the direct effect of environmental temperature.

This settles the second difficulty: the thought that if species richness is supposed to depend on atmospheric temperature, then clearly dry deserts, which are relatively species-poor for their latitude, must constitute a decisive counter-example (Huston 1979). It is indeed possible for water-deficit to affect species richness. Clearly the chief environmental gradient, on a global scale, is of temperature, but there are large deviations from this gradient in some parts of the Earth where water becomes limiting. Similarly there may be areas where nutrient supply is ecologically limiting; but these areas would seem to be relatively small and difficult to perceive on a gross geographical scale.

The third difficulty arises in understanding the link between biomass or population size and the number of species. For this, it is necessary to consider a stochastic model of the persistence of local populations. It is now widely believed that local populations of most, indeed probably of all species, are subject to perpetual turnover, becoming stochastically extinct after an ecologically comparatively short period of time, to be recolonised usually, at some later date, by immigrants from some other population (e.g. Levins 1970, Dempster 1983, Maurer 1990, Hanski 1991, Thomas 1991, Harrison 1991, Thomas and Harrison 1992, Hanski and Gilpin 1991). G. Beven for example observed a constant turnover of this kind among the bird species present in an English woodland over a period of years (in Williamson 1981). The number of species remained roughly constant, but the actual species composition changed markedly during the period of observation. There is still serious doubt about the applicability of this type of stochastic model, and some workers consider this theory to be no theory at all (for the general question of stochastic modelling, Krüger et al. 1987; for controversies over stochastic population modelling, Turner 1992). Rohde (1992) in a review of the empirical correlation between diversity and temperature, uses—and then rejects—a deterministic model of habitat saturation to link the two.

Given that population persistence is a positive function of population size (Williamson 1981, Lande 1987, table 14.1 of Williamson 1989), it is a matter of no great difficulty to see that, with the same constant loss and replacement rates of populations in any two areas, the one which contains large populations will have more species at dynamic equilibrium than the area in which the populations are small. Or in even simpler terms, when populations of all species become smaller, the abundant species become scarcer, the rarer species become very sparse indeed, and the rarest species vanish (Hutchinson 1959, Ricklefs 1973). This pattern will not be destroyed if species have different rates of recolonisation.

Thus given this model of species turnover, there is a comparatively direct link between the supply of resources coming into an ecosystem and the equilibrium number of species it will contain (Fig. 11.1(c)). On the global scale this means that species numbers will be rather directly related to thermal energy in the form of average temperatures, with deviations from the overall global gradient in arid or nutrient-poor zones. Wright et al. (1993) develop this argument into a full stochastic model connecting individual consumption through to species richness.

Adaptation and range limitation

We have found that some critics believe that the theory lacks adequate physiological foundations. The distribution of organisms must indeed depend on their physiological responses to the climate. But it does not follow that everything about their distribution can be predicted from physiological information alone. Indeed, Carter and Prince (1981), in a seminal and somewhat neglected paper showed that the limit to the distribution of a species would occur well short of the point at which actual survival or reproduction of individuals became physiologically impossible: the discrepancy arises because there are processes involving the propagation of whole populations which cannot be predicted solely from a knowledge of individual physiology. As usual, reductionism is only a partly effective strategy in science.

On the other hand, it is clear that any theory of populations must not attempt explanations which require physiological processes that are known to be untrue: to predict that populations would become larger as the energy supply decreased might require metabolism to decline with rising temperature, and hence would be ridiculous. There is indeed a theoretical limit to the relationship when the environment is hot enough to denature biological molecules. However, these conditions do not normally apply on a planetary scale. More seriously, metabolism will decline with increasing temperature beyond some critical value, for although chemical reactions will increase in rate with temperature, the functioning of biological macromolecules may not. Clearly, for an Antarctic Icefish, the temperature of the temperate zone would be lethal. Perhaps each species has a range of temperature in which it functions, and this determines its latitudinal range (with the discrepancy described by Carter and Prince 1981), and that is all there is to be said.

The problem with this neat solution is that it does not explain why there are so many more species adapted to the tropics than to the temperate zone and to the temperate zone than to the Arctic. Is it that there are fewer species in the temperate zone because fewer species are adapted to lower temperatures, or that the number of species at this latitude is somehow regulated (Wallace hypotheses would suggest by speciation rates; the energy theory would suggest by population dynamics), and that these species become adapted to the zone in which they live? Or is there some kind of feedback loop?

A currently canvassed version of the Wallace hypothesis, suggests that the rate of speciation (or clade generation) is higher in the tropics (Jablonski 1993, Latham and Ricklefs 1993, 1994). This might arise from an increased mutation rate: Stehli *et al.* (1969) suggested increased ultraviolet radiation, and Rohde (1992) proposed higher tropical temperatures as the direct cause. However, it is not clear that speciation rates are at all tied to mutation rates, and the effect would most readily arise from either Rohde's alternative suggestion—shorter tropical generation-times—or, if the long term rate of species formation is indeed proportional to land-surface area (or some equivalent measure for marine systems), from the greater area of the planet at low latitudes, that is, there is simply more land (or sea) in general, between the equator and five degrees, than in a five degree belt in the temperate zone (planetary surface area is a direct function of the cosine of latitude). One might think that the greater latitudinal range of the species at high latitudes (Rapoport's rule, see Stevens 1989) would offset this effect: however, the important point is not the number of daughter species which any one species produces during its lifetime, but the number of species which it produces with their points of origin at a particular latitude. The effect of land area will ensure that over a long enough time, whether from one widespread species, or from the total ensemble of species, more will be produced in this way 'adapted' to lower latitudes. The theory then implies certain limitations on not only the rate of global extinction of species (which must be latitudinally distributed in such a way as to maintain the gradient—for instance, by being much the same at all latitudes and certainly not biased toward the tropics), but on the rate at which species expand from their points or areas of origin— if the rate of expansion is too fast, then the gradient will be substantially destroyed. But why do species continue to maintain the particular adaptations of their area of origin, rather than moving bodily, or spreading outwards to occupy a large part of the latitudinal range of their continent? There are four major hypotheses:

1. *External restraint on evolutionary change*, for instance from competition with species in adjacent latitudinal belts; according to the structural theory, tropical species should have difficulty invading higher latitudes because the broad-niched species already there will have saturated the ecological space.

2. *Shortage of time*: we know little of the rate at which species adapt to new environments and spread, although some actual cases are extremely rapid (Hudson 1965, 1972); in general we expect adaptation to environmental gradients to be gradual and steady, with newly arising genetic variation in the form of mutations of relatively small phenotypic effect ('polygenes') allowing ever more extensive adaptation to more extreme climates and other factors. Perhaps this process is so slow that it cannot be completed across the full range of global latitudes within the average life-span of a species.

3. *Internal restraint*, postulated in many versions of the punctuated equilibrium theory (Gould 1980), but in a more plausible version as the problem of adaptation to major environmental steps, which may require several rather large, simultaneous mutations, or the invention of a new adaptive suite, and which will therefore occur rarely and sporadically. Latham and Ricklefs (1993, 1994) suggest that frost constitutes one such barrier: only rarely do individual species solve the problem of adapting not merely to a change in their metabolic rate with lowering temperature or increasing seasonality, but

to a change in the state of water. This, they suggest, traps large numbers of species within the tropical belt.

4. *External restraints imposed by the dynamic process* of Fig. 11.1(c), which prevents species from entering environments at the margin of their range for sufficiently long for them to adapt: the appropriate mutations (whether of the small kind that permit continuous slow adaptation, or of the large kind that deal with major environmental challenges) may be available in the genome, but the species may be in the new environment for such short periods that the chance of their occurring when they are favoured by natural selection is substantially reduced.

Tests of the energy/dynamic theory

Many studies have shown that species richness indeed correlates, on a global or local scale, with the energetic element of climate: temperature or evapotranspiration (e.g. Richerson and Lum 1980, Wright 1983, Currie and Paquin 1987, Turner *et al.* 1987, Adams and Woodward 1989); Wright *et al.* (1994) (see also Rohde 1992) give a comprehensive tabulation and summary of these and other studies. This abundant evidence unfortunately does not confirm the species–energy hypothesis, but merely rediscovers the observation on which it was based: that species richness is correlated with latitude, and that latitude is correlated with temperature (an appropriately smoothed surface of either of them defines the planet's equator with surprising accuracy—Stehli 1968). The findings make no critical separation between temperature and latitude as the 'causes' of the gradient, the only way out being to argue that latitudinal position cannot by itself cause anything. However, that argument leaves the field wide open for other correlatives of latitude, such as area.

It is here that the strength of the species–energy hypothesis becomes apparent: it makes a large number of subsidiary predictions, which in the Popperian tradition can be used in attempts to refute it.

The microclimate effect

If there is a direct influence of thermal energy on species diversity, then the relevant measure of energy is that experienced by the organism. Meteorological records are taken in screens placed at a standard height of 1.4 m over mown grass, and as far as possible from the influences of objects that might affect their microclimate. The temperatures they record are therefore more relevant to larger and more exposed organisms and less relevant to those that are smaller and less exposed to the general atmosphere. Hence, it is predicted that

(1) meteorological records should be good predictors for organisms such as birds;

(2) solar ectotherms (e.g. butterflies and grasshoppers) and those invertebrates which occupy the boundary layer (e.g. phytophagous insect larvae) should be further predicted by the hours of bright sunshine (the total hours when there is enough infrared to burn when focused—a fair predictor of the heating of the boundary layer), but negatively related to wind-speed (which tends to destroy the boundary layer—e.g., Grace 1981);

(3) soil organisms (e.g. centipedes) should correlate more with soil temperature than with air temperature; and

(4) organisms which spend most of their life in fresh water, for which no extensive meteorological data exists (e.g. dragonflies, water beetles) should correlate poorly with all available climatic data; and

(5) low growing plants should show boundary layer effects; emergent trees, whose temperature is rather closely coupled to that of the air (Grace *et al.* 1989) should show less effect of the microclimate.

The aridity effect

When aridity limits species richness we would expect that, if the dynamic theory is valid, either temperature and water availability (predicted with reasonable reliability on the macroclimatic scale by rainfall) will be significant predictors of species richness, or species richness will be well predicted by actual evapotranspiration, the combined function of temperature and water availability (Rosenzweig 1968). It would not be a particularly critical test to examine these relationships in continental deserts, as a positive result would tell us only what we already know, that the species richness of arid deserts is low for their temperature or latitude. However, the test can be carried out in areas which few people would think of as arid, such as the British Isles (which are indeed proverbially rainy), by examining the distribution of ferns and bryophytes, plants which are heavily dependent on liquid water for reproduction, and comparing this with the majority of flowering plants, which are not.

The topography effect

The climate experienced by organisms may differ from the broader meteorological readings because of the aspect and slope of the ground; in England the microclimate of south facing slopes, particularly of short turf, can be up to 11°C warmer on sunny days than the opposite north facing slope (reverse for southern hemisphere) (Rorison *et al.* 1986). Thus the species richness of the south slopes should be considerably greater, not merely than that of the facing north slopes, but greater also than that of a level plain at the same latitude and altitude. According to the energy theory then, the south slopes will add species to the local richness, while the north slopes, although species-poorer, will not actually subtract any species from the regional list. We therefore expect that dissected topographies, particularly those with valleys running predominantly east–west, will have more species than smooth landscapes under the same macroclimate.

It is a classic finding of Simpson (1964) that the diversity of North American mammals is greatest in topographically rough areas (the mountains of the West). Plant diversity in California similarly increases with topographic roughness (Richerson and Lum 1980). Simpson's interpretation, favoured also by Pagel *et al.* (1991), is that this results from increased habitat diversity. In the extreme, topographically varied areas can contain habitats from the Arctic to the humid tropical within a few dozen kilometres: critical analysis will be required to distinguish this effect from the climatic effect proposed here.

The winter effect

The energy theory predicts that only groups which are active in the winter will show a positive relationship with winter thermal energy; dormant or diapausing organisms should show no correlation or, because warm winters raise metabolic rates and reduce fat reserves, a negative correlation. We began our research in this field by testing this effect in British butterflies and moths. The hypothesis survived the test: their species richness correlated positively with the summer temperature and negatively with the temperature in winter (Turner *et al.* 1987).

The body-size effect

If the primary cause of the species–climate relationship is mortality, etc. produced by the energetic relations of individuals, then certain testable consequences follow for warm-blooded endotherms: on account of their surface:volume ratios, their weights (averaged across species) should increase in colder climates. Cousins (1989) has shown that this indeed applies to British birds. But according to the species–energy hypothesis, this effect will occur only when heat-loss is critical for survival. Thus, except for the smallest species, even quite far north in the temperate zone this increase in weight should be far less marked for birds that visit only during the summer, when temperature is less critical than in winter (Root 1988a,b). In winter, resident and visiting birds of most size-classes should show a strong weight–temperature relationship. Eventually such theories will have to be integrated with the totally independent work on the partitioning of energy resources according to body size (reviewed by Lawton 1990).

The seasons effect

If the species–energy hypothesis is valid, then species richness should correlate with climate only in the season when a taxonomic group is present in the area under study. This can be tested with migratory and non-migratory birds, as migratory species are present only for part of the year. Thus, in a temperate area, the summer visitors should correlate with the summer climate only; the winter visitors should correlate with the winter climate; and year-round residents, with some complications depending on how much they change their range between the two seasons, should correlate both with summer and winter temperatures.

Areas such as western Europe, including in microcosm the British Isles, are ideal for carrying out this test, as the overall gradient of temperature switches through almost a right angle between summer and winter; contrary to one's naive expectation, summer and winter temperatures in this area are not highly correlated (Turner and Lennon in Greenwood 1992). Turner *et al.* (1988) made use of this right-angle switch in the temperature of Britain to perform the seasons test on the distributions of a subset of the British birds (54 species of 'small' insectivores, that is smaller than a Starling), divided into summer visitors, winter visitors and year-round residents. The species richness of the winter visitors was highly correlated with the winter temperature and not at all with the summer, and the residents in winter correlated with both the winter and the summer temperature. In the summer the results were somewhat ambiguous: the visitors seemed to correlate well with the summer temperature (and not at all with the winter), but the distribution of breeding residents seemed not to be satisfactorily explained by the climate at all. We suggested that this provided qualified support for the energy/dynamic hypothesis.

Birds and the seasons test

We (Turner, Lennon and Greenwood, in preparation) have now extended this preliminary study to include the whole of the British bird fauna (Sharrock 1976, Lack 1986). In addition to dividing the birds into the four above classes by their migratory status, we divided them into three weight classes, 'light, medium and heavy' (Table 11.1). This allowed us to test not only the seasons effect, but effectively the body-size effect as well. The birds were further divided into cross-classified subclasses according to their habitat (Table 11.1 footnote) and their diet (to search for systematic differences according to the birds' ecology). Association of species richness with the average temperature of the four seasons (winter being December through February, summer being April through September, and spring and autumn the intervening months) was tested by step-backward multiple regression, using the species richness in quadrats of 20 km per side centred on 84 widely distributed climate recording stations (map in Turner *et al.* 1988).

Table 11.1 shows the significant remaining variables after backward elimination from the full set, showing the most prominent association of each subgroup of birds with the climate of the four seasons. Thus, for example, the species richness of the lightest weight winter residents in all habitat and diet classes combined (WRAA) is significantly correlated with the

Table 1 Associations of species richness with seasonal temperature in British birds

(a) Light birds (<50 g)

Summer distributions				Winter distributions			
Group	r^2	Temperature of	N	Group	r^2	Temperature of	N
Residents				Residents			
SRAA	0.326	+SMR	45	**WRAA**	**0.708**	+WIN+SMR–SPR	41
SRAI	0.355	+SMR	28	**WRAI**	**0.662**	+SMR	15
SRAO	0.000		13	**WRAO**	**0.620**	+WIN+SMR–SPR	9
				WRAV	0.319	+SMR	16
SRLA	0.264	+SMR	43	**WRLA**	**0.686**	+WIN+SMR–SPR	39
SRLI	0.372	+SMR	27	**WRLI**	**0.689**	+WIN+SMR–SPR	14
Visitors				Visitors			
SVAA	**0.499**	+SMR–SPR	37	**WVAA**	**0.422**	+WIN+*SMR*–SPR	34
SVAI	**0.492**	+SMR–SPR	34	**WVAI**	**0.569**	+WIN+*SMR*–SPR	18
SVLA	**0.522**	+SMR–SPR	33	WVAO	0.120	+AUT	7
SVLI	**0.513**	+SMR–SPR	30	WVAV	0.139	+*SMR*	8

Table 1 *Continued*

(b) Medium weight birds (50–350 g)

Summer distributions

Group	r^2	Temperature of	N
Residents			
SRAA	0.000		38
SRAI	0.185	*–SMR*	15
SRAO	**0.587**	+SMR–SPR	11
SRAR	0.079	–SPR	7
SRLA	0.000		35
SRLO	**0.572**	+SMR–AUT	9
Visitors			
SVAA	0.140	*–SMR*+AUT	27
SVAI	0.303	*–SMR*+AUT	19
SVAP	0.136	*–SMR*+SPR	6
SVLA	0.000		19
SVLI	0.221	*–SMR*	13

Winter distributions

Group	r^2	Temperature of	N
Residents			
WRAA	**0.463**	+SMR	29
WRAI	0.00		15
WRAO	**0.524**	+SMR–SPR	11
WRAR	0.347	+SMR	7
WRAV	0.265	+SMR	6
WRLA	0.442	+SMR	27
WRLI	0.000		6
WRLO	**0.525**	+SMR–SPR	8
Visitors			
WVAA	**0.527**	+WIN–SPR+AUT	47
WVAI	**0.536**	+WIN–SPR+AUT	37
WVAO	0.111	*+SMR*	8
WVLA	**0.542**	+WIN+*SMR*–SPR	28
WVLI	**0.566**	+WIN+*SMR*–SPR	21
WVLO	0.000		6
WVSA	**0.485**	+WIN–SPR+AUT	17

Table 1 *Continued*

(c) Heavy birds (>350 g)

	Summer distributions				Winter distributions			
	Group	r^2	Temperature of	N	Group	r^2	Temperature of	N
	Residents				Residents			
	SRAA	**0.418**	$-SMR$ +SPR	67	**WRAA**	0.366	+ AUT	48
	SRAI	**0.540**	$-SMR$ +**WIN**	12	WRAI	**0.447**	$-SMR$+AUT	6
	SRAO	0.273	+**SMR** $-$SPR	21	WRAO	**0.473**	+**SMR**–AUT	9
	SRAP	**0.558**	$-SMR$ +SPR + AUT	14	WRAP	**0.444**	$-SMR$+AUT	12
	SRAR	**0.545**	$-SMR$ + SPR	12	WRAR	0.195	+**WIN**$-SMR$	10
	SRAV	0.000		10	WRAV	0.396	+**SMR**–SPR	13
	SRFA	0.000		22	**WRFA**	**0.641**	+**SMR**	10
	SRFO	0.227	+**SMR**	9	**WRLA**	0.000		27
	SRFP	0.226	$-SMR-WIN$+SPR	8	WRLR	0.195	+**WIN**$-SMR$	10
	SRLA	**0.405**	$-SMR$ + SPR	35	WRLV	0.392	+**SMR**–SPR	11
	SRLO	0.238	+**SMR** $-$SPR	13	**WRMA**	**0.446**	$-SMR$+AUT	11
	SRLR	**0.545**	$-SMR$ + SPR	12	**WRMP**	**0.446**	+**WIN**$-SMR$	9
	SRLV	0.127	$-$AUT					
	SRMA	**0.643**	$-SMR$+**WIN**	13	Visitors			
	SRMP	**0.590**	$-SMR$+**WIN**	9	**WVAA**	**0.406**	+ AUT	63
					WVAI	0.295	+**WIN**	22
	Visitors				WVAO	**0.537**	+*SMR*	10
	SVAA	0.221	$-SMR$ +SPR	9	WVAP	0.355	$-SMR$+AUT	16
					WVAR	0.346	+**WIN**	10
					WVAV	0.316	+ AUT	15
					WVFA	**0.513**	+*SMR*	21
					WVFI	0.244	+**WIN**	6

Table 1 *Continued*

				N
WVFO	**0.537**	*+SMR*		8
WVFV	0.384	*+SMR*		6
WVLA	0.316	+AUT		28
WVLR	0.346		+WIN	10
WVLV	0.186	+AUT		11
WVMA	0.306		+WIN	19
WVMI	0.203		+WIN	11
WVMP	0.348	−SMR+AUT	+WIN	10
WVSA	0.398		+WIN	10

The predictors of species richness are the temperatures of those of the four seasons (SPR, spring; SMR, summer; AUT, autumn; WIN, winter) which remain after backward elimination in a multiple regression. The sign shows the direction of the association. Bird groups are coded by: migratory status (S, summer; W, winter; R, resident; V, visitor); habitats (A, all combined; F, fresh water; L, land; M, marine; S, shore); diets (A, all combined; I, insects and other invertebrates; O, omnivores; P, piscivores; R, mammals and birds (raptors); V, vegetarians). Groups with fewer than six species are omitted. *N* is the number of species in the category. For the significance of bold and italics, see text.

temperature of summer, winter and (inversely) spring. Associations which are predicted by the species–energy hypothesis are printed in bold type; major anomalies are in italics. The coefficient of determination (r^2) of each model is also given; high values are bold (as are groups with more than 20 species); low values tend to indicate unsatisfactory models.

The significant pattern is the way in which summer and winter temperature correlate with the four migration classes (summer and winter, visitors and residents) *within* each weight class (it is not legitimate to take a head-count of the number of times a regressor appears in any part of the table, as the groups are nested and also overlap a little—some species may be included in two habitats or diet classes).

Winter temperature appears often as a significant regressor in models for winter birds, both residents and visitors. It therefore appears that the species richness of wintering birds is indeed strongly influenced by (or at least cor-related with) the ambient winter temperature. Conversely, winter temperature appears seldom in the models for summer distributions (and only in the heaviest class). Again, consistent with the energy hypothesis, particularly for visitors, winter temperature is not influencing summer distributions to any extent.

What of bird distributions in summer? Only in the light-weight birds does *summer temperature* appear as a frequent significant regressor: among the middle and heavy-weight birds it appears inconsistently and more often than not with a negative sign. This indicates that the ambient temperature during summer influences the species richness of birds (both residents and visitors) in summer quite strongly, but only if the birds are small. The ambient tem-perature does not detectably influence the species richness of heavier birds (above 50 grams in weight).

This constitutes rather good support for the energy hypothesis, when the body-size effect is also taken into account. It is reasonable to suppose that birds of all sizes are under temperature stress during the British winter. Long periods of cold, when food is restricted by seasonal dearth, or by snow and frost, cause the birds to burn their fat reserves in order to maintain their body temperature; cold winter nights can literally cause them to drop dead off their perches (Dobinson and Richards 1964, Newton 1972, Elkins 1983, Clarke *et al.* 1993). Hence in accord with the energy hypothesis, the species richness of all birds in winter is correlated with the winter temperature. In summer birds have much less of a problem, since the air temperature spends most of its time above the critical threshold (Root 1988*a,b*). During the summer it is therefore reasonable to expect that large and middle size birds suffer little thermal stress and should show little correlation with the temperature in that season. On the other hand, the smallest birds may still suffer some thermal stress especially during the cool of the night. Hence the species richness only of these lightest birds is influenced by temperature during the summer.

Or to put it another way, if body size is taken into account, the species energy hypothesis has passed the seasons test, which predicts that summer bird distributions should be correlated with summer temperature, and winter distributions with winter temperature; the lightest birds do indeed switch their distributions between seasons in just this way. The heavier birds also switch in a predictable way, from having their species richness strongly correlated with temperature in the winter, to showing no particular correlation with climate during the summer. All these findings apply to both residents and visitors.

The effect of weight is still seen, rather attenuated, in the winter distributions. Winter temperature is a consistent regressor for light birds, is rather erratic for the heaviest, and has a striking and unexplained association in the medium class, correlating strongly with winter visitors and not at all with winter residents.

Among the anomalous correlations, spring and autumn temperatures give little cause for concern. Autumn appears chiefly alongside winter temperature as a predictor of the richness of winter visitors, which is fully compatible with the energy hypothesis in view of the facts that these species arrive during these months (October and November), and that winter surveying started mid-way through November. Spring seldom appears by itself, but only as an adjunct to summer temperature, when it tends to appear with a negative coefficient (or positive if summer is negative). As the *simple* regression (not tabulated) of species richness on spring temperature is usually positive, spring temperature is clearly taking on the role of a 'spoiler' variable in the multiple regression models, which are using it with reverse sign in an attempt to fit residuals from the main regression on summer temperature.

There is, however, one further outstanding anomaly: a frequent correlation, for all weights of birds, of *winter* distributions with *summer* temperatures. For residents this could be attributed to the summer distributions casting a 'shadow' over the winter: as most resident species only partly shift their range between the two seasons, one might expect that some of the correlation with summer temperature would remain the year round. But the effect is shown also, frequently, by winter *visitors*. This is not expected on the simple energy hypothesis. But, if we recall that for an endotherm the energy theory can be rephrased as the proposal that decreasing environmental temperatures increase the requirement of the animal for nutrient energy, a compatible explanation becomes plain: that we are seeing here an effect of the stored largess of the productivity of summer. Food resources, both vegetable and invertebrate, build up in areas with favourable summer climates and remain banked in diapause and dormancy during the winter, when they are the main food supply for many overwintering birds. The ability of birds to withstand cold, even quite extreme cold, is remarkable, provided the bird has a large enough food supply during the day to build up fat reserves which can be metabolised to maintain body temperature during the cold of the night (Newton 1972, Elkins 1983). It is therefore likely in terms of the energy theory that summer temperatures can affect the species richness of birds in winter via this stored food supply.

Conclusion

So far, the species–energy theory has survived our attempts to destroy it: the winter test is confirmed by British lepidoptera, and the seasons and body-size tests (with an interesting complication from stored summer harvests) by British birds; the distribution of North American mammals and Californian plants conforms *prima facie* with the topography test, but critical work is needed to distinguish this climatic effect from increased habitat diversity.

A research project which thoroughly falsified these and the other predictions we have made would put the energy hypothesis firmly out of court. Validation of the species–energy theory by survival of the tests would not necessarily, on the other hand, mean that other theories were invalid. Both the species–energy theory, and the tropical speciation version of the Wallace theory, suggest that the gradient of species richness is set up by a dynamic equilibrium between species birth and species extinction. The theories differ in that one invokes global birth (speciation) and global extinction, whereas the other invokes only local birth (colonisation) and local extinction. It is a good working hypothesis that a greater rate of speciation in the tropics produces the long term and large scale latitudinal gradient of species richness on the planet. But it is also a good working hypothesis that the process outlined in Fig. 11.1(b)–(c) is inevitable, and will produce a considerable redistribution of species richness along regional gradients in climate and perhaps other resources. The tests outlined above should, when completed, give us a very good idea whether it does. If the tests fail the hypothesis, indicating that the dynamic process does not redistribute species richness in this way, then the urgent question will be '*why not?*'

Acknowledgements

We acknowledge the financial support provided by NERC through its TIGER (Terrestrial Initiative in Global Environmental Research) programme, award number GST/02647. Initial work on bird distributions was carried out with the support of SERC (graduate studentship) and NERC (small research grant GR9/531), and with a generous grant for computing equipment from the Royal Society.

References

Adams, J. M. and Woodward, F. I. (1989). Patterns in tree species richness as a test of the glacial extinction hypothesis. *Nature*, **339**, 699–701.
Begon, M., Harper, J. L. and Townsend, C. R. (1986). *Ecology. Individuals, populations and communities*. Blackwell Scientific, Oxford.
Beirne, B. P. (1952). *The origin and history of the British fauna*. Methuen, London.
Brown, J. H. and Gibson, A. C. (1983). *Biogeography*. Mosby, St Louis.

Carter, S. D. and Prince, R. N. (1981). Epidemic models used to explain biogeo-graphical distribution limits. *Nature*, **293**, 644–645.

Clarke, J. A., Baillie, S. R., Clarke, N. A. and Langston R. H. W. (1993). Estuary water capacity following severe weather mortality. (ESTU TID 4113). BTO Research Report no. 103, p. 166. British Trust for Ornithology, Thetford, Norfolk.

Colinvaux, P. (1993). *Ecology 2*. Wiley, New York.

Connell, J. H. and Orias, E. (1964). The ecological regulation of species diversity. *American Naturalist*, **98**, 399–414.

Cousins, S. H. (1989). Species richness and the energy theory. *Nature*, **340**, 350–351.

Currie, D. J. (1991). Energy and large-scale patterns of animal- and plant-species richness. *American Naturalist*, **137**, 27–49.

Currie, D. J. and Paquin, V. (1987). Large-scale biogeographical patterns of species richness in trees. *Nature*, **329**, 326–327.

Dempster, J. P. (1983). The natural control of populations of butterflies and moths. *Biological Reviews of the Cambridge Philosophical Society*, **58**, 461–481.

Dennis, R. L. H. (1977). *The British butterflies. Their origin and establishment*. E. W. Classey, Faringdon, Oxfordshire.

Dobinson, H. M. and Richards, A. J. (1964). The effect of the severe winter of 1962/63 on British bird populations. *British Birds* **57**, 373–434.

Elkins, N. (1983). *Weather and bird behaviour*. T. and A. D. Poyser, Calton, Staffordshire.

Ford, E. B. (1945). *Butterflies*. Collins, London.

Gould, S. J. (1980). Is a new and general theory of evolution emerging? *Paleobiology*, **6**, 119–130.

Grace, J. (1981). Some effects of wind on plants. In *Plants and their atmospheric environment* (eds. J. Grace, E. D. Ford and P. G. Jarvis), pp. 125–130. Blackwell Scientific, Oxford.

Grace, J., Allen, S. J. and Wilson, C. (1989). Climate and the meristem temperatures of plant communities near the tree line. *Oecologia*, **79**, 198–204.

Gray, A. (1878) Forest geography and archaeology: a lecture delivered before the Harvard University Natural History Society, April 18, 1878. *American Journal of Science and Arts [third series]*, **16**, 85–94, 183–196.

Greenwood, J. J. D. (1992). Understanding bird distributions. *Trends in Ecology and Evolution*, **7**, 252–253.

Hanski, I. (1991). Single-species metapopulation dynamics: concepts, models and observations. *Biological Journal of the Linnean Society of London*, **42**, 17–38.

Hanski, I. and Gilpin, M. (1991). Metapopulation dynamics: brief history and conceptual domain. *Biological Journal of the Linnean Society*, **42**, 3–16.

Harrison Mathews, L. (1952). *British mammals*. Collins, London.

Harrison, S. (1991). Local extinction in a metapopulation context: an empirical evaluation. *Biological Journal of the Linnean Society*, **42**, 73–88.

Hawkins, B. A. (1990). Global patterns of parasitoid assemblage size. *Journal of animal Ecology*, **59**, 57–72.

Hudson, R. (1965). The spread of the Collared Dove in Britain and Ireland. *British Birds*, **58**, 105–139.

Hudson, R. (1972). Collared Doves in Britain and Ireland during 1965–1970. *British Birds*, **65**, 139–155.

Huston, M. (1979). A general hypothesis of species diversity. *American Naturalist*, **113**, 81–101.

Hutchinson, G. E. (1959). Homage to Santa Rosalia *or* Why are there so many kinds of animals? *American Naturalist*, **93**, 145–159.

Jablonski, D. (1993). The tropics as a source of evolutionary novelty through geological time. *Nature*, **364**, 142–144.

Janzen, D. H. (1970). Herbivores and the number of tree species in tropical forests. *American Naturalist*, **104**, 501–528.

Krüger, L., Daston, L. J. and Heidelberger, M. (eds.) (1987). *The probabilistic revolution. Volume 1: Ideas in history.* MIT Press, Cambridge, MA.

Lack, P. (1986). *The atlas of wintering birds in Britain and Ireland.* T. & A. D. Poyser, Calton, Staffs.

Lande, R. (1987). Extinction thresholds in demographic models of territorial populations. *American Naturalist*, **130**, 624–635.

Latham, R. E. and Ricklefs, R. E. (1993). Global patterns of tree species richness in moist forests: energy-diversity theory does not account for variation in species richness. *Oikos*, **67**, 325–333.

Latham, R. E. and Ricklefs, R. E. (1994). Continental comparisons of temperate-zone tree species diversity. In *Species diversity in ecological communities: historical and geographical perspectives* (eds. R. E. Ricklefs and D. Schluter), pp. 294–314. University of Chicago Press.

Lawton, J. H. (1990). Species richness and population dynamics of animal assemblages. Patterns in body size:abundance space. *Philosophical Transactions of the Royal Society of London series B*, **330**, 283–291.

Levins, R. (1970). Extinction. In *Some mathematical problems in biology* (ed.) M. Gerstenhaber. pp. 77–107. Mathematical Society, Providence, RI.

MacArthur, J. W. (1975). Environmental fluctuations and species diversity. In *Ecology and evolution of communities* (eds. M. L. Cody and J. M. Diamond), pp. 74–80. Belknap Press, Cambridge, Mass.

MacArthur, R. H. and MacArthur, J. W. (1961). On bird species diversity. *Ecology*, **42**, 594–598.

Mathews, J. R. (1955). *Origin and distribution of the British flora.* Hutchinson, London.

Maurer, B. A. (1990). The relationship between distribution and abundance in a patchy environment. *Oikos*, **58**, 181–189.

May, R. M. (1973). *Stability and complexity in model ecosystems.* Princeton University Press.

May, R. M. (1990). How many species? *Philosophical Transactions of the Royal Society of London series B*, **330**, 293–304.

Newton, I. (1972). *Finches.* Collins, London.

Pagel, R. M. D., May, R. M. and Collie, A. (1991). Ecological aspects of the geographic distribution and diversity of mammal species. *American Naturalist*, **137**, 791–815.

Pennington, W. (1969). *The history of British vegetation.* English Universities Press, London.

Pianka, E. R. (1974). *Evolutionary ecology.* Harper and Row, New York.

Ragge, D. R. (1988). The distribution and history of the British Orthoptera. In *Grasshoppers and allied insects of Great Britain and Ireland* (eds. J. A. Marshall and E. C. M. Haes), pp. 25–33. Harley Books, Great Horkesley, Colchester, Essex.

Richerson, P. J. and Lum, K-L. (1980). Patterns of plant species diversity in California: relation to weather and topography. *American Naturalist*, **116**, 504–536.

Ricklefs, R. E. (1973). *Ecology.* Chiron Press, Newton, MA.

Rohde, K. (1992). Latitudinal gradients in species diversity: the search for the primary cause. *Oikos*, **65**, 514–527.

Root, T. (1988a). Environmental factors associated with avian distributional boundaries. *Journal of Biogeography*, **15**, 489–505.

Root, T. (1988b). Energy constraints on avian distributions and abundances. *Ecology*, **69**, 330–339.

Rorison, I. H., Sutton, F. and Hunt, R. (1986). Local climate, topography and plant growth in Lathkill Dale NNR. I. A twelve-year summary of solar radiation and temperature. *Plant, Cell and Environment*, **9**, 49–56.

Rosenzweig, M. L. (1968). Net primary productivity of terrestrial communities: predictions from climatological data. *American Naturalist*, **116**, 504–536.

Silvertown, J. (1985). History of a latitudinal diversity gradient: woody plants in Europe 13,000–1000 years BP. *Journal of Biogeography*, **12**, 519–525.

Simpson, G. G. (1964). Species densities of North American mammals. *Systematic Zoology*, **13**, 361–389.

Sharrock, J. T. R. (1976). *The atlas of breeding birds in Britain and Ireland.* British Trust for Ornithology, Tring, Herts.

Stehli, F. G. (1968). Taxonomic diversity gradients in pole location: the recent model. In *Evolution and environment* (ed. E. T. Drake), pp. 163–227.Yale University Press, New Haven, CT.

Stehli, F. G., Douglas, R. G. and Newall, N. D. (1969). Generation and maintenance of gradients in taxonomic diversity. *Science*, **164**, 947–949.

Stevens, G. C. (1989). The latitudinal gradient in geographical range: how so many species coexist in the tropics. *American Naturalist*, **133**, 240–256.

Thomas, C. D. (1991). Spatial and temporal variability in a butterfly population. *Oecologia*, **87**, 577–580.

Thomas, C. D. and Harrison, S. (1992). Spatial dynamics of a patchily distributed butterfly species. *Journal of Animal Ecology*, **61**, 437–441.

Turner, J. R. G. (1992). Stochastic processes in populations: the horse behind the cart? In *Genes in ecology* (eds. R. J. Berry, T. J. Crawford, and G. M. Hewitt), pp. 29–53. Blackwell Scientific, Oxford.

Turner, J. R. G., Gatehouse, C. M. and Corey, C. A. (1987). Does solar energy control organic diversity? Butterflies, moths and the British climate. *Oikos*, **48**, 195–203.

Turner, J. R. G., Lennon, J. J. and Lawrenson, J. A. (1988). British bird species distributions and the energy theory. *Nature*, **335**, 539–541.

Turner, J. R. G., Lennon, J. J. and Greenwood, J. J. D. *in prep*. Testing the dynamic theory of biodiversity: 2. British birds and the seasons.

Wallace, A. R. (1878). *Tropical nature, and other essays.* Macmillan, London.

Williamson, M. H. (1981). *Island populations.* Oxford University Press.

Williamson, M. H. (1989). Mathematical models of invasion. In *Biological invasions: a global perspective* (eds. J. A. Drake, and others), pp. 329–350. Wiley, London and New York.

Wright, D. H. (1983). Species-energy theory: an extension of species-area theory. *Oikos*, **41**, 496–506.

Wright, D. H., Currie, D. J. and Maurer, B. A. (1994). Energy supply and patterns of species richness on local and regional scales. In *Species diversity in ecological communities: historical and geographical perspectives* (eds. R. E. Ricklefs and D. Schluter), pp. 64–74. University of Chicago Press.

12

Spatial covariance in the species richness of higher taxa

Kevin J. Gaston

Introduction

As a field of study, biodiversity engenders a variety of responses. In some quarters it retains the status of a 'band wagon', and is argued to embody nothing distinctive. In others it is viewed as an important new area of work with a clear agenda of its own. Whatever one's position, it is evident that development of the concept of biodiversity has served to highlight, afresh or for the first time, a number of important, yet largely unresolved, questions about the natural world. These include issues such as how many species there are (e.g. May 1988, 1990, Gaston 1991, Hawksworth 1991, Grassle and Maciolek 1992, Hammond 1992, Stork 1993), present levels of extinction (e.g. Whitmore and Sayer 1992, Smith *et al.* 1993*a*, 1993*b*, Gaston 1994), and the importance of individual species in maintaining ecosystem function and integrity (e.g. Walker 1992, Schulze and Mooney 1993). An additional issue, with implications for all these, is that of how numbers of species in different higher taxa covary in space. This issue forms the central topic of this paper.

Background

Relationships between the numbers of species in different groups in different areas (Fig. 12.1) attract attention in theoretical and applied contexts. In the arena of theory, their study has primarily focused on what determines the numbers of species in an area. Although scattered, a moderate sized literature has sought to document and explore patterns of covariance, often comparing the variation in the species richness of one higher taxon potentially explained by the richness of another with that explained by environmental variables (e.g. Kiester 1971, Power 1972, Rogers 1976, Schall and Pianka 1977, 1978, Gilbert and Smiley 1978, Pianka and Schall 1981, Fernandes and Price 1988, 1991, Currie 1991, Compton and Hawkins 1992). Topics of particular emphasis have included interactions of plant and insect species richness (e.g. Kostrowicki 1969, Murdoch *et al.* 1972, Abbott 1974, Lindroth *et al.* 1973, Otte 1976, Slansky 1976, Eastop 1978, Hockin 1981, Reed 1982, Williams 1982, Itamies 1983, Rust *et al.* 1985, Danks 1986, 1993, Dixon *et al.* 1987,

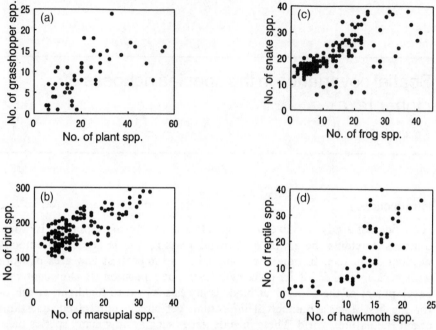

Fig. 12.1 Relationships between the numbers of species of various pairs of higher taxa. (a) grasshopper and plant species for 50 × 50 m plots in North and South American grasslands (from data in Otte 1976), (b) bird and marsupial species in 150 × 150 mile (approx. 240 km) squares across Australia (from data in Pianka and Schall 1981), (c) snake and frog species in 150 × 150 mile (approx. 240 km) squares across Australia (from data in Pianka and Schall 1981), and (d) reptile and hawkmoth species in 152,000 km^2 squares across Europe (K.J. Gaston and R. David, unpublished data).

S.G. Nilsson *et al.* 1988, Brown and Opler 1990, Becker 1992, Gaston 1992, Samways 1994), changes in species richness with succession (e.g. Southwood *et al.* 1979, Brown and Hyman 1986, Brown and Southwood 1987, Brown and Gange 1989) and with latitude (e.g. Kiester 1971, Gauld 1986, Gauld *et al.* 1992), and effects of resource fragmentation (e.g. Dixon *et al.* 1987, Gauld 1991, Gaston 1992, Gauld and Gaston, 1994).

In the applied arena, relationships between numbers of species in different higher taxa are important to effective biodiversity conservation. Information about numbers of species present in different areas may facilitate their prioritisation for determining the resources which should be put into their conservation and enable them to be monitored for the success or failure of that conservation; more broadly, this information may facilitate monitoring of biotic responses to environmental change. As is well-documented, recent years have seen increased emphasis on the need to conserve wholesale biodiversity (a large part of overall biodiversity; Williams and Gaston 1994) and not simply its more conspicuous or popular elements (Myers 1979, Wilson

1987). Given the finite effort available, one sensible way to generate species richness estimates is to identify a higher taxon (or set of higher taxa), an 'indicator group' or 'predictor set' (Kitching 1993), whose richness is both more readily assessed and correlated with the richness of the higher taxon of primary interest (this use of the term 'indicator group' should be distinguished from others).

Until recently, remarkably little evidence had been provided to support claims for the indicator status of any higher taxa. Too often, supposed indicator groups seemed simply to be those which their advocates studied. This problem has been overcome to some extent, with several studies exploring the value of the richness of one or more higher taxa in predicting either patterns of richness in others, or which sets of areas should be conserved to the best benefit of all groups concerned (Emberson 1985, Thomas and Mallorie 1985, Smith and Theberge 1986, Usher 1986, 1992, Yen 1987, Burbidge *et al.* 1992, Kremen 1992, Pearson and Cassola 1992, Ryti 1992, Dinerstein and Wikramanayake 1993, Oliver 1993, Peterson *et al.* 1993, Prendergast *et al.* 1993a, Saetersdal *et al.* 1993, Usher and Thompson 1993).

General conclusions are not forthcoming from this work, both theoretical and applied, for several reasons. First, the data are extremely heterogeneous, particularly in features such as quality and spatial scale (Anderson and Marcus 1993). Second, reported results are often difficult to interpret, due to the ambiguity of the analyses performed. Third, no formal model has been proposed against which results can be compared. Finally, where studies address mechanisms, they often fail to consider the full range of possibilities. In the following sections I discuss these various problems.

Analytical considerations

Hotspots versus all spots

Analyses of spatial covariance in species richness can be divided into two broad groups. These are concerned with whether there is (1) coincidence between areas in which particular higher taxa are most speciose (so-called 'hotspots'; Mittermeier 1988, Mittermeier and Werner 1990, Prendergast *et al.* 1993a,b, Gaston and David in press), or (2) an overall relationship between the numbers of species of two higher taxa in different areas. These two groups should not be confused. Demonstration that hotspots do or do not coincide provides limited information about overall relationships. Likewise, broad relationships between the richness of different higher taxa need not result in coincidence of areas where they are most speciose. Although my primary concern in this chapter is with overall relationships between numbers of species of different higher taxa, and not with the coincidence of hotspots, some of the same analytical considerations apply.

Endemism and rarity versus species richness

In measuring biodiversity, the assumption is often tacitly made that areas which differ in observed levels of endemism, or numbers of rare species tend to differ in overall species richness in a similar way. This would suggest that spatial covariance in levels of endemism, or rarity of pairs of higher taxa provides information about covariance in numbers of species and vice-versa. There are, however, no general patterns in the relationship between endemism or rarity and species richness. Examples are known where patterns of endemism or rarity either do or do not correlate with overall species richness, although most *published* studies report significant correlations (Pearson 1977, Dzwonko and Loster 1989, Järvinen 1982*a*, White *et al.* 1984, Thomas and Mallorie 1985, C. Nilsson *et al.* 1988, Wheeler 1988, Peterson *et al.* 1993, Prendergast *et al.* 1993*a*).

Sampling artefacts

Differential patterns of recording species from any two higher taxa may lead to spurious patterns in covariance. For most pairs of higher taxa it is virtually impossible to ensure equivalent levels of recording. Where serious differences do exist they will, however, usually be known, and both can and need to be accommodated. More generally, spatial variation in recording intensity can seriously frustrate attempts to establish real patterns of relationship between the richness of two higher taxa. Methods are increasingly being developed to reduce such problems (e.g. Prendergast *et al.* 1993*b*), but they often remain severe, especially at smaller spatial scales.

Spatial scale

It has been widely observed that the spatial scale of observation has important consequences for the patterns documented in many ecological and biogeographical phenomena (e.g. Dayton and Tegner 1984, Giller and Gee 1987, Levin 1992). The same considerations are necessary in the context of spatial covariance in the species richness of different higher taxa. Several aspects of spatial scale are liable to be important, including (1) the unit area size used for richness contrasts and (2) the size of the region in which all of these areas are encompassed.

Relationships between the numbers of species of different higher taxa across areas have been explored for areas of a wide spectrum of sizes (Fig. 12.2). At one extreme, Gaston and Hudson (1994) documented covariance in numbers of species of various higher taxa for entire biogeographic regions. At an intermediate scale, Currie (1991) explored relationships between the richness of different higher taxa in North America across grid squares of (latitude × longitude) $2.5° × 2.5°$ and $5° × 2.5°$. At a yet smaller scale, Reed (1982) reported the relationship between the numbers of butterfly and plant species on British and Irish islands.

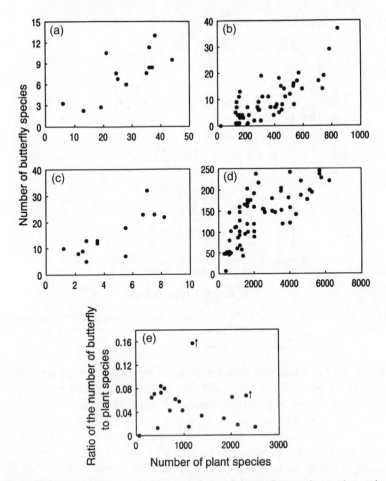

Fig. 12.2 Relationships between the numbers of butterfly species and numbers of plant species. (a) Butterfly species (mean values for 500 m transects) and plant species (mean in 2 m quadrats) for localities in the Moroccan Atlas mountains (from Thomas and Mallorie 1985), (b) butterfly and plant species on British and Irish islands (from data in Reed 1982), (c) butterfly species and the average floral richness (number of types of plants in flower) for transects through forest in Madagascar (from Kremen 1992), (d) Papilionoidea species and vascular plant species in different provinces and regions of the Palaearctic (from Kostrowicki 1969), and (e) the ratio of the numbers of butterfly to plant species in different countries and the number of plant species per 1000 km^2 (from Dixon et al. 1987).

Whilst it is possible to find documented relationships between the numbers of species in the same two higher taxa at various scales, as far as I am aware no concerted effort has been made to explore these differences in the same region using data collected and analysed in a similar way. Figure 12.3 provides such an example in which the relationship is roughly the same at two scales.

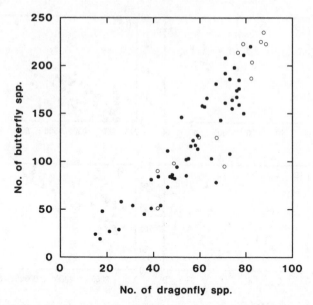

Fig. 12.3 Relationship between the numbers of butterfly species and the numbers of dragonfly species in grid squares of two different sizes across Europe. Solid circles—approx. 152,000 km^2, open circles—approx. 610,000 km^2.

The difficulty of obtaining data for areas of similar sizes may be manifested in a variety of ways. Even for grid-based studies it remains a concern. The unit areas are seldom equivalent, especially when the effects of topography are taken into account; moreover, typically it is difficult to determine what the relevant area actually is. Ocean areas frequently are excluded in studies of terrestrial organisms but, although the areas of other habitat types which the organisms do not occupy (and perhaps actively avoid) might equally be removed, this is rarely done. Substantial differences in the sizes of areas may also result because species occurrences and levels of species richness are commonly plotted on Mercator (and related) map projections, where relative land areas at different latitudes are not represented correctly: land areas appear larger the further they are from the equator. Udvardy (1981) provides some figures for real land areas at different latitudes to illustrate the problem, and Williams (1994) provides entries into the literature on equal-area grid systems. As well as resulting from data limitations, the problem of differing sized areas may arise simply because it is useful to explore patterns across areas of differing size (e.g. islands, biogeographic regions, national parks).

It may be desirable to remove the effects of area from analyses of spatial covariance in species numbers. Information is seldom available as to how species numbers and area scale within each of the areas considered (species–area relationships determined across all of the areas, rather than separately for each one, may be potentially misleading; see Holt 1993). It therefore becomes necessary to make assumptions about these relationships. Dixon *et al.* (1987) took such an approach in exploring how the richness of aphids scaled with that of plants across different countries and larger regions. They assumed that both aphid and plant richness in the various areas scaled as $s = x/a^{0.25}$, where s is the number of species per unit area, x the number of species in an area, and a the size of the area.

Species richness distributions

The frequency distribution of species richness of a higher taxon in different areas has been little explored (but see Brown and Kurzius 1987, Ouborg 1993, Väisänen and Heliövaara 1994). Perusal of available data for areas of only very approximately similar size reveals, however, that this distribution can take a wide range of forms. For example, analyses of a number of groups across Europe reveal distributions which are approximately uniform, approximately normal, and strongly skewed (Fig. 12.4). It seems likely that within any one region the basic shape of the distribution observed may be a function of the size (and hence on a grid system the number) of each area.

Such variation in the shape of species richness distributions suggests that consideration of the appropriate transformations may need to be made on a

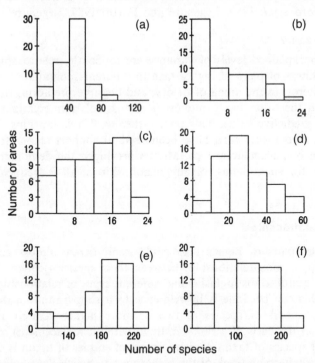

Fig. 12.4 Frequency distributions of the numbers of areas (152,000 km^2 squares across Europe) containing different numbers of species, for (a) freshwater fish, (b) gymnosperms, (c) amphibians, (d) noctuine moths, (e) birds, and (f) butterflies. See Gaston and David (in press) for data sources.

case by case basis, and that leaving the data untransformed or log-transformed may be inappropriate.

Spatial autocorrelation

Interpretation of patterns of spatial covariation in the richness of different taxa can be complicated seriously by spatial autocorrelation in the data, although studies have almost universally ignored this problem. Similarity in the species richness of a higher taxon in two areas is likely to be a negative function of the distance which separates them. The further they are apart, the more dissimilar the environmental conditions they experience are likely to be (the 'reddened spectrum'; Williamson 1987), and the more different the levels of species richness. The existence of spatial autocorrelation in species numbers means that in analyses of spatial covariance in the richness of two higher taxa, individual data points will not be strictly independent.

The literature associated with the importance of spatial dependence, particularly in ecological studies, is growing rapidly (e.g. Palmer 1988, Legendre and Fortin 1989, Borcard et al. 1992, Rossi et al. 1992, Legendre 1993). Techniques for analysing bivariate relationships where both variables may be spatially autocorrelated are becoming available (e.g. Clifford and Richardson 1985, Clifford et al. 1989, Legendre and Fortin 1989, Legendre 1993).

Predictive value

Even when explained levels of variance are moderate, relationships between species richness of pairs of higher taxa in different geographic areas may be of some interest in the arena of theory, and beg interpretation. In an applied context, on the other hand, interest is almost solely in relationships which have high predictive value. This said, rather weak relationships between the richness of two higher taxa have been used to support claims for the indicator value of particular groups. More worryingly, the demonstration, as a foundation for such claims, of any relationship at all is frequently entirely lacking.

Types of relationships

In principle, pairs of higher taxa could demonstrate almost any form of relationship, or non-relationship, between their species richness in different areas. It is useful to distinguish some different types of relationship. One way in which this can be done is by reference to a simple random draw model. Expected numbers of species of two or more higher taxa in an area are generated on the basis of a random draw (without replacement) of their sum (number of species of taxon A plus number of species of taxon B etc.) from a regional pool, keeping track of the higher taxa to which the randomly chosen species belong. The regional pool comprises the sum of the total number of species of the higher taxa considered in the region within which the area is embedded. Although not generally applied explicitly in the present context, such a model has been used in addressing related issues. The primary complication to its use in exploring patterns of spatial covariance in richness lies in the delineation of the regional species pool. In island situations this may be comparatively straightforward, but is almost invariably more difficult in continental situations (Cornell and Lawton 1992, Warren and Gaston 1992, Eriksson 1993), although some attempt at establishing approximate estimates of the size and structure of the pool can usually be made.

Examples of relationships between the numbers of species in two higher taxa, compared with predictions from the random draw model, are given in Fig. 12.5. For some data sets the model provides a remarkably good overall fit. For others, whilst it fails to predict the correct values, it predicts a relationship of approximately the form observed. In yet other cases the model is in a direct sense rather less informative, but remains a reference point against which relationships can be compared. The richness of higher taxon A may, for example, be seen to increase at a faster or a slower rate with respect to the numbers of higher taxon B than predicted.

The random draw model predicts positive patterns of spatial covariance in the species numbers of two higher taxa. This limitation is not as severe as it might seem, both because the majority of interest seems to lie in positive relationships, and because strong inverse correlations are documented rather seldom (for examples see Kiester 1971, Schall and Pianka 1977, 1978, Pianka and Schall 1981, Dixon et al. 1987, Kouki et al. 1994). Negative relationships have been termed 'complementary', or as 'exhibiting complementarity', in that when the richness of one of the groups is high that of the other is low, and vice-versa (Darlington 1957, Schall and Pianka 1977, Pianka and Schall 1981). This use of the term 'complementarity' should be distinguished from applications in the context of conservation prioritisation (Vane-Wright et al. 1991, Pressey et al. 1993), although the two are not unrelated.

The random draw model can be used not only in some sense as a null model for the shape of the relationship between the richness of higher taxa, but also for the level of correlation. A distribution of predicted correlation coefficients can be generated by drawing separately from the regional pool the sum of the number of species of the two higher taxa recorded for each area, calculating the correlation coefficient between the numbers of species drawn for each of the higher taxa, and repeating the exercise n times. Expected correlation coefficients may be high when the spread of the sums is also high and a moderate to large proportion of the regional pool occurs in one or more areas.

Mechanisms

Species richness may be influenced by a number of factors, such as available energy or production (Brown and Davidson 1977, Schall and Pianka 1978, Abramsky and Rosenzweig 1983, Wright 1983, Turner et al. 1987, 1988, Currie and Paquin 1987, Owen 1988, 1989, Adams and Woodward 1989,

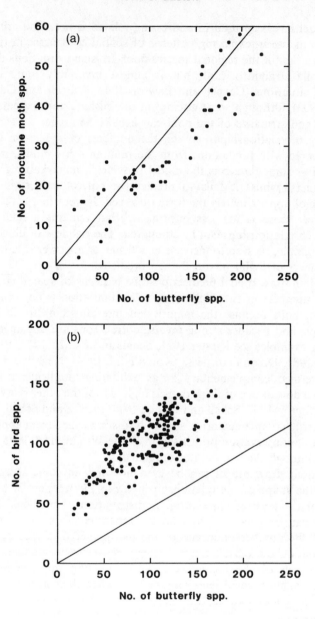

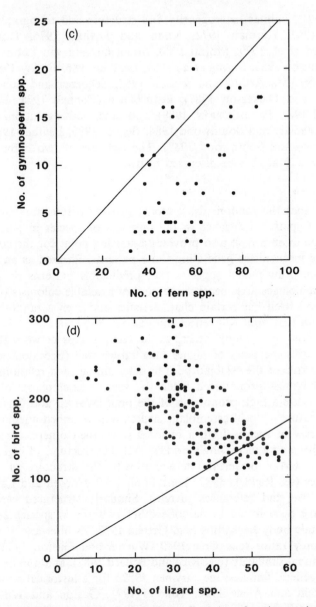

Fig. 12.5 Relationships between the numbers of species of various pairs of higher taxa, compared with the predictions of a random draw model based on the numbers of species of the two higher taxa in the regional species pool: (a) noctuine moth and butterfly species in 152,000 km^2 squares across Europe, (b) bird and butterfly species in 275 × 275 km squares across North America (from data in Pearson and Cassola 1992, and D.L. Pearson personal communication), (c) gymnosperm and fern species in 152,000 km^2 squares across Europe, (d) bird and lizard (exclusive of non-*Ctenotus* skinks) species in 150 × 150 mile (approx. 240 km) squares across Australia (from data in Pianka and Schall 1981).

Currie 1991), habitat heterogeneity and structure and topographic diversity (Pianka 1967, Harman 1972, Kohn and Levitan 1976, Holland 1978, Southwood *et al.* 1979, Moran 1980, Boomsma and van Loon 1982, Tonn and Magnuson 1982, Strong *et al.* 1984, Leather 1986, Friday 1987), rates of immigration (MacArthur and Wilson 1967, Schoener and Schoener 1983, Robinson and Dickerson 1987), disturbance (Connell 1978, Hildrew and Townsend 1987, Petraitis *et al.* 1989), and time, both evolutionary and ecological (Kennedy and Southwood 1984, Beaver 1985, Leather 1986, Ricklefs 1987, Brown and Southwood 1987). The richness of two higher taxa may covary for several reasons discussed below.

Random draw

Not only does the random draw model provide a reference against which patterns of spatial covariance in the numbers of species in pairs of higher taxa can be examined, it also provides a starting point for the consideration of possible mechanisms generating those patterns. Phrased as an hypothesis, the random draw model suggests that predictable patterns of structure in species assemblages arise because the pool of available colonists (the regional species pool) itself has certain characteristics and even a random sample of species from that pool will reflect this.

The random draw model describes the relationships between the numbers of species of some pairs of higher taxa rather well (Beccaloni and Gaston 1995). The role of the regional pool in determining such relationships seems likely to be most pronounced when the species assemblages of individual areas constitute a high proportion of the pool (Warren and Gaston 1992).

If one feature of local assemblage structure is described adequately by the random draw of species from a regional pool, then others are likely to be also. It is becoming increasingly evident that the structures of regional species pools can exert a powerful interaction with the structure of more local assemblages (see Ricklefs 1987, Tonn *et al.* 1990, Cornell and Lawton 1992, Eriksson 1993 and references therein). Similarly structured random draw models have been found to describe other patterns in species assemblages. These include body size ratios (e.g. Greene 1987, Zwölfer and Brandl 1989), predator–prey ratios (e.g. Cole 1980, Warren and Gaston 1992), levels of floral or faunal similarity (e.g. Rice and Bellard 1982), and numbers of higher taxa (e.g. genera, families; see Järvinen 1982*b* for a historical review; Rogers 1976, Gotelli and Abele 1982, Tokeshi 1991, Gaston and Williams 1993, Williams and Gaston 1994). In this sense, these patterns are not independent.

With reference to spatial covariance in species richness, the connection is readily seen with patterns in numbers of higher taxa. If the numbers of higher taxa in different areas are similar to those predicted by a random draw of species, then the proportions of species in different higher taxa in those areas will tend to remain approximately constant. If those proportions are approximately constant, then there will be strong relationships between the species richness of pairs of higher taxa in different areas.

The random draw model in some sense simply removes the level of explanation away from the assemblages in the different areas to the regional pool, raising the question of why the pool has the structure that it does (van Valen 1982). However, as well as providing a starting point for exploring spatial covariance in richness, the random draw model may be modified to explore more complicated scenarios. For example, the probability of choosing each species from the regional pool could be altered to account for the relative size of its geographic distribution (Harvey *et al.* 1983, and Crowley 1992 provide entries to the literature on this topic).

Interactions

The most frequently proffered explanations for why the species richness of two higher taxa covary concern some form of direct interaction. This interaction may be trophic, competitive, or mutualistic. It may also take other forms. For example, documented relationships between the numbers of species of higher taxa of animals and of plants have often been interpreted as resulting in part from the greater architectural complexity and the greater habitat diversity associated with high plant species richness, as well as from the greater diversity of food resources provided by more speciose plant assemblages (e.g. Murdoch *et al.* 1972, Power 1972). Determining the respective contributions of these processes presents a formidable challenge.

Relationships resulting from interactions between the species of two higher taxa may be linear, but probably more often they will be curvilinear, particularly when documented over a wide range of values of species richness. In the context of trophic interactions, curvilinear relationships are of particular interest as possible expressions of resource fragmentation, especially when species of one taxon specialise in the consumption of one or a few species of the other taxon. The expected negative relationship between species richness and population densities means that as resource species richness increases, each resource species should be harder for consumers to find; this, in turn, may result in a reduction of the ratio of consumer to resource species. Such a mechanism has been postulated to account for geographical patterns of covariance in the species numbers of some groups of parasitoids (Janzen 1981), insect herbivores (Dixon *et al.* 1987), and their respective hosts (i.e. insect herbivores and plants). The pressure of resource fragmentation may be or have been a major force on the ecologies of organisms in the tropics, where such fragmentation appears often to be greatest.

Common determinants of diversity

Pairs of higher taxa may covary in their species richness, not because of any interdependence, but because both richnesses depend on the same third set of factors (see Warren and Gaston 1992). This dependence need not be of similar form, and might thus lead to negative as well as to positive patterns of covariance.

Unravelling the roles of common determinants of diversity may be difficult. Broad generalisations as to which factors favour richness need not translate into spatial covariance in their richness. It is axiomatic that most higher taxa are species-poor in arctic regions, at high altitudes, and in arid environments, and are species-rich at low latitudes, at low to middle elevations, and in forest environments. There is, nonetheless, substantial variation in the richness of any one group at any point on its environmental spectrum, and there are major differences between groups in the particulars of how richness changes with environmental conditions.

Spatial covariance in different factors

Covariance in species richness may occur through no shared factors, but because the factors which determine the species richness of two higher taxa themselves show spatial covariance. This seems particularly likely, because many environmental variables are known to covary in space. This is an acknowledged difficulty with attempting to impute some causal basis to correspondence between levels of species richness and patterns of environmental variables such as temperature and rainfall; environmental variables show high levels of correlation.

Discriminating between these various mechanisms is not straightforward for two particular reasons. First, the numbers of species of two higher taxa may frequently be correlated in a similar way with a number of variables. Second, many studies concern moderate to large spatial scales, at which it is either difficult or impossible to use an experimental approach, with the result that discussions of causation centre on patterns of correlation.

In sum, different mechanisms are likely to be most relevant in different situations (depending on such factors as spatial scale, geographic region, season etc.), but in many instances it may be wrong to try and impute causation solely to one mechanism.

Conclusion

Spatial covariance in the species richness of different higher taxa has received remarkably little detailed consideration. Whilst numbers of analyses have been performed, these have been largely superficial. Nonetheless, this issue seems central to an understanding of the structure of species assemblages, and is of importance to the practice of conservation.

The way ahead seems clear. Future studies need to consider the variety of factors which could be contributing to observed patterns of covariance, and to attempt to dissect their effects. Once artefactual problems have, where possible, been addressed, a broad view needs to be taken of potential underlying mechanisms.

The imperatives of conservation provide little time for the collection and analysis of new and detailed data sets with which to examine the issue of spatial covariance in species richness. What lessons can be drawn from our present understanding? First, there seem to be few grounds for simply assuming that conservation strategies centred on particular higher taxa will be equally effective in conserving the species of other higher taxa. Much will depend on the particulars of the groups of organisms, region, and spatial scales. Second, even strong positive relationships between the richness of several pairs of higher taxa need not mean that the richness of one or more of those higher taxa is strongly correlated with the overall species richness (summed across all taxonomic groups) of the areas concerned. Third, even where strong relationships between the species richness of pairs of higher taxa apparently exist, they may be complex, making assumptions of linearity hazardous. Regrettably, despite such warnings, there is for most regions of the world little alternative to basing the prioritisation of sites on one or at most a few higher taxa. The extent to which this strategy serves to conserve species of other higher taxa may have a profound effect on how future generations view the success of present-day conservation efforts.

One reason why development of the concept of biodiversity has served to highlight a number of fundamental issues, either for the first time or afresh, is that they often lie at the interface of two or more disciplines. As such, only some of their aspects tend to receive attention. The multi-disciplinary approach to the study of biodiversity yields an opportunity to explore these issues more thoroughly. In the case of spatial covariance in the species richness of different higher taxa this could perhaps most usefully result in a closer synthesis of theoretical and applied treatments.

Acknowledgements

This work benefited greatly from discussions with Phil Warren and Paul Williams. Rhian David, Phil Warren and Paul Williams kindly commented on the manuscript. Rhian David assisted with the collation of information. Financial support for the work was provided by the Natural Environment Research Council.

References

Abbott, I. (1974). Numbers of plant, insect and landbird species on nineteen remote islands in the southern hemisphere. *Biological Journal of the Linnean Society*, **6**, 143–152.

Abramsky, Z. and Rosenzweig, M. L. (1983). Tilman's predicted productivity–diversity relationship shown by desert rodents. *Nature*, **309**, 150–151.

Adams, J. M. and Woodward, F. I. (1989). Patterns in tree species richness as a test of the glacial extinction hypothesis. *Nature*, **339**, 699–701.

Anderson, S. and Marcus, L. F. (1993). Effect of quadrat size on measurements of species density. *Journal of Biogeography*, **20**, 421–428.

Beaver, R. A. (1985). Geographical variation in food web structure in *Nepenthes* pitcher plants. *Ecological Entomology*, **10**, 241–248.

Beccaloni, G. W. and Gaston, K. J. (1995). The species richness of Neotropical forest butterflies: indicators and Ithomiines. *Biological Conservation*, **71**, 77–86.

Becker, P. (1992). Colonization of islands by carnivorous and herbivorous Heteroptera and Coleoptera: effects of island area, plant species richness, and 'extinction' rates. *Journal of Biogeography*, **19**, 163–171.

Boomsma, J. J. and van Loon, A. J. (1982). Structure and diversity of ant communities in successive coastal dune valleys. *Journal of Animal Ecology*, **51**, 957–974.

Borcard, D., Legendre, P. and Drapeau, P. (1992). Partialling out the spatial component of ecological variation. *Ecology*, **73**, 1045–1055.

Brown, J. H. and Davidson, D. W. (1977). Competition between seed-eating rodents and ants in desert ecosystems. *Science*, **196**, 880–882.

Brown, J. H. and Kurzius, M. A. (1987). Composition of desert rodent faunas: combinations of coexisting species. *Annales Zoologici Fennici*, **24**, 227–237.

Brown, J. W. and Opler, P. A. (1990). Patterns of butterfly species density in peninsula Florida. *Journal of Biogeography*, **17**, 615–622.

Brown, V. K. and Gange, A. C. (1989). Herbivory by soil-dwelling insects depresses plant species richness. *Functional Ecology*, **3**, 667–671.

Brown, V. K. and Hyman, P. S. (1986). Successional communities of plants and phytophagous Coleoptera. *Journal of Ecology*, **74**, 963–975.

Brown, V. K. and Southwood, T. R. E. (1987). Secondary succession: patterns and strategies. In *Colonization, succession and stability* (eds. A. J. Gray, M. J. Crawley and P. J. Edwards), pp. 315–337. Blackwell Scientific, Oxford.

Burbidge, A. H., Leichester, K., McDavitt, S. and Majer, J. D. (1992). Ants as indicators of disturbance at Yanchep National Park, Western Australia. *Journal of the Royal Society of Western Australia*, **75**, 89–95.

Clifford, P. and Richardson, S. (1985). Testing the association between two spatial processes. *Statistics and Decisions*, Suppl Issue **2**, 155–160.

Clifford, P., Richardson, S. and Hémon, D. (1989). Assessing the significance of the correlation between two spatial processes. *Biometrics*, **45**, 123–134.

Cole, B. J. (1980). Trophic structure of a grassland insect community. *Nature*, **288**, 76–77.

Compton, S. G. and Hawkins, B. A. (1992). Determinants of species richness in southern African fig wasp assemblages. *Oecologia*, **91**, 68–74.

Connell, J. H. (1978). Diversity in tropical rain forests and coral reefs. *Science*, **199**, 1302–1310.

Cornell, H. V. and Lawton, J. H. (1992). Species interactions, local and regional processes, and limits to the richness of ecological communities: a theoretical perspective. *Journal of Animal Ecology*, **61**, 1–12.

Crowley, P. H. (1992). Resampling methods for computation-intensive data analysis in ecology and evolution. *Annual Review of Ecology and Systematics*, **23**, 405–447.

Currie, D. J. (1991). Energy and large-scale patterns of animal and plant species richness. *American Naturalist*, **137**, 27–49

Currie, D. J. and Paquin, V. (1987). Large-scale biogeographical patterns of species richness of trees. *Nature*, **329**, 326–327.

Danks, H. V. (1986). Insect plant interactions in arctic regions. *Revue d'Entomologie du Quebec*, **31**, 52–75.

Danks, H. V. (1993). Patterns of diversity in the Canadian insect fauna. *Memoirs of the Entomological Society of Canada*, **165**, 51–74.

Darlington, P. J. (1957). *Zoogeography*. Wiley, New York.

Dayton, P. K. and Tegner, M. J. (1984). The importance of scale in community ecology: A kelp forest example with terrestrial analogs. In *A new ecology: novel approaches to interactive systems* (eds. P. W. Price, C. N. Slobodchikoff and W. S. Gaud), pp. 457–481. Wiley, New York.

Dinerstein, E. and Wikramanayake, E. D. (1993). Beyond 'hotspots': how to prioritize investments to conserve biodiversity in the Indo-Pacific region. *Conservation Biology*, **7**, 53–65.

Dixon, A. F. G., Kindlmann, P., Leps, J. and Holman, J. (1987). Why there are so few species of aphids, especially in the tropics. *American Naturalist*, **129**, 580–592.

Dzwonko, Z. and Loster, S. (1989). Distribution of vascular plant species in small woodlands on the Western Carpathian foothills. *Oikos*, **56**, 77–86.

Eastop, V. F. (1978). Diversity of the Sternorrhyncha within major climatic zones. In *Diversity of insect faunas* (eds. L. A. Mound and N. Waloff), pp. 71–88. Blackwell Scientific, Oxford.

Emberson, R. M. (1985). Comparisons of site conservation value using plant and soil arthropod species. *Bulletin of the British Ecological Society*, **16**, 16–17.

Eriksson, O. (1993). The species-pool hypothesis and plant community diversity. *Oikos*, **68**, 371–374.

Fernandes, G. W. and Price, P. W. (1988). Biogeographical gradients in galling species richness: tests of hypotheses. *Oecologia*, **76**, 161–167.

Fernandes, G. W. and Price, P. W. (1991). Comparison of tropical and temperate galling species richness: the roles of environmental harshness and plant nutrient status. In *Plant–animal interactions: evolutionary ecology in tropical and temperate regions* (eds. P. W. Price, T. M. Lewinsohn, G. W. Fernandes and W. W. Benson), pp. 91–115. John Wiley, New York.

Friday, L. E. (1987). The diversity of macroinvertebrate and macrophyte communities in ponds. *Freshwater Biology*, **12**, 87–104.

Gaston, K. J. (1991). The global magnitude of insect species richness. *Conservation Biology*, **5**, 283–296.

Gaston, K. J. (1992). Regional numbers of insect and plant species. *Functional Ecology*, **6**, 243–247.

Gaston, K. J. (1994). Estimating extinction rates: Joseph Banks' legacy. *Trends in Ecology and Evolution*, **9**, 1–3.

Gaston, K. J. and David, R. Hotspots across Europe. *Biodiversity Letters*, in press.

Gaston, K. J. and Hudson, E. (1994). Regional patterns of diversity and estimates of global insect species richness *Biodiversity and Conservation*, **3**, 493–500.

Gaston, K. J. and Williams, P. H. (1993). Mapping the world's species—the higher taxon approach. *Biodiversity Letters*, **1**, 2–8.

Gauld, I. D. (1986). Latitudinal gradients in ichneumonid species-richness in Australia. *Ecological Entomology*, **11**, 155–161.

Gauld, I. D. (1991). The Ichneumonidae of Costa Rica, 1: Introduction, keys to subfamilies, and keys to the species of the lower pimpliform subfamilies Rhyssinae, Pimplinae, Poemeniinae, Acaenitinae and Cyllocerinae. *Memoirs of the American Entomological Institute* No. 47.

Gauld, I. D. and Gaston, K. J. (1994). The taste of enemy-free space: parasitoids and nasty hosts. In *Parasitoid community ecology* (eds. B. A. Hawkins and W. Sheehan), pp. 279–299. Oxford University Press.

Gauld, I. D., Gaston, K. J. and Janzen, D. H. (1992). Plant allelochemicals, tritrophic interactions and the anomalous diversity of tropical parasitoids: the 'nasty' host hypothesis. *Oikos*, **65**, 353–357.

Gilbert, L. E. and Smiley, J. T. (1978). Determinants of local diversity and phytophagous insects: host specialists in tropical environments. In *Diversity of insect faunas* (eds. L. A. Mound and N. Waloff), pp. 89–104. Blackwell Scientific, Oxford.

Giller, P. S. and Gee, J. H. R. (1987). The analysis of community organisation: the influence of equilibrium, scale and terminology. In *Organisation of communities: past and present* (eds. J. H. R. Gee and P. S. Giller), pp. 519–542. Blackwell Scientific, Oxford.

Gotelli, N. J. and Abele, L. G. (1982). Statistical distributions of West Indian land bird families. *Journal of Biogeography*, **9**, 421–435.

Grassle, J. F. and Maciolek, N. J. (1992). Deep-sea species richness: regional and local diversity estimates from quantitative bottom samples. *American Naturalist*, **139**, 313–341.

Greene, E. (1987). Sizing up size ratios. *Trends in Ecology and Evolution*, **2**, 79–81.

Hammond, P. M. (1992). Species inventory. In *Global biodiversity: status of the earth's living resources* (ed. B. Groombridge), pp. 17–39. Chapman and Hall, London.

Harman, W. (1972). Benthic substrates: their effect on freshwater molluscs. *Ecology*, 53, 271–272.

Harvey, P. H., Colwell, R. K., Silvertown, J. W. and May, R. M. (1983). Null models in ecology. *Annual Reviews of Ecology and Systematics*, **14**, 189–211.

Hawksworth, D. L. (1991). The fungal dimension of biodiversity: magnitude, significance, and conservation. *Mycological Research*, **95**, 641–655.

Hildrew, A. G. and Townsend, C. R. (1987). Organisation in freshwater benthic communities. In *Organisation of communities: past and present* (eds. J. H. R. Gee and P. S. Giller), pp. 347–371. Blackwell Scientific, Oxford.

Hockin. D. C. (1981). The environmental determinants of the insular butterfly faunas of the British Isles. *Biological Journal of the Linnean Society*, **16**, 63–70

Holland, P. G. (1978). An evolutionary biogeography of the genus *Aloe*. *Journal of Biogeography*, **5**, 213–226.

Holt, R. D. (1993) Ecology at the mesoscale: the influence of regional processes on local communities. In *Species diversity in ecological communities: historical and geographical perspectives* (eds. R. E. Ricklefs and D. Schluter), pp. 77–88. Chicago University Press.

Itamies. J. (1983). Factors contributing to the succession of plants and Lepidoptera on the islands off Rauma, SW Finland. *Acta Universitatis Oulensis*, A 142.1983. Biol 18.

Janzen, D. H. (1981). The peak in North American ichneumonid species richness lies between 38° and 42° N. *Ecology*, **62**, 532–537.

Järvinen, O. (1982a). Conservation of endangered plant populations: single large or several small reserves. *Oikos*, **38**, 301–307.

Järvinen, O. (1982b). Species-to-genus ratios in biogeography: a historical note. *Journal of Biogeography*, **9**, 363–370.

Kennedy, C. E. J. and Southwood, T. R. E. (1984). The number of species of insects associated with British trees: a re-analysis. *Journal of Animal Ecology*, **53**, 455–478.

Kiester, A. R. (1971). Species density of North American amphibians and reptiles. *Systematic Zoology*, **20**, 127–137.

Kitching, R. L. (1993). Towards rapid biodiversity assessment—lessons following studies of arthropods of rainforest canopies. In *Rapid biodiversity assessment* (ed. A. J. Beattie), pp. 26–30. Research Unit for Biodiversity and Bioresources, Macquarie University.

Kohn, A. J. and Levitan, P. J. (1976). Effect of habitat complexity on population density and species richness in tropical intertidal predatory gastropod assemblages. *Oecologia*, **25**, 199–210.

Kostrowicki, A. S. (1969). Geography of the palaearctic Papilionoidea (Lepidoptera). *Zaklad Zoologiii Systematycznej, Polskiej Akademii Nauk. Paustwowe Wydawnictwo Naukowe*, 1–314.

Kouki, J., Niemelä, P. and Viitasaari, M. (1994). Reversed latitudinal gradient in species richness of sawflies (Hymenoptera, Symphyta). *Annales Zoologici Fennici*, **31**, 83–88.

Kremen, C. (1992). Assessing the indicator properties of species assemblages for natural areas monitoring. *Ecological Applications*, **2**, 203–217.

Leather, S. R. (1986). Insect species richness of the British Rosaceae: the importance of host range, plant architecture, age of establishment, taxonomic isolation and species area relationships. *Journal of Animal Ecology*, **55**, 841–860.

Legendre. P. (1993). Spatial autocorrelation: trouble or new paradigm? *Ecology*, **74**, 1659–1673.

Legendre, P. and Fortin, M-J. (1989). Spatial pattern and ecological analysis. *Vegetatio*, **80**, 107–138.

Levin, S. A. (1992). The problem of pattern and scale in ecology. *Ecology*, **73**, 1943–1967.

Lindroth, C. H., Andersson, H., Bodvarsson, H. and Richter, S. H. (1973). Surtsey, Iceland: The development of a new fauna, 1963–1970: Terrestrial invertebrates. *Entomologica Scandinavica Supplementum*, 5.

MacArthur, R. H. and Wilson, E. O. (1967). *The theory of island biogeography*. Princeton University Press, Princeton, New Jersey.

May, R. M. (1988). How many species are there on earth? *Science*, **241**, 1441–1449.

May, R. M. (1990). How many species? *Philosophical Transactions of the Royal Society, London*, **B330**, 293–304.

Mittermeier, R. A. (1988). Primate diversity and the tropical forest: case studies from Brazil and Madagascar and the importance of the megadiversity countries. In *Biodiversity* (ed. E. O. Wilson), pp. 145–154. National Academy Press, Washington.

Mittermeier, R. A. and Werner, T. B. (1990). Wealth of plants and animals unites 'megadiversity' countries. *Tropicus*, **4**, 1, 4–5.

Moran, V. C. (1980). Interactions between phytophagous insects and their *Opuntia* hosts. *Ecological Entomology*, **5**, 153–164.

Murdoch, W. W., Evans, F. C. and Peterson, C. H. (1972). Diversity and pattern in plants and insects. *Ecology*, **53**, 819–828

Myers, N. (1979). *The sinking ark, a new look at the problem of disappearing species*. Pergamon, Oxford.

Nilsson, C., Grelsson, G., Johansson, M. and Sperens, U. (1988). Can rarity and diversity be predicted in vegetation along river banks? *Biological Conservation*, **44**, 201–212.

Nilsson, S. G., Bengtsson, J. and As, S. (1988). Habitat diversity or area *per se?* Species richness of woody plants, carabid beetles and land snails on islands. *Journal of Animal Ecology*, **57**, 685–704.

Oliver, I. (1993). Rapid biodiversity assessment and its application to fauna conservation in production forests. In *Rapid biodiversity assessment* (ed. A. J. Beattie), pp. 31–34. Research Unit for Biodiversity and Bioresources, Macquarie University.

Otte, D. (1976). Species richness patterns of New World desert grasshoppers in relation to plant diversity. *Journal of Biogeography*, **3**, 197–209.

Ouborg, N. J. (1993). Isolation, population size and extinction: the classical and metapopulation approaches applied to vascular plants along the Dutch Rhine-system. *Oikos*, **66**, 298–308.

Owen, J. G. (1988). On productivity as a predictor of rodent and carnivore diversity. *Ecology*, **69**, 1161–1165.

Owen, J. G. (1989). Patterns on herpetofaunal species richness: relation to temperature, precipitation, and variance in elevation. *Journal of Biogeography*, **16**, 141–150.

Palmer, M. W. (1988). Fractal geometry: a tool for describing patterns of plant communities. *Vegetatio*, **75**, 91–102.

Pearson, D. L. (1977). A pantropical comparison of bird community structure on six lowland forest sites. *Condor*, **79**, 232–244.

Pearson, D. L. and Cassola, F. (1992). World-wide species richness patterns of tiger beetles (Coleoptera: Cicindelidae): indicator taxon for biodiversity and conservation studies. *Conservation Biology*, **6**, 376–391.

Peterson, A. T., Flores-Villela, O. A., León-Paniagua, L. S., Llorente-Bousquets, J. E., Luis-Martinez, M. A., Navarro-Sigüenza, A. G., Torres-Chávez, M. G. and Vargas-Fernández, I. (1993). Conservation priorities in Mexico: moving up in the world. *Biodiversity Letters*, **1**, 33–38.

Petraitis, P. S., Latham, R. E. and Niesenbaum, R. A. (1989). The maintenance of species diversity by disturbance. *Quarterly Review of Biology*, **64**, 393–418.

Pianka, E. R. (1967). On lizard species diversity: North American flatland deserts. *Ecology*, **48**, 333–351.

Pianka, E. R. and Schall, J. J. (1981). Species densities of Australian vertebrates. In *Ecological biogeography of Australia* (ed. A. Keast), pp. 1675–1694. Dr W. Junk, The Hague.

Power, D. M. (1972). Numbers of bird species on the California islands. *Evolution*, **26**, 451–463.

Prendergast, J. R., Quinn, R. M., Lawton, J. H., Eversham, B. C. and Gibbons, D. W. (1993a). Rare species, the coincidence of diversity hotspots and conservation strategies. *Nature*, **365**, 335–337.

Prendergast, J. R., Wood, S. N., Lawton, J. H. and Eversham, B. C. (1993b). Correcting for variation in recording effort in analyses of diversity hotspots. *Biodiversity Letters*, **1**, 39-53.

Pressey, R. L., Humphries, C. J., Margules, C. R., Vane-Wright, R. I. and Williams, P. H. (1993). Beyond opportunism: key principles for systematic reserve selection. *Trends in Ecology and Evolution*, **8**, 124–128.

Reed, T. M. (1982). The number of butterfly species on British islands. *Proceedings of the 3rd Congress of European Lepidopterists*, Cambridge 146–152.

Rice, J. and Bellard, R. J. (1982). A simulation study of moss floras using Jaccard's coefficient of similarity. *Journal of Biogeography*, **9**, 411–419.

Ricklefs, R. E. (1987). Community diversity: relative roles of local and regional processes. *Science*, **235**, 167–171.

Robinson, J. V. and Dickerson, J. E. (1987). Does invasion sequence affect community structure? *Ecology*, **68**, 587–595.

Rogers, J. S. (1976). Species density and taxonomic diversity of Texas amphibians and reptiles. *Systematic Zoology*, **25**, 26-40.

Rossi, R. E., Mulla, D. J., Journel, A. G. and Franz, E. H. (1992). Geostatistical tools for modelling and interpreting ecological spatial dependence. *Ecological Monographs*, **62**, 277–314.

Rust, R, Menke, A. and Miller, D. (1985). A biogeographic comparison of the bees, sphecid wasps, and mealybugs of the California channel islands (Hymenoptera,

Homoptera). In *Entomology of the California channel islands: proceedings of the first symposium* (eds. A.S Menke and D. R. Miller), pp. 29–59. Santa Barbara Museum of Natural History, Santa Barbara, California.

Ryti, R. T. (1992). Effect of the focal taxon on the selection of nature reserves. *Ecological Applications*, **2**, 404–410.

Saetersdal, M., Line, J. M. and Birks, H. J. B. (1993). How to maximize biological diversity in nature reserve selection: vascular plants and breeding birds in deciduous woodlands, western Norway. *Biological Conservation*, **66**, 131–138.

Samways, M. J. (1994). *Insect conservation biology*. Chapman and Hall, London.

Schall, J. J. and Pianka, E. R. (1977). Species densities of reptiles and amphibians on the Iberian Peninsula. *Acta Vertebrata Doñana*, **4**, 27–34.

Schall, J. J. and Pianka, E. R. (1978). Geographical trends in numbers of species. *Science*, **201**, 679–686.

Schoener, T. W. and Schoener, A. (1983). Distribution of vertebrates on some very small islands. II. Patterns in species number. *Journal of Animal Ecology*, **52**, 237–262.

Schulze, E-D. and Mooney, H. A. (eds.) (1993). *Biodiversity and ecosystem function*. Springer-Verlag, Berlin.

Slansky, F. (1976). Phagism relationships amongst butterflies. *Journal of the New York Entomological Society*, **84**, 91–105.

Smith, F. D. M., May, R. M., Pellew, R. Johnson, T. H. and Walter, K. S. (1993*a*). Estimating extinction rates. *Nature*, **364**, 494–496.

Smith, F. D. M., May, R. M., Pellew, R., Johnson, T. H. and Walter, K. S. (1993*b*). How much do we know about the current extinction rate? *Trends in Ecology and Evolution*, **8**, 375–378.

Smith, P. G. R. and Theberge, J. B. (1986). Evaluating biotic diversity in environmentally significant areas in the northwest territories of Canada. *Biological Conservation*, **36**, 1–18.

Southwood, T. R. E., Brown, V. K. and Reader, P. M. (1979). The relationships of plant and insect diversities in succession. *Biological Journal of the Linnean Society*, **12**, 327–348.

Stork, N. E. (1993). How many species are there? *Biodiversity and Conservation*, **2**, 215–232.

Strong, D. R., Lawton, J. H. and Southwood, T. R. E. (1984). *Insects on plants: community patterns and mechanisms*. Blackwell Scientific, Oxford.

Thomas, C. D. and Mallorie, H. C. (1985). Rarity, species richness and conservation: butterflies of the Atlas mountains in Morocco. *Biological Conservation*, **33**, 95–117.

Tokeshi, M. (1991). Faunal assembly in chironomids (Diptera): generic association and spread. *Biological Journal of the Linnean Society*, **44**, 353–367.

Tonn, W. M. and Magnuson, J. J. (1982). Patterns in the species composition and richness of fish assemblages in N. Wisconsin lakes. *Ecology*, **63**, 1149–1166.

Tonn, W. M., Magnuson, J. J., Rask, M. and Toivonen, J. (1990). Intercontinental comparison of small-lake fish assemblages: the balance between local and regional processes. *American Naturalist*, **136**, 345–375.

Turner, J. R. G., Gatehouse, C. M. and Corey, C. A. (1987). Does solar energy control organic diversity? Butterflies, moths and the British climate. *Oikos*, **48**, 195–205.

Turner, J. R. G., Lennon, J. J. and Lawrenson, J. A. (1988). British bird species distributions and the energy theory. *Nature*, **335**, 539–541.

Udvardy, M. D. F. (1981). An overview of grid-based atlas works in ornithology. *Studies in Avian Biology*, **6**, 103–109.

Usher, M. B. (1986). Insect conservation: the relevance of population and community ecology and of biogeography. In *Proceedings of the 3rd European Congress of Entomology. Part 3* (ed. H. H. W. Velthuis), pp. 387–398. Nederlandse Entomologische Vereniging, Amsterdam.

Usher, M. B. (1992). Management and diversity of arthropoda in *Calluna* heathland. *Biodiversity and Conservation*, **1**, 63–79.

Usher, M. B. and Thompson, D. B. A. (1993). Variation in the upland heathlands of Great Britain: conservation importance. *Biological Conservation*, **66**, 69–81.

Väisänen, R. and Heliövaara, K. (1994). Hot-spots of insect diversity in northern Europe. *Annales Zoologici Fennici*, **31**, 71–81.

Vane-Wright, R. I., Humphries, C. J. and Williams, P. H. (1991). What to protect?— systematics and the agony of choice. *Biological Conservation*, **55**, 235–254.

van Valen, L. M. (1982). A pitfall in random sampling. Nature **295**, 171.

Walker, B. H. (1992). Biodiversity and ecological redundancy. *Conservation Biology*, **6**, 18–23.

Warren, P. H. and Gaston, K. J. (1992). Predator–prey ratios: a special case of a general pattern? *Philosophical Transactions of the Royal Society, London*, **B338**, 113–130.

Wheeler, B. D. (1988). Species richness, species rarity and conservation evaluation of rich-fen vegetation in lowland England and Wales. *Journal of Applied Ecology*, **25**, 331–352.

White, P. S., Miller, R. I. and Ramseur, G. S. (1984). The species-area relationship of the southern Appalachian high peaks: Vascular plant richness and rare plant distributions. *Castanea*, **49**, 47–61.

Whitmore, T.C, and Sayer, J. A. (eds.) (1992). *Tropical deforestation and species extinction*. Chapman and Hall, London.

Williams, G. R. (1982). Species-area and similar relationships of insects and vascular plants on the southern outlying islands of New Zealand. *New Zealand Journal of Ecology*, **5**, 86–96.

Williams, P. H. (1994). Choosing conservation areas: using taxonomy to measure more of biodiversity. In *Biodiversity and conservation* (ed. T-Y. Moon). Seoul. In press.

Williams, P. H. and Gaston, K. J. (1994). Measuring more of biodiversity: can higher taxon richness predict wholesale species richness? *Biological Conservation*, **67**, 211–217.

Williamson, M. H. (1987). Are communities ever stable? In *Colonisation, succession and stability* (eds A. J. Gray, M. J. Crawley and P. J. Edwards), pp. 352–371. Blackwell Scientific, Oxford.

Wilson, E. O. (1987). The little things that run the world (the importance and conservation of invertebrates). *Conservation Biology*, **1**, 344–346.

Wright, D. H. (1983). Species–energy theory: an extension of species–area theory. *Oikos*, **41**, 496–506.

Yen, A. L. (1987). A preliminary assessment of the correlation between plant, vertebrate and Coleoptera communities in the Victorian mallee. In *The role of invertebrates in conservation and biological survey* (ed. J. D. Majer), pp. 73–88. Dept. Conservation and Land Management, Western Australia.

Zwölfer, H. and Brandl, R. (1989). Niches and size relationships in Coleoptera associated with Cardueae host plants: adaptations to resource gradients. *Oecologia*, **78**, 60–68.

13

Biodiversity of parasites in relation to their life-cycles

Serge Morand

Introduction

Parasites with simple life-cycles (SLC) require a single host individual in their development, whereas complex life-cycle (CLC) parasites need several hosts. Several groups of parasites are composed entirely of complex life-cycle species (CLC) with at least one intermediate host and one definitive host, as in the case of digenous trematodes, cestodes and acanthocephalans. Although the intermediate host can serve as the site of parasite reproduction, it can also shelter infesting stages and be a prey for the definitive host, with the parasite then becoming an integral part of a predator–prey network.

Few studies have tried to understand the evolution and maintenance of complex life-cycles (Dobson 1989; Dobson and Merenlender 1991; Esch and Fernandez 1993), or the influence of CLC parasites on the dynamics of a multi-species system. Recently, Freeland and Boulton (1992) argued that CLC parasites may predominate parasitic associations and that they may stabilize multi-species systems.

The role of CLC parasites in host–parasite associations should be examined from both evolutionary and ecological perspectives. From the few studies available on the ecology of these species, I will compare cross-species patterns, attempting to answer the following questions: Do CLC parasites predominate in host–parasite associations? Are CLC parasites more taxonomically diverse than SLC species? What role can a CLC parasite play in host community ecology? Finally, using the answers to these questions, I will defend the hypothesis that the evolution of complex life-cycles has had a stabilizing effect on multi-species systems.

Complex life-cycles predominate in host–parasite associations

Freeland and Boulton (1992) classified the incidence of CLC parasites in domestic terrestrial vertebrates, finding that more than 60% of parasite species have CLCs (based on data of Soulsby 1982). However, whereas CLCs are rare in protozoans (33%), they are the rule in cestodes, trematodes and acanthocephalans (100%). In nematodes, they occur in 65% of species. Likewise, data collected for wild mammals (Table 13.1) show that CLC parasites predominate, constituting more than 70% of the helminth species.

Table 13.1   Life-cycles of helminths parasitizing some wildlife terrestrial mammals

Host (Number of helminth species)	SLC species (%)	CLC species (%)	References
Coyote (29)	21	69	Pence and Meinzer 1978
Black bear (13)	46	54	Pence 1990
Bobcat (22)	9	86	Stone and Pence 1978
Bobcat (14)	21	79	Tiekotter 1985
Racoon (15)	13	87	Snyder 1985
Ground squirrel (7)	29	71	Rosenberg and Pence 1978
Gray squirel (8)	50	50	Conti et al. 1984
Black rat (19)	32	68	Hasegawa et al. 1994
Red fox (13)	38	62	Borgsteedde 1984
Red fox (10)	40	60	Miguel 1993
Red fox (8)	13	83	Dible et al. 1983
Bank vole (9)	44	56	Kisielewska 1970
Bank vole (6)	50	50	Haukisalmi et al. 1988
Grasshopper rat (5)	0	100	Pfaffenberger et al. 1985
Round-tailed muskrat (6)	83	17	Forrester et al. 1987
Cotton Rat (8)	13	87	Rosenberg and Pence 1978
Wood mouse (9)	33	67	Montgomery and Montgomery 1988
Cottontail rabbit(9)	56	44	Lepitzki 1992
Total	28	72	

Differences between CLC and SLC parasites

Background

According to Price (1980), because of the complexity of their life-cycles, CLC parasites should be found in situations with high probabilities of extinction and low probabilities of colonization. At least two long-term studies show that CLC parasites persist in their host populations. The acanthocephalan *Pomphorhynchus laevis* is described by Kennedy and Rumpus (1977) as stable over a nine-year period in the populations of its intermediate and definitive hosts. A long term study carried out by Théron et al. (1992) showed that the population dynamics of the trematode *Schistosoma mansoni*, parasitizing the black rat in a marshy forest of Guadeloupe, were stable over an eight-year period.

CLC and SLC parasites differ in their genetic variability, disease severity, and the ability to invade host communities. Genetic variability, expressed in terms of mean heterozygosity per locus, the proportion of polymorphic loci and the average number of alleles per locus, is higher in CLC than in SLC ascaridoid nematodes (Bullini et al., 1986). These authors suggest that the increase in polymorphism is linked to biochemical adaptations which are necessary for survival in poikilothermal and homeothermal hosts. Polymorphism evolves in CLC ascaridoids within poikilothermic hosts (e.g. *Anisakis* in fish) and monomorphism in SLC ascaridoids parasitizing homeothermic hosts (e.g. *Toxocara* in mammals).

With respect to disease severity, Ewald (1983) showed that CLC microparasites (viruses, bacteria and protozoans) are more pathogenic for man than are SLC species. Indeed, almost 70% of microparasites causing mortalities exceeding 1% in man have CLCs, and CLC species represent less than 20% of pathologies causing less than 1% mortality. To my knowledge, there are no analogous data concerning macroparasites.

Dobson and May (1986) analysed the patterns of invasion of pathogens and parasites into communities. They suggested that invasion is more difficult for CLC parasites than for SLC parasites. For example, the percentage of CLC parasite species accompanying transcontinental fish transfers is lower (19%) than the percentage of CLC species in British (64%) or Canadian faunas (69%).

Other differences between SCL and CLC parasites

CLC and host specificity   Host specificity (or host range) refers to the number of host species infected by a given parasite species. Some parasites are restricted to one species of host (high specificity) whereas others are found on a wide range of hosts (low specificity).

Poulin (1992) recently analysed the host range characteristics of freshwater fish parasites. Among the macroparasites, nematodes and anthocephalans are the least specific; the monogenes are the most specific (Fig. 13.1(a)). Although, on the whole, a wider host range is observed in CLC parasites, there are some exceptions. Copepods and monogenes have very similar cycles (both groups are ectoparasites with free-living larvae) and yet the copepods have wider host ranges than do CLC trematodes.

I have compiled data concerning the specificity of two groups of nematodes which parasitize vertebrates: the oxyuroids (SLC species) and the ascaridoids (SLC or CLC species). There are significant differences between the average numbers of known hosts between groups (Mann–Whitney test, $p < 0.01$) (Fig. 13.1(b)). In contrast, there is no significant difference between the average number of known hosts in SLC and CLC ascaridoids (Mann–Whitney test, $p = 0.916$). This latter result highlights how phylogenetic effects can predominate over life-cycle characteristics.

Life-history traits   As for other organisms, a parasite's main life-history traits are its lifespan, size, and reproduction. Certain life-history traits may be modified as the parasite's development becomes increasingly complex. This point has not yet been systematically explored either theoretically or empirically. Some data for parasitic nematodes are shown in Table 13.2. Since this group has both SLC and CLC representatives, we can test the

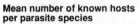

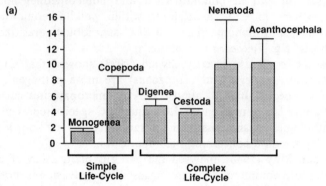

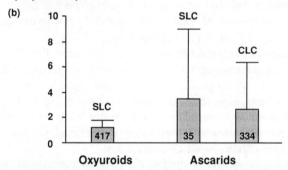

Fig. 13.1 (a) Host specificity (mean number of hosts ± s.d.) of 176 species of freshwater fish parasites (adapted from Poulin, 1992). (b) Host specificity (mean number of hosts ± s.d.) of oxyuroid (SLC) and ascaridoid (SLC or CLC) parasites of vertebrates (data from Skrjabin *et al.* 1960; Mosgovoy 1953; Sprent 1983) (the numbers in boxes refer to the numbers of species).

incidence of cycle complexity in life-history traits, and in particular the relationship between female fecundity and size. A linear relationship links fecundity (expressed as the logarithm of the total number of offspring produced per female) and female size (expressed as the log of body length) (Fig. 13.2). This relation holds for both CLC and SLC parasites (ANOVA, $p = 0.85$). There is no effect of type of life-cycle on maturation time or lifespan of either infesting stages or adult stages (Games Howell test, $p > 0.05$; Day and Quinn 1989).

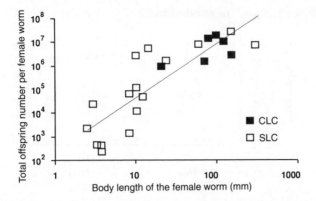

Fig. 13.2 Relationship between fecundity (in terms of the logarithm of the total number of offspring per female) and female length (in terms of the logarithm of female body length). The following relationship is found: ln(offspring) = 1.48 ln(female length in mm) + 8.2 (R^2 = 0.38, p = 0.026). There is no effect of complex life-cycles on this relationship (ANOVA, p = 0.85).

These results suggest that, whatever the type of life-cycle, parasites invest their energy in egg production in a similar manner, supporting the remark of Jennings and Calow (1975) that

entoparasitism provides a continuous, superabundant, and easily obtained food supply and a relatively predictable environment so that environment limitations on fecundity are released and the parent can produce and provision eggs without strain and without risk of over-expenditure.

Prevalence of infection   Based on data from Table 13.2, CLC parasites are less prevalent than SLC parasites (Fig. 13.3(a)). This trend disappears when examining one group of parasites, nematodes, in which both life-cycles occur (Fig. 13.3(b)).

Taxonomic diversification   Parasitism by nematodes accompanies a rise in both the average number of genera per family (Mann–Whitney, p = 0.01) and the average number of species per genus (Mann–Whitney, p = 0.01) (Fig. 13.4). Further, CLC nematode families outnumber SLCs by 15 to 10 (Anderson 1992). At a lower level of taxonomy, life-cycle complexity is not correlated with taxonomic diversification either at the level of the genus (case of vertebrate parasites, $p > 0.05$) or the species (case of oxyurids and ascarids, $p > 0.05$) (Fig. 13.4).

Advantages of being a CLC parasite

Increased reproductive success

One would intuitively expect that, somehow, the evolution of CLCs should be accompanied, or even motivated, by increases in reproductive success.

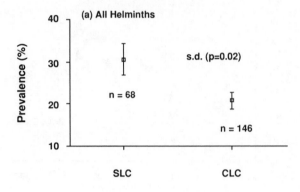

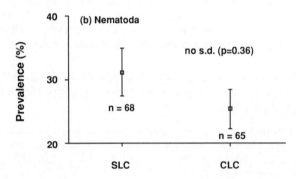

Fig. 13.3 (a) Mean prevalence (± s.e.) of CLC and SLC for helminth parasites of vertebrates (data from Table 13.2, arcsine transformation of data, in Zar (1984)). CLC parasites are less frequent in terms of prevalence than SLCs ($t = 2.34$, $p = 0.02$; n refers to number of cases). (b) Mean prevalence (± s.e.) of CLC and SLC nematode parasites of vertebrate hosts (data from Table 13.2, arcsine transformation of data, in Zar (1984)). No differences between CLC parasites and SLC parasites ($t = 0.91$, $p = 0.36$; n refers to number of cases).

What scant evidence exists is provided by the work of Robert and colleagues (Robert *et al.* 1988, 1990).

These authors studied the biology of two cestode species which are parasites of two sympatric flat fish species. One of the cestodes (*Bothriocephalus barbatus*) has a copepod as its intermediate host, and the brill (*Scophtalmus rhombus*), a predator of copepods during its early life, as its definitive host. The second cestode (*Bothriocephalus gregarius*) has the same intermediate host and a different definitive host: the turbot (*Psetta maxima*). However, for this latter parasite, a non-obligatory (paratenic) third host may be involved: the goby, *Pommatoschistus marmoratus*, a fish which becomes infested when consuming copepods. The cestode's larvae concentrate in the goby's intestine and retain their capacity to infest the turbot; the latter becomes infested in consuming these gobies (Fig. 13.5). Infestation measurements in the field

Table 2 Values of life-history traits of some nematodes

Species	Egg or larva production (female^{-1} day^{-1})	Life expectancies of adults	Prepatent period	Life expectancies of free stages	Length of adult female (mm)	Length of egg (μm)	Complex life cycle
Ascaris lumbricoïdes	7300-22700	1-2y	50-80 d	28-84 d	200-410	50-75	no
Parascaris equorum	100000	270 d	94 d	—	150	100	no
Toxocara canis	200000	32	14 d	—	150	75-80	no-yes
Toxocara cati	19000	39	90 d	—	70	70-75	no-yes
Necator americanus	15000	3-10 y	42 d	3-10 d	9-11	70-40	no
Ancylostoma duodenale	10000/25000	4-5y	28-50	3-10 d	10-18	60-40	no
Nematospiroides dubia	200/1000	12 w	9-11 d	73 w	12	—	no
Amidostomum anseris	1850	200 d	50 d	—	—	—	no
Trichonstrongylus tenuis	356	1 y	7.5 d	7-13 y	10	46-75	no
Ostertagia ostertagi	200	1 y	21 d	—	8.5	65-80	no
Nematospiroides dubius	200-1000	84 d	10 d	—	12	—	no
Nippostrongylus brasiliensis	1000-1500	15 d	6 d	—	33-58	58-33	no
Pseudoterranova decipiens	7000	2-3 w	25 d	15 d	52-64	46-52	yes
Trichinella spiralis	100-1500	—	7 d	—	2-4	—	no
Trichuris trichuira	3000-7000	1-3 y	5-9 w	11-30 d	40-80	70-90	no
Trichuris muris	4000-8000	5-6 w	2 w	—	10	—	no
Enterobius vermicularis	10000-12000 (total per F)	1-3 m	16-28 d	14-56 d	10	57	no
Passalurus ambiguus	1000-2000 (total per F)	—	25-27 d	—	8-9	95	no
Syphacia muris	480 (total per F)	—	7 d	—	3.4	90	no
Aspiculuris tetraptera	450 (total per F)	—	24 d	—	3.8	95	no

Strongyloides stercolaris	50	56 d	2–3 w	—	2–2.5	50–30	yes, FLG
Loa loa	—	15 y	52 w	10–14 d	70	—	yes
Acanthocheilonema vittae	7000	450 d	50 d	—	—	170–205 L	yes
Wuchereria bancrofti	10000 L	5–4 y	52 w	12 d	100	275–8 L	yes
Litosomoides carinii	15 000 L	1 y	10–11 w	—	100–140	90–7 L	yes
Haemonchus contortus	5000	10 m	3 w	—	18–30	85–44	no
Muellerius capillaris	560 L	4–7 y	6 w	4 w	19–23	—	yes
Capillaria hepatica	—	59 d	16 d	—	98	62–37	no
Trichinella spiralis	100–1500	30 d	7 d	—	3	100 L	no

d, day; w, week; m, month; y, year; L, larvae; FLG, free–living generation
Data compiled from various sources: see Mozgovoy, 1953; Skrjabin *et al*. 1960; Anderson, 1982, 1992; Mehlorn, 1988; DesClers, 1990.

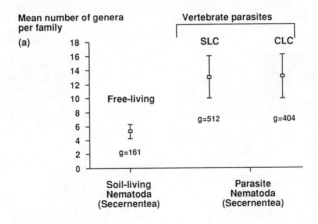

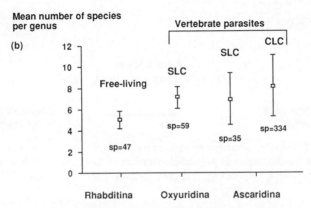

Fig. 13.4 Taxonomic diversification within the Nematoda as a function of life-cycle characteristics: (a) mean number of genera per family ($\pm$ s.e.); (b) mean number of species per genus ($\pm$ s.e.). Data compiled from various sources (see Goodey 1951; Mozgovoy 1953; Andrassy 1983; Sidiqui 1986; Anderson 1992) (g = number of genera; sp = number of species).

show that the infestation rate of *Psetta maxima* is much higher (prevalence: 79%, mean number per host: 49) than that of *Scophtalmus rhombus* (prevalence: 36%, mean number per host: 0.57). Indeed, if the two hosts are infested during the first years of their planktivorous life via infected copepods, the infestation can occur in older *Psetta maxima* simply by consuming infected gobies. The hypothesis of increased reproductive success in CLCs could be tested by comparing reproductive success in communities with and without the goby.

Organization of host communities

The time-delays in the development and reproduction of parasites are among the factors which tend to destabilize host–parasite equilibria (Anderson and

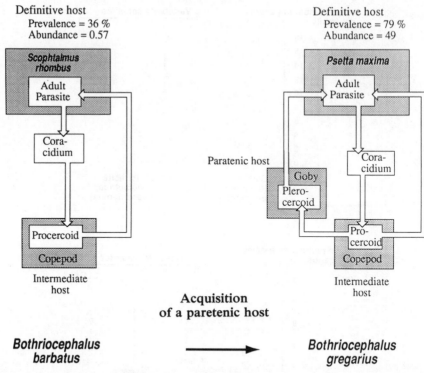

Fig. 13.5 Life-cycle, abundance and prevalence of two cestodes *Bothriocephalus barbatus* and *Bothriocephalus gregarius*, parasites of *Scophthalmus rhombus* and *Psetta maxima*, respectively (adapted from Robert *et al.* 1988, 1990).

May 1978). Transmission delays can be reduced if the parasite modifies host behaviour in such a way that the passage of the parasite from one stage to another is easier and quicker (Holmes and Bethel 1972; Moore 1984; Combes 1991). The next section describes a theoretical model of a CLC parasite whose cycle is inserted into a predator–prey relationship.

A simple parasite–predator–prey model   The model is a modified version of one presented by Dobson and Keymer (1985) (Fig. 13.6), differing from Dobson and Keymer by taking into consideration population changes in definitive hosts and transmission resulting from both predation and free-living parasites. Five differential equations describe the changes in the density of the definitive host predator (H_d), the intermediate host prey (H_1), the parasite's adult stage (P_d), the parasite's larval stage (P_1), and the parasite's free living stage (W). The intermediate host (H_1) is infected (β) by the free-living stages of the parasite (W). Intermediate hosts are preyed upon by the definitive host (ε) and the presence of the parasites increases the susceptibility to predation (ρ). The parasite induces host mortality in its definitive host (α_p);

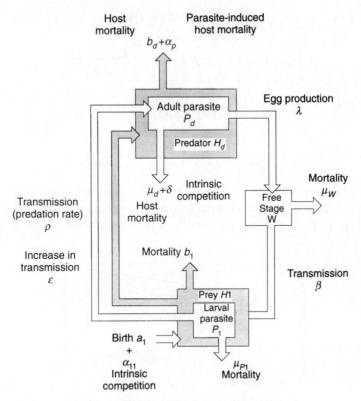

Fig. 13.6 Hypothetical prey–parasite–predator relationship with variables and parameters of the model (see text).

intraspecific competition leading to population reductions occurs between the parasite's adult stages (δ). Since the increase of predation (ρ) is proportional to the number of parasites sheltered by the intermediate host, the distribution of parasites within the host population must be taken into account, and this is done using a negative binomial distribution with clumping parameter k_1. The parasite's distribution in the definitive host also follows a negative binomial law, with parameter k_d. Finally, regulation of the prey population occurs by intraspecific competion (α_{11}).

The equations are

$$\frac{dW}{dt} = \lambda P_d - \mu_w W - \beta H_1 W$$

$$\frac{dP_1}{dt} = \beta H_1 W - (b_1 + \mu_1)P_1 - \varepsilon H_d P_1 - \rho \varepsilon H_d P_1 \left[1 + \frac{P_1(k_1 + 1)}{H_1 k_1} \right]$$

$$\frac{dP_d}{dt} = \varepsilon H_d P_l + \rho \varepsilon H_d P_l \left[1 + \frac{P_l(k_l + 1)}{H_l k_l} \right] -$$

$$P_d \left[\mu_d + b_d + (\delta + \alpha_p)) + (\delta + \alpha_p) \frac{P_d(k_d + 1)}{H_d k_d} \right]$$

$$\frac{dH_d}{dt} = a_d (\rho \varepsilon H_d P_l + \varepsilon H_d H_l) - b_d H_d - \alpha_p P_d$$

$$\frac{dH_l}{dt} = (a_1 - b_1) H_1 - \alpha_{11} H_1^2 - \rho \varepsilon H_d P_l - \varepsilon H_d H_l.$$

The equilibrium values of the larval and adult stages of parasites are affected by the influence of the parasite on its own transmission via the intermediate host (Fig. 13.7). Increasing ρ raises the equilibrium value of the number of

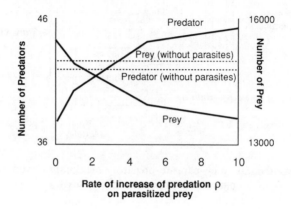

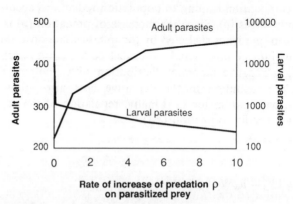

Fig. 13.7 The effect of ρ (rate of increase of predation due to parasite manipulation) on predator and prey population sizes at equilibrium (values of other parameters: $\lambda = 2000$; $\mu_w = 0.99$; $\mu_1 = 0$; $\mu_d = 0.20$; $\beta = 10^{-7}$; $b_1 = 0.3$; $b_d = 0.15$; $k_1 = 8$; $k_d = 1$; $\varepsilon = 0.01$; $\delta = 0.1$; $\alpha_p = 0.001$; $\alpha_1 = 1.9$; $a_d = 0.001$; $\alpha_{11} = 0.0008$).

predators; there is a threshold value of ρ beyond which the number of predators is higher than the equilibrium value without parasites. The opposite is true for the prey, which are less numerous than in a parasite-free situation.

Under the assumption that the mortality induced by the parasite (α_p) is minimized by competition between parasites to occupy optimum sites in the definitive host (i.e., $\delta > \alpha_p$, $\alpha_p > 0$) (Esh and Fernandez 1993), the introduction of a parasite can decrease the equilibrium number of predators and increase that of the prey (when the latter is not affected by the larval parasites, or $\rho = 0$) (Fig. 13.7).

The results of this model are in agreement the predictions of energy-budget models describing why parasites exploit predators and prey as definitive and intermediate hosts, respectively (Lafferty 1992).

Transmission rates and system resilience   The return-time to equilibrium (an inverse measure of resilience) is shorter when the parasite increases transmission efficiency via effects on the intermediate host (Fig. 13.8). Thus, by its action on the prey, the parasite increases the resilience and stability of a three-variable system.

CLC parasites: a possible role in community stability

Parasitism is now acknowledged as an important agent in population regulation (e.g. Dobson 1988, 1989; Scott 1988; Minchella and Scott 1991). Both CLC and SLC parasites are known to have the potential of regulating their host populations (Gibson *et al.* 1972; Hudson and Dobson 1989).

Freeland (1983) developed a different view of parasitism. Far from denying its regulatory effect on host populations, he integrated this into a multi-species system. He postulated that parasites allow the co-existence of species by lowering their coefficient of inter-specific competition (Freeland 1983; see also Holt 1977; Holt and Pickering, 1985). Following Holt and Pickering (1985), Freeland and Boulton (1992) have shown by simulation that parasites can stabilize multi-species systems, suggesting trophic networks with parasites are more feasible than those without parasites. However, these theoretical results have not been confirmed since no food web has yet incorporated parasites (Freeland and Boulton 1992).

Conclusion

At least for microparasites of man we can draw some conclusions about parasites in relation to their life-cycles. First, CLC species are more numerous and are more pathogenic than SLC species (Ewald 1983). Second, there is undoubtedly a cost associated with complexity, in particular the necessity to adapt to hosts that are taxonomically remote, but this cost has not yet been measured in detail; nonetheless, increased complexity may be beneficial to

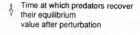

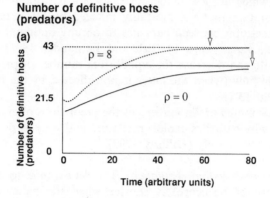

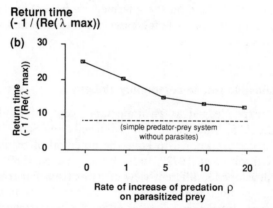

Fig. 13.8 (a) Numerical simulations showing the effect of system perturbation on the return time. The return time decreases as ρ increases (ρ = rate of increase of prediction due to parasite manipulation). (b) Return times in relation to ρ (the eigenvalues λ are calculated at each equilibrium point using the Jacobian matrix of the system, numerical values are given in Fig. 13.6). The return time is equal to $-1/Re(\lambda max)$, where $Re(\lambda max)$ is the largest real part of the eigenvalue (May 1974). The introduction of a parasite has a destabilizing effect on the predator–prey system by increasing return time. By manipulating prey (increasing ρ), the parasite reduces its destabilizing influence.

parasites if it leads to higher transmission rates (Dobson 1989). Third, since CLC species have more difficultly invading host communities as compared to SLC species (Dobson and May 1986), one may expect that the former has either lower extinction rates or higher speciation rates than the latter. At least with regards to speciation in CLC nematodes, one fails to find greater taxonomic diversification at the level of either species or genera. So then what could explain the high proportion of CLC parasites in natural communities? Recall from our simple model that CLC parasites can:

(1) produce a density increase in the definitive host at equilibrium, despite the parasite's impact on survival (at equilibrium, prey density is lower than in the absence of parasites) and

(2) counter-act the destabilizing effect of a directly-transmitted parasite in a community (by increasing time delays or decreasing resilience and by changing the behaviour of intermediate hosts).

To explain the larger number of CLC compared to SLC species I suggest we reconsider the hypothesis of Freeland and Boulton (1992) that complex life-cycles promote system persistence. Community destabilization and local extinctions are more likely to be induced by SLC parasites because of their greater ability to invade a community and their more focused effect on single species. Research into evolution towards complexity should concentrate on factors which increase transmission rates, whereas research into the predominance of CLC parasites should focus on factors which increase community stability.

Acknowledgements

I am grateful to C. Combes, A. Théron, E. Faliex for their help during the preparation of this work. I also thank anonymous reviewers for their comments which greatly improved the original manuscript.

References

Anderson, R. C. (1992). *Nematode parasites of vertebrates: their development and transmission.* C. A. B. International, Cambridge.

Anderson, R. M. (1982). The population dynamics and control of hookworm and roundworm infections. In *The population dynamics of infectious diseases: theory and applications* (ed. R. M. Anderson), pp. 67–108. Chapman and Hall, London.

Anderson, R. M. and May, R. M. (1978). Population biology of infectious diseases, Part I. *Nature*, **280**, 361–67.

Andrassy, I. (1983). *A taxonomic review of the suborder Rhabditina (Nematoda: Secernentia).* Orstom, Paris.

Borgsteede, F. M. M. (1984). Helminth parasites of wild foxes (*Vulpes vulpes* L.) in The Netherlands. *Zeitschrift für Parasitenkunde,* **70**, 281–85.

Bullini, G., Nascetti, L., Paggi, P., Orecchia, Mattiucci, S. and Berland, B. (1986). Genetic variation of ascaroid worms with different life-cycles. *Evolution*, **40**, 437–40.

Combes, C. (1991). Ethological aspects of parasite transmission. *The American Naturalist*, **138**, 866–80.

Conti, J. A., Forrester D. J., and Frohlich, R. K. (1984). Helminths of urban gray squirrels in Florida. *The Journal of Parasitology*, **70**, 143–44.

Day, R. W. and Quinn, G. P. (1989). Comparisons of treatments after an analysis of variance. *Ecological Monographs*, **59**, 433–63.

DesClers, S. (1990). Modelling the life-cycle of the sealworm (*Pseudoterranova decipiens*) in Scottish Waters. *Canadian Bulletin of Fisheries and Aquatic Science*, **222**, 273–88.

Dible, E. D., Font, W. F. and Wittrock, D. D. (1988). Helminths of the red fox, *Vulpes vulpes* L., in West Central Wisconsin. *The Journal of Parasitology*, **69**, 1170–72.

Dobson, A. P. (1988). The population biology of parasite-induced changes in host behavior. *The Quaterly Review of Biology*, **63**, 139–65.

Dobson, A. P. (1989). The population biology of parasitic helminths in animal populations. In *Applied mathematical ecology* (eds. S. A. Levin, T. G. Hallam and L. J. Gross), pp. 145–75. Springer-Verlag, London.

Dobson, A. P. and Keymer, A. E. (1985). Life-history models. In *Acantocephalan biology* (eds. D. W. T. Crompton and B. B. Nickol), pp. 347–84. Cambridge University Press.

Dobson, A. P. and May, R. M. (1986). Patterns of invasion by pathogens and parasites. In *Ecology of biological invasions of North America and Hawaii* (eds. H. A. Mooney and J. A. Drake), pp. 58–76. Springer-Verlag, New York.

Dobson, A. P. and Merenlender, A. (1991). Coevolution of macroparasites and their hosts. In *Parasite-host associations, coexistence or conflict?* (eds. C. A. Toft, A. E. Aeschlimann and L. Bolis), pp. 83–101. Oxford University Press.

Esch, G. W. and Fernandez, J. C. (1993). *A functional biology of parasitism*. Chapman and Hall, London.

Ewald, P. E. (1983). Host–parasite relations, vectors, and the evolution of disease severity. *Annual Review of Ecology and Sytematics*, **14**, 465–85.

Forrester, D. J., Pence, D. B., Bush, A. O., Lee, D. M. and Holler, N. R. (1987). Ecological analysis of the helminths of round-tailed muskrats (*Neofiber alleni* True) in southern Florida. *Canadian Journal of Zoology*, **65**, 2976–79.

Freeland, W. J. (1983). Parasites and the coexistence of animal host species. *The American Naturalist*, **121**, 223–36.

Freeland, W. J. and Boulton, W. J. (1992). Coevolution of food webs: parasites, predators and plant secondary compounds. *Biotropica*, **24**, 309–17.

Gibson, C. G., Broughton, E. and Choquette, L. P. E. (1972). Waterfowl mortality caused by *Cyathocotyle bushiensis* Khan, 1962, St. Lawurence River, Québec. *Canadian Journal of Zoology*, **50**, 1351–56.

Goodey, T. (1951). *Soil and freshwater nematodes*. London, Methuen.

Hasegawa, H. J., Kobayashi, J. and Otsuru, M. (1994). Helminth parasites collected from *Rattus rattus* on Lanyu, Taiwan. *The Journal of the Helminthological Society of Washington*, **61**, 95–102.

Haukisalmi, V., Henttonen, H. and Tenora, F. (1988). Population dynamics of common and rare helminths in cyclic vole populations. *Journal of Animal Ecology*, **57**, 807–25.

Holmes, J. C. and Bethel, W. M. (1972). Modification of intermediate host behaviour by parasites. In *Behavioural aspects of parasite transmission* (eds. E.U Canning and C. Wright), pp. 123–49. Academic Press, London.

Holt, R. D. (1977). Predation, apparent competition, and the structure of prey communities. *Theoretical Population Biology*, **2**, 197–229.

Holt, R. D. and Pickering, J. (1985). Infectious disease and species coexistence: a model of Lotka–Volterra form. *The American Naturalist*, **126**, 196–211.

Hudson, P. J. and Dobson, A. P. (1989). Population biology of Trichonstrongylus tenuis, a parasite of economic importance for red grouse management. *Parasitology Today*, **5**, 283–91.

Jennings, J. B. and Calow, P. C. (1975). The relationship between high fecundity and the evolution of entoparasitism. *Oecologia*, **21**, 109–15.

Kennedy, C.R and Rumpus, A. (1977). Long term changes in the size of the *Pomphorhynchus laevis* (Acanthocephala) population in the River Avon. *Journal of Fish Biology*, **10**, 35–45.

Kisielewska, K. (1970). Ecological organization of intestinal helminth groupings in *Clethrionomys glareolus* (Schreb.) (Rodentia). I. Structure and seasonal dynamics of helminth groupings in a host population in the Biolowieza National Park. *Acta Parasitologica Polonica*, **13**, 121–47.

Lafferty, K. D. (1992). Foraging on prey that are modified by parasites. *The American Naturalist*, **140**, 854–67.

Lepitzki, D. A. W., Woolf, A. and Bunn, B. M. (1992). Parasites of cottontail rabbits of Southern Illinois. *The Journal of Parasitology*, **78**, 1082–83.

May, R. M. (1974). *Stability and complexity in model ecosystems*. Princeton University Press.

Mehlhorn, H. (1988). *Parasitology in focus*. Springer-Verlag, Berlin.

Miguel, J. (1993). Contribución al conocimiento de la helmintofauna de los carnivoros silvestres de Cataluña. Unpublished thesis, University of Barcelona.

Minchella, D. J. and Scott, M. E. (1991). Parasitism: A cryptic determinant of animal community structure. *Trends in Ecology and Evolution*, **6**, 250–53.

Montgomery, S. S. J. and Montgomery, W. I. (1988). Cyclic and non-cyclic dynamics in populations of the helminth parasites of wood mice, *Apodemus sylvaticus*. *Journal of Helminthology*, **62**, 78–90.

Moore, J. (1984). Parasites that change the behavior of their host. *Scientific American*, **68**, 1174–76.

Mozgovoy, A. A. (1953). 'Essentials of Nematology. Vol. II. Ascaridita of animals and man and the diseases they causes' (in Russian). Parts 1 & 2. Izdatel'stvo Akademii Nauk SSSR, Moscow.

Pence, D. B. (1990). Helminth community of mammalian hosts: concepts at the infracommunity, component and compound community levels. In *Parasite communities: patterns and processes* (eds. G. Esch, A. Bush and J. Aho), pp. 233–61. Chapman and Hall, London.

Pence, D. B. and Meinzer, W. P. (1978). Helminth parasitism in the coyote, *Canis latrans*, from the rolling plains of Texas. *The Journal of Parasitology*, **6**, 339–44.

Pfaffenberger, G. S., Kemether, K. and de Bruin, D. (1985). Helminths of sympatric population of Kangaroo rats (*Dipodomys ordii*) and Grasshopper mice (*Onychomys leucogaster*) from the high plains of eastern New Mexico. *The Journal of Parasitology*, **71**, 592–595.

Poulin, R. (1992). Determinants of host-specificity in parasites of freshwater fishes. *International Journal for Parasitology*, **22**, 753–58.

Price, P. W. (1980). *Evolutionary biology of parasites*. Princeton University Press.

Robert, F., Boy, V. and Gabrion, C. (1990). Biology of parasite populations: population dynamics of bothriocephalids (Cestoda-Pseudophyllidae) in teleostean fish. *Journal of Fish Biology*, **37**, 327–42.

Robert, F., Renaud, F. Mathieu, E. and Gabrion, C. (1988). Importance of the paratenic host in the biology of *Botriocephalus gregarius* (Cestoda, Pseudophyllidea), a parasite of the turbot. *International Journal of Parasitology*, **18**, 611–21.

Rosenberg, G. W. and Pence, D. B. (1978). Circulation of helminth species in a rodent population from the high plains of Texas. *Occasional Paper of the Museum, Texas Technology University*, **56**, 1–10.

Scott, M. E. (1988). The impact of infection and disease on animal populations: Implications for conservation biology. *Conservation Biology*, **2**, 40–56.

Sidiqi, M. R. (1986). *Tylenchida, parasites of plants and insects*. Commonwealth Institute of Parasitology, Farnham Royal, Slough.

Skrjabin, K. I., Shikhobalova, N. P. and Lagodovskaya, E. A. (1960). *Oxyurata of Animals and Man. Izdatel'stvo Akademii Nauk SSSR. Moksva* (Translated by the Israel Program for Scientific Translations, Jerusalem, 1974).

Snyder, D. E. (1985). Helminth parasites from Illinois racoon *Procyon lator*. *The Journal of Parasitology*, **71**, 274–78.

Snyder, D. E. and Fitzgerald, P. R. (1985). The relationship of *Baylisacaris procyonis* to Illinois raccoons (*Procyon lotor*). *The Journal of Parasitology*, **71**, 596–598.

Soulsby, E. J. L. (1982). *Helminths, arthropods and protozoa of domestic animals*. Bailliére Tindall, East Sussex, England.

Sprent, J. F. A. (1983). Observations on the systematics of ascaridoid nematodes. In *Concepts in nematode systematics* (eds. A. R. Stone, H. M. Platt and L. F. Khalil), pp. 309–19. Academic Press, London.

Stone, J. E. and Pence, D. B. (1978). Ecology of helminth parasitism in the bobcat of West texas. *The Journal of Parasitology*, **64**, 295–302.

Théron, A., Pointier, J. P., Morand, S., Imber-Establet, D. and Borel, G. (1992). Long-term dynamics of natural populations of *Schistosoma mansoni* among *Rattus rattus* in patchy environment. *Parasitology*, **104**, 291–98.

Tiekotter, K. L. (1985). Helminth species diversity and biology in the bobcat, *Lynx rufus* (Schreber), from Nebraska. *The Journal of Parasitology*, **71**, 227–34.

Zar, J. H. (1984). *Biostatistical analysis*. Prentice Hall, Englewood Cliffs, New Jersey.

14

Biodiversity value and taxonomic relatedness

Paul H. Williams

Introduction

The threat of exinction hangs over not only particular species, but also entire faunas and floras. The very immensity of the variety of all life dictates that most of this richness will only ever be conserved *in situ*. Therefore conservationists need to know which areas contibute most to overall biodiversity. This, in turn, requires an ability to measure 'biodiversity value' (McNeely *et al.* 1990; Reid *et al.* 1992; Groombridge 1992; Wilson 1992).

Biodiversity conservation is not necessarily equivalent to species conservation. Species conservation, when justified as part of biodiversity conservation, appears to assume the possibility of 'collecting the set' of all species for protection. Under these circumstances the most threatened species are often seen as the most immediate priority (a reactive strategy), and other species are assumed to be safe for the time being. But, if it is not possible to protect everything (which is perhaps realistic at least when dealing with tropical biodiversity), and particularly if it is not possible to foresee threats or act in time for the species that were assumed safe, then a lot of resources could be invested in a few species, while elsewhere, through some unexpected change in land use, the majority might be lost. A more appropriate (proactive) strategy might be to 'salvage the most', because it is then that the areas with the greatest contribution to total biodiversity acquire the greatest priority. It follows that, in certain circumstances, threatened species or areas might even be seen as a poor investment, particularly if equivalent, lower risk choices for biodiversity value were available (Williams *et al.* 1994).

Measuring biodiversity as the variety of all life will usually be impractical or too expensive (May 1990; Ehrlich 1992). Consequently we need indirect, approximate measures for a more realistic approach (e.g. Currie 1991; Bibby *et al.* 1992; Scott *et al.* 1993; Williams *et al.* 1994). This does not mean that measures based on detailed information about species are irrelevant. In contrast, taxonomic measures should be valuable because they can provide a baseline measure of biodiversity value for assessing which of the more rapid and low-cost surrogate approaches may be most effective. Higher taxa constitute one of these potentially low-cost surrogates.

Here I summarise recent discussions of taxonomic or phylogenetic measures of biodiversity, and particularly how they might contribute to measuring biodiversity value. This is not intended as a general review of taxonomic measures (e.g. Pielou 1967; Altschul and Lipman 1990; May 1990; Vane-Wright *et al.* 1991; Crozier 1992; Faith 1992*a,b*; Nixon and Wheeler 1992; Weitzman 1992, 1993; Solow *et al.* 1993; Crozier and Kusmierski in press), but instead describes one perspective based on Williams *et al.* (1994) and work in progress (in Humphries *et al.*, in prep.).

Where does value lie in biodiversity?

Biodiversity can be seen as the irreducible complexity of all life. As such, there is no single objective definition or measure of biodiversity, only measures appropriate for restricted purposes (Norton 1994). Consequently, conservationists should begin by specifying where in biodiversity lies the value that is in need of conservation.

One example is to focus on biodiversity's (rather vaguely defined) 'option value' for the future (IUCN *et al.* 1980). Reid (1994, p. 10) describes this as maximising the human capacity to adapt to changing ecological conditions which, in turn, requires maximising the rest of life's capacity to adapt to change. Naturally, other approaches are possible and, like any single formulation of biodiversity value, IUCN *et al.*'s approach cannot include all of them. However, an interpretation of option value can provide a basis from which to compare the consequences and trade-offs between one kind of value and any others.

The notion of option value redirects the focus of biodiversity value from the level of species (or higher) down to the level of species' characters, attributes or features (used here for particular states of homologous characters—the differences between species). These characters are seen as the fundamental units with option value for the future (Faith 1992*a*). Crucially, option value gives any included characters equal value because of the inevitable ignorance or uncertainty of precise needs for the future. Biodiversity conservation would then focus on maximising the amount of 'currency' (counted as the number of different, and valued, biological characters) to be held within the protection system 'bank' (the set of protected species, ecosystems or areas). Thus, the paradoxical consequence of equal value for characters as units of currency is that their owners, the individuals, species or areas, may have different value because they contribute different numbers of complementary or novel characters for representation in the protection system.

Do all characters of organisms have option value?

Option value may be considered to reside in all characters of species. But if more information is available, some classes of characters may be considered less valuable, so that not all characters would be counted as part of the currency of biodiversity. The problem is that increased selection by conservationists for richness in one class of characters may in effect deselect for another to some extent, even if scores for the two are highly correlated among individuals. Not only does selecting a currency require taking responsibility for the choice of what is of most value, but it also implies accepting a lesser value for the remaining characters.

The consequences of currency choice for conservation decisions may diverge because different kinds of characters are distributed among species in different ways. These different patterns of character ownership can be predicted to some extent by models of evolutionary change in characters. The simplest of these is the 'empirical' model that assumes a character sample to be representative of all unsampled characters. Alternatively, 'anagenetic' and 'cladogenetic' models require that contentious assumptions about 'clock-like' patterns of character change be explicitly accepted or rejected. In reality, currencies may yet prove to be highly correlated among species, so that any direct diversity measurement could present an approximate surrogate for any other, although this is a bold assumption that remains to be confirmed.

Williams *et al.* (in press) conclude that although molecular biologists have understandably adopted genetic characters (particularly in the sense of differences in base-pair sequences, e.g. Fig. 14.1(a)) in preference to other classes of characters as the preferred currency for conservation, there is no special theoretical justification for this. Phenotypic characters were suggested as possibly closer to the currency of popular value. A greater value for phenotypic characters might be justified not so much because it is the most directly perceived currency (and so perhaps linked to any emotional response), but because it may be more directly useful to a wider audience and, perhaps more directly than some 'genetic' characters, may be the actual stuff from which future options (for evolution or utility) could be maintained.

Currencies: genetic characters

All of the more usual justifications for conservation (e.g. moral, aesthetic, cultural) can be applied at the level of genetic characters, but these provide no special theoretical grounds for valuing them over and above any other currency. Generally, because phenotypic characters or traits are expected to be (to a large extent) expressions of parts of the genetic code, genetic characters have been seen as a more fundamental currency of diversity among organisms. Thus, in a recent book on biodiversity by Wilson (1992), although the biological species is identified as the 'pivotal unit' for the description of biodiversity, genetics is referred to for the ultimate sources of this diversity.

Utilitarian justifications for valuing intra-specific genetic characters may have originated from selective breeding programmes involving single species. Value has been extended to interspecific genetic characters for their potential insertion in transgenic organisms and for pharmaceutical prospecting among little-known species. In a recent review of biodiversity by Groombridge

(a)

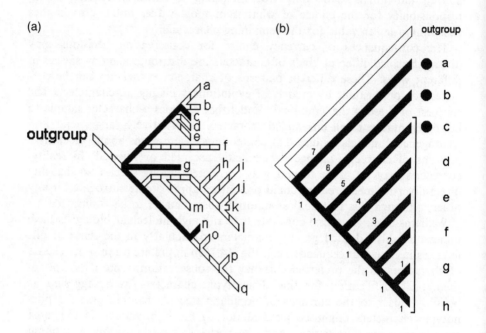

(b)

outgroup

(c)

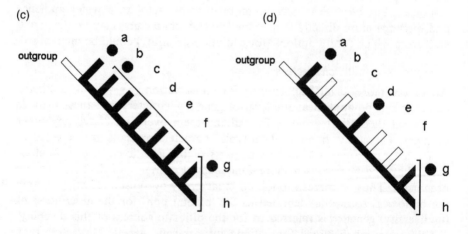

(d)

(1992), most references to the value of interspecific genetic diversity are also connected with plant breeding or with pharmaceutical prospecting. However, it is arguably the phenotypic products of whole genes, including molecules, that are actually most valued. To the extent that phenotypic characters are expressions of genetic characters at the level of entire transcribed genes, phenotypic diversity and genetic diversity can be seen as equivalent views of the level at which option value resides. The only purpose for the distinction here is that phenotypic diversity deliberately excludes variety at lower levels of organisation, such as within DNA base sequences where these differences have no effect on phenotype.

Currencies: phenotypic characters

It is perhaps phenotypic, and especially morphological, diversity that people respond to first and foremost. For many today, just as for Linnaeus (1758, pp. 12–13), the diversity of life is divided into mammals, birds, reptiles and amphibians, fish, insects, 'worms' or 'animals' (a category which excludes those previously listed), and plants. Thus when Groombridge (1992, p. iv) reviewed species diversity, it was divided into microoganisms, lower plants, higher plants, nematodes, deep-sea invertebrates, soil macrofauna, fishes, higher vertebrates and island species. While habitat criteria clearly influenced

---

Legend to facing page

Fig. 14.1 Genealogical trees used to infer the distribution of character changes and biodiversity value. (a) Inferring genetic diversity from a sample of DNA, using an empirical metric (distance) tree for populations (here labelled a–q) of crested newts (*Triturus cristatus* 'superspecies') and spanning sub-tree length, re-drawn from Fig. 4 in Faith (1992*a*) (each step along the branches represents a change at a single site in mitochondrial DNA; the spanning subtree for three geographically central populations, c, g, and n, is shown in black). (b) Using a genealogical classification of eight modern species, a–h, to infer diversity in order to choose the set of three species with the most diversity value, if a correlation between numbers of character changes and duration in time is assumed to yield an ultrametric tree, so that character richness may be maximised (species a + b + c/d/e/f/g/h) by using oldest-taxon richness or spanning-subtree length. (c) If no such correlation with time is assumed, so that numbers of character changes associated with each internode are not predictable from the neighbouring node's position (or alternatively character changes are associated primarily with nodes and are independent of time), then character richness may be maximised (species a + b/c/d/e/f + g/h) by using a spanning sub-tree node count. (d) However, if diversity value is maximised as richness of character combinations and (for example) character changes are assumed to occur as for case c, then value is maximised by even or regular representation of tree shape (species a + d + g/h) using cladistic dispersion or p-median. Outgroups are treated as unprotected (Williams and Humphries 1994). Species essential to any maximally-scoring set are shown by spots at the end of black branches, whereas species that remain equivalent as choices are shown bracketed across the ends of alternative grey branches, with branches to species contributing lower value to these sets shown in white.

this classification, morphological similarity is predominant, and genetic similarity (monophyly) is conspicuously lacking.

The most important justification for a preference for valuing phenotypic characters (or their underlying genetic characters in the sense of whole, transcribed genes) may be that it is the currency of most direct use to people. The utilitarian justifications for using whole-gene characters can all be applied to the phenotypic products of genes, including molecules.

Currencies: functional characters

Functional diversity among species is expected to have a premium value for preserving the integrity of ecosystems, a particularly important goal for conservationists (IUCN *et al.* 1980; Walker 1992; Schulze and Mooney 1993).

Unfortunately, it is particularly unclear how a currency based on functional characters might be measured (Gaston 1994; Williams *et al.* 1994). At least part of the conservation value of morphological characters may lie in their relationship to functional characters. When defined broadly, individual functional groups tend to comprise entire higher taxa, or at least it has often been found convenient to categorise species in this way (Moran and Southwood 1982; Stork 1987). Thus, there is probably some relationship between functional-group richness of an area and its higher-taxon richness, as counted across a particular rank of the nomenclatural hierarchy (Gaston 1994). The relationship is not simple because even very disparate organisms may sometimes belong to the same guild (Eggleton and Gaston 1990; Hochberg and Lawton 1990). In that sense, organisms may be functionally equivalent, and yet they could be of very different (and not consistent) 'desirability' among conservationists (e.g. nematode and hymenopteran parasitoids, hawkmoths and hummingbirds, seed-eating ants and rodents).

Predicting character distributions: the role of taxonomy

If ideal, complete knowledge of character differences were available, then, in principle, this could be used directly in diversity measures (Williams *et al.* 1991; Weitzman 1992, 1993; Solow *et al.* 1993) without recourse to evolutionary models. It is much more usual for overall character differences to be extrapolated from data for small samples of characters. In this case, cladistic methods can be used to distinguish the divergence component of this information in order to build trees (classifications) as estimates of genealogical pattern. These trees should then be more generally predictive of all of the other kinds of character difference (Farris 1983). However, this does require the selection of an explicit, special evolutionary model for the way character changes are distributed across trees. The possibilities include an 'empirical' model, an anagenetic 'clock' model, and a cladogenetic model. The choice of model will be governed by which currency of biodiversity value has been selected, and by how these particular characters are believed to change with time (Williams *et al.* 1994).

The 'empirical' model of character change, as employed here, uses counts of character differences as estimates of the branch lengths of a metric tree (Fig. 14.1(a)). That is, even though the vast majority of characters remain unsampled (e.g. Patterson *et al.* 1993), branch-length estimates from differences within a sample of characters used to reconstruct the tree must be accepted as representative of the overall distribution of character changes (Faith 1992*a*). This requires that the sample of characters be unbiased, not only in the sense that sampling effort should be even among branches (Williams and Humphries 1994), but also in that patterns of both sampled and all unsampled characters should behave as though they obeyed the same evolutionary model. This could present difficulties, for example, if selectively neutral DNA sequence data were used to predict the distribution of characters of functional morphology that are selectively constrained.

Another model of evolutionary change in characters that has been particularly popular with molecular biologists is the 'clock' model, usually associated with character changes along branches of trees (anagenesis). The regularity of this molecular clock remains contentious (Scherer 1990; Gillespie 1991), but perhaps because a substantial proportion of the genetic code has been considered to be selectively neutral in evolution (Kimura 1983), it is more often accepted that numbers of changes in these genetic characters, rather than in other kinds of characters, are related to duration in time (as opposed to changes merely accumulating with time at widely varying rates) and so to the age of lineages. Clock models are assumed by some tree-building methods (UPGMA and some distance methods), and if necessary, even the nomenclatural hierarchy can be used to infer branch-length information. For example, one interpretation might be to place all modern species on trees at an equal distance from the root (e.g. converting a cladogram into an ultrametric tree diagram) not only on a time axis, but also in terms of numbers of character changes (Fig. 14.1(b)).

In contrast, if no assumptions about clock-like correlations between numbers of character changes and time duration are made, then it may not be straightforward to predict from the position of a branching point on a tree how many character changes are likely to be associated either with that branching event (cladogenetic changes) or with the neighbouring branch lengths (anagenetic changes) (Fig. 14.1(c),(d)). However, it is still possible (for morphological diversity in particular) that the predictive value of the tree's structure may be high, particularly if there is a correlation between numbers of character changes and numbers of branching points along a lineage.

For example, one key difference from molecular sequence characters in the interpretation of classifications for morphological characters has been that a relationship between numbers of character changes and duration in time is generally less widely accepted than for molecular characters. There are cases where such a correlation has been found (Smith *et al.* 1992), and others where it has been rejected (Cloutier 1991). Thus the coelacanth *Latimeria chalumnae* among vertebrates, and the maidenhair tree *Ginkgo biloba* among seed plants, may both have diverged from their closest surviving relatives a long time ago, and may both have undergone great molecular change (*Latimeria*: Hillis *et al.* 1991; *Ginkgo*: Hamby and Zimmer 1992), but morphological or ecological change does not appear to have been commensurately great (*Latimeria*: Forey in prep.; *Ginkgo*: Crane 1988). One explanation for this pattern could be that most morphological changes may be associated with cladogenesis (changing at the branching of lineages, with speciation) (Gould and Eldredge 1977, 1993; Vrba 1980; Ax 1987). This is not to say that there need be a constant number of morphological changes at cladogenesis, but it would be useful to know whether there is a correlation between the numbers of character changes and the numbers of cladogenetic events among lineages within a particular group. Of course, any assessment of whether this pattern were general would be greatly strengthened if it were possible to take account of any evidence for 'hidden' cladogenetic events involving extinct branches.

Conserving option value

After selecting an appropriate currency of biodiversity value and an evolutionary model that will predict its distribution over genealogical trees, the major remaining choice is whether the greatest option value is seen as lying with individual characters, or with their integration as combinations of characters.

Maximising character richness

If value is placed on different characters independent of one another, such as might be the case in pharmaceutical prospecting, then the aim of a biodiversity measure would be to give the highest score to the set of species that maximises the number of different character states represented (character richness).

Overall character richness for subsets of species can be maximised on metric trees with branch lengths proportional to relative numbers of character changes by selecting the set of species (often a local fauna or flora) with the maximum total branch length within its minimum spanning subtree (the shortest part of the tree joining the set of chosen species, Fig. 14.1(a)–(c)) (Faith 1992a,b). Ultrametric trees resulting from purely clock-like character change are a special case, for which diversity is maximised by selecting the richest set of species in terms of the earliest diverging taxa (Fig. 14.1(b)). A measure of higher-taxon richness is available to give greater weight to these sets when branch-length information is lacking (Williams *et al.* 1991; Williams 1992). Even for this case, spanning-subtree length remains a consistent approach (Faith 1992a,b; Faith and Walker 1993), but only if the branch lengths implied by ultrametric trees are explicitly added.

Maximising character combinations

Much of the foregoing discussion has been based on the premise that the greatest option value of biodiversity will result from maximising richness in individual character states. However, it may be broader combinations of states among different characters that are seen to have greater value. These combinations may be required as integrated, functional suites of characters to perform ecosystem services (e.g. a requirement for a dispersive, para-sitoid of forest caterpillar 'X'). An alternative approach to option value would therefore be to maximise richness in different combinations of characters.

Choosing species that are regularly spaced across asymmetric trees (Fig. 14.1(d)) samples most evenly the overall cladistic topology of the tree, in the sense that it evenly represents taxa at all levels in the hierarchy, and therefore represents the combinations of characters (as owned by terminal taxa) defining the taxa at all of these levels (Williams 1993a; Williams and Humphries 1994). A measure of regularity based on Poisson's work on rare events (Greig-Smith 1983) was used as the basis for a cladistic dispersion measure, which favours sets of species with the largest and most even numbers of nodes between them (Williams *et al.* 1991). The initial formula-tion had mathematical inconsistencies, but, more importantly, had no clearly justified balance between regularity and numbers of species or sub-tree length (Williams *et al.* 1991; Faith 1992b). A more consistent, revised formulation employed regularity as an optional and subsidiary criterion to spanning sub-tree node count, so that it is used only to break ties (Williams 1992, 1993a; Williams and Humphries 1994).

Recently, Faith and Walker (1993) have drawn on a family of p-median procedures from the operations research literature (e.g. Love *et al.* 1988) which also attempts to locate objects in a regular pattern on a network (such as a metric tree). Using the continuous p-median, they seek to represent all possible character combinations occurring notionally along the branches of a tree by minimising the sum or average distances on the tree from every point along the branches to the nearest protected species. Faith (personal com-munication) has preferred this continuous p-median over another form, the discrete p-median, because several similar species in unresolved 'bushes' (polytomies) contribute less to the score with the continuous p-median. However, this bias of the discrete p-median towards 'bushes' of species is also shown by the continuous p-median if the number of species in the bush is sufficiently high. Furthermore, this preference has no apparent theoretical justification from the character-pattern model. In contrast, Williams *et al.* (1994) suggest that it may be more appropriate to use the discrete p-median, because the conservation objective may be to seek representation of only those character combinations actually owned by extant (terminal) species on trees. The discrete p-median diversity measure minimises the sum or average distances on the tree from every unprotected (terminal) species to the nearest

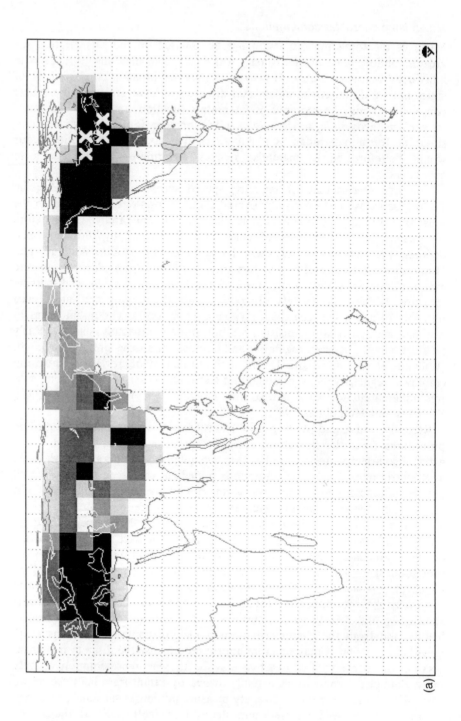

(a)

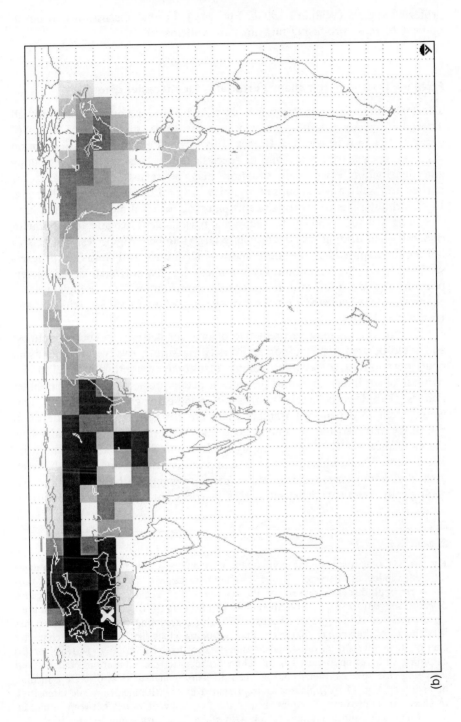

(b)

protected species (Williams 1993*b*; Fig. 14.2). Further discussion and development of these promising measures are anticipated.

From idealism to pragmatism: surrogates for character diversity

The taxonomic diversity approach described above recognises that not all areas and species can be protected in some sense, and that the value of organisms cannot be counted directly. It is formulated for maximising option value within a taxonomic group of species for which estimates of genealogy and distribution exist. This situation may be realistic for planning for particularly well known groups of organisms (e.g. Mickleburgh *et al.* 1992; Fig. 14.2), although this is not the same as planning for total biodiversity (all life).

Species richness is likely to become a reasonable surrogate for character richness when dealing with increasingly large numbers of species. For this to fail would require that the more species-rich faunas and floras be progressively more highly taxonomically clumped (i.e. hypo-diverse or character-poor in the sense of additional species contributing few complementary characters) with increasing numbers of species (Williams 1993*a*; Williams and Humphries 1994).

Indicator groups may be predictive of broader patterns in species richness and could be used with taxonomic diversity measures. However, it remains an open question as to whether adding taxonomic weighting to the species richness of an indicator group provides a better predictor of total character richness than does the species richness of an indicator group on its own. This would seem to depend on whether there are strongly repeating patterns in genealogy and distribution among taxa. Such patterns could be explained by the vicariance model of biogeography (Nelson and Platnick 1981; Humphries and Parenti 1986), which deals with common histories of divergence between areas and populations (without subsequent dispersal). Perhaps unfortunately, geologically frequent dispersal across barriers, or spread of populations

---

Legend to pages 270 and 271

Fig. 14.2 Comparison of the consequences of different value of characters and of different models of character change for the distribution of biodiversity value. Maps show regional diversity scores for the 29 species of cuckoo bumble bees (genus *Bombus*, subgenus *Psithyrus*) on an equal-area grid map (grid-cell area *c.* 611 000 km^2, for intervals of 10° longitude) (Williams 1993*a*): (a) assuming value of characters lies in richness of individual characters and assuming clock-like patterns of character change, then diversity is measured as older-taxon diversity (cf. Fig. 14.1(b)); or (b) assuming value of characters lies in richness of character combinations, but without assuming clock-like character change, then diversity is measured as discrete p-median diversity (cf. Fig. 14.1(d)). Scores are represented by logarithmic grey-scale intensities, in classes of approximately equal size by the frequency of values between minimum (light grey) and maximum (black with white 'X'), with white for no data.

through new bridges of suitable habitat, may disrupt the generality of these patterns and reduce the predictiveness of this level of indicator approach.

Higher taxa can provide a surrogate for species that reduces the massive extrapolation between groups of the indicator approach, while avoiding the simplifying assumptions of environmental diversity ('ED' of Faith and Walker 1993), these being (1) uniform random distribution of species in niche space (or simple graded transformations thereof) and (2) equilibrium distribution of organisms among patches of suitable habitat. The principle behind using higher taxa is that mapping 1000 genera or families represents more of total biodiversity than mapping 1000 species without a commensurate increase in costs. One interpretation of higher taxa is to consider higher-taxon richness as a surrogate for species richness (Gaston and Williams 1993; Williams and Gaston 1994; Williams *et al.* 1994). It may also be possible to use higher taxa with tree-based measures, at least when taxa are monophyletic (Williams 1993*a*). This interpretation is likely to be most robust for the character richness approach with the clock model (because the diversity measure is most dependent on the highest-level, most ancient relationships), and least robust for the character combination approach with the cladogenetic model (because the diversity measure is more dependent on complete sampling of the lowest-level taxa, which are obscured within higher taxa).

A particular strength of the higher-taxon surrogate for species data is that because it retains direct information on the identity of taxa within each area, there is some knowledge of the spatial turnover of taxa among areas (Williams 1993*a*; Williams *et al.* 1994). This allows taxa from selected priority areas to be removed from further consideration, in order to search for the next areas that are richest in complementary taxa (Kirkpatrick 1983; Ackery and Vane-Wright 1984; Burley 1988; Margules *et al.* 1988; Rebelo and Siegfried 1990; Vane-Wright *et al.* 1991). Existing protected areas can be taken into account or not, as required. This use of complementarity among floras or faunas has great potential for increasing the efficiency and flexibility of reserve selection methods for conservation (Pressey *et al.* 1993).

Conclusion

The conclusion drawn from the present interpretation of option value in terms of characters is not that biodiversity must always be scored using taxonomic measures, but rather that it can add further justification to other, less direct approaches that use more practical surrogates. The consequence of IUCN *et al.* (1980) identifying biodiversity value with option value could be seen as placing the traditional three levels of biodiversity (genes, species and ecosystem; e.g. Groombridge, 1992) within a scale of surrogacy for measuring overall biodiversity, arranged from the more direct (e.g. taxonomic measures) to the more remote (e.g. landscape measures). But in opposition to this scale, approaches at these different levels have other crucial strengths, such that they may become more practical for measuring overall biodiversity (and more relevant to the processes that maintain ecosystem viability) at the progressively higher scales of characters, species, higher taxa, and classifications or ordinations of species assemblages, ecosystems, land patches and landscapes.

Acknowledgements

My thanks to Dan Faith, Kevin Gaston, Paul Harvey, Chris Humphries, Chris Margules, Sandra Mitchell, Mark Pagel and Dick Vane-Wright for stimulating discussion, although they do not necessarily share the views expressed here, and to the referees for helpful comments.

References

Ackery, P. R. and Vane-Wright, R. I. (1984) *Milkweed butterflies, their cladistics and biology, being an account of the natural history of the Danainae, a subfamily of the Lepidoptera*. British Museum (Natural History) and Cornell University Press, London.

Altschul, S. F. and Lipman, D. J. (1990) Equal animals. *Nature*, London, **348**, 493–494.

Ax, P. (1987) *The phylogenetic system*. Wiley, Chichester.

Bibby, C. J., Collar, N. J., Crosby, M. J., Heath, M. F., Imboden, C., Johnson, T. H., Long, A. J., Stattersfield, A. J. and Thirgood, S. J. (1992) *Putting biodiversity on the map: priority areas for global conservation*. ICBP, Cambridge.

Burley, F. W. (1988) Monitoring biological diversity for setting priorities in conservation. In *Biodiversity* (ed. E. O. Wilson), pp. 227–230. National Academy Press, Washington DC.

Cloutier, R. (1991) Patterns, trends, and rates of evolution within the Actinistia. *Environmental Biology of Fishes*, **32**, 23–58.

Crane, P. R. (1988) Major clades and relationships in the 'higher' gymnosperms. In *Origin and evolution of gymnosperms* (ed. by C. B. Beck), pp. 218–272. Columbia University Press, New York.

Crozier, R. H. (1992) Genetic diversity and the agony of choice. *Biological Conservation*, **61**, 11–15.

Crozier, R. H. and Kusmierski, R. M. (in press) Genetic distances and the setting of conservation priorities. *Biological Conservation*.

Currie, D. J. (1991) Energy and large-scale patterns of animal- and plant-species richness. *American Naturalist*, **137**, 27–49.

Eggleton, P. and Gaston, K. J. (1990) 'Parasitoid' species and assemblages: convenient definitions or misleading compromises? *Oikos*, **59**, 417–421.

Ehrlich, P. R. (1992) Population biology of checkerspot butterflies and the preservation of global biodiversity. *Oikos*, **63**, 6–12.

Faith, D. P. (1992a) Conservation evaluation and phylogenetic diversity. *Biological Conservation*, **61**, 1–10.

Faith, D. P. (1992b) Systematics and conservation: on predicting the feature diversity of subsets of taxa. *Cladistics*, **8**, 361–373.

Faith, D. P. and Walker, P. A. (1993) *DIVERSITY: a software package for sampling phylogenetic and environmental diversity, reference and user's guide*. Privately distributed, Lyneham.

Farris, J. S. (1983) The logical basis of phylogenetic analysis. In *Advances in cladistics II* (ed. by N. I. Platnick and V. I. Funk), pp. 7-36. Columbia University Press, New York.

Gaston, K. J. (1994) Biodiversity measurement. *Progress in Physical Geography*, **18**, 565–574.

Gaston, K. J. and Williams, P. H. (1993) Mapping the world's species—the higher taxon approach. *Biodiversity Letters*, **1**, 2–8.

Gillespie, J. H. (1991) *The causes of molecular evolution*. Oxford University Press.

Gould, S. J. and Eldredge, N. (1977). Punctuated equilibria: the tempo and mode of evolution reconsidered. *Paleobiology*, **3**, 115–151.

Gould, S. J. and Eldredge, N. (1993). Punctuated equilibrium comes of age. *Nature*, London, **366**, 223–227.

Greig-Smith, P. (1983) *Quantitative plant ecology*. Blackwell Scientific Publications, Oxford.

Groombridge, B. (ed.) (1992) *Global biodiversity, status of the Earth's living resources*. Chapman and Hall, London.

Hamby, R. K. and Zimmer, E. A. (1992) Ribosomal RNA as a phylogenetic tool in plant systematics. In *Molecular systematics in plants* (ed. by P. S. Soltis, D. E. Soltis and J. J. Doyle), pp. 50–91. Chapman and Hall, London.

Hillis D. M., Dixon, M. T. and Ammerman, L. K. (1991) The relationships of the coelacanth *Latimeria chalumnae*: evidence from sequences of vertebrate 28S ribosomal RNA genes. *Environmental Biology of Fishes*, **32**, 119–131.

Hochberg, M. E. and Lawton, J. H. (1990) Competition between kingdoms. *Trends in Ecology and Evolution*, **5**, 367–371.

Humphries, C. J. and Parenti, L. R. (1986) *Cladistic biogeography*. Clarendon Press, Oxford.

IUCN, UNEP and WWF (1980) *World conservation strategy, living resource conservation for sustainable development*. IUCN, UNEP and WWF, Gland.

Kimura, M. (1983) *The neutral theory of molecular evolution*. Cambridge University Press.

Kirkpatrick, J. B. (1983) An iterative method for establishing priorities for the selection of nature reserves: an example from Tasmania. *Biological Conservation*, **25**, 127–134.

Linnaeus, C. (1758) *Systema naturae per regna tria naturae, secundum classes, ordines, genera, species, cum characteribus, differentiis, synonymis, locis* (Vol. 1, edn. 10). Holmiae.

Love, R. F., Morris, J. G. and Wesolowsky, G. O. (1988) *Facilities location. Models and methods*. North-Holland, London.

McNeely, J. A., Miller, K. R., Reid, W. V., Mittermeier, R. A. and Werner, T. B. (1990) *Conserving the world's biological diversity*. IUCN, WRI, CI, WWF and World Bank, Washington D.C.

Margules, C. R., Nicholls, A. O. and Pressey, R. L. (1988) Selecting networks of reserves to maximise biological diversity. *Biological Conservation*, **43**, 63–76.

May, R. M. (1990) Taxonomy as destiny. *Nature*, London, **347**, 129–130.

Mickleburgh, S. P., Hutson, A. M. and Racey, P. A. (1992) *Old World fruit bats, an action plan for their conservation*. IUCN, Gland.

Moran, V. C. and Southwood, T. R. E. (1982) The guild composition of arthropod communities in trees. *Journal of Animal Ecology*, **51**, 289–306.

Nelson, G. and Platnick, N. I. (1981) *Systematics and biogeography: cladistics and vicariance*. Columbia University Press, New York.

Nixon, K. C. and Wheeler, Q. D. (1992) Measures of phylogenetic diversity. In *Extinction and phylogeny* (ed. by M. J. Novacek and Q. D. Wheeler), pp. 216–234. Columbia University Press, New York.

Norton, B. G. (1994) On what we should save: the role of culture in determining conservation targets. In *Systematics and conservation evaluation* (ed. by P. L. Forey, C. J. Humphries and R. I. Vane-Wright), pp. 23–39. Oxford University Press.

Patterson, C., Williams, D. M. and Humphries, C. J. (1993) Congruence between molecular and morphological phylogenies. *Annual Review of Ecology and Systematics*, **24**, 153–188.

Pielou, E. C. (1967) The use of information theory in the study of the diversity of biological populations. *Proceedings of the 5th Berkeley Symposium on Mathematics and Statistical Probability*, **4**, 163–177.

Pressey, R.L., Humphries, C.J., Margules, C.R., Vane-Wright, R.I. and Williams, P.H. (1993) Beyond opportunism: key principles for systematic reserve selection. *Trends in Ecology and Evolution*, **8**, 124–128.

Rebelo, A. G. and Siegfried, W. R. (1990) Protection of fynbos vegetation: ideal and real-world options. *Biological Conservation*, **54**, 15–31.

Reid, W. V. (1994) Setting objectives for conservation evaluation. In *Systematics and conservation evaluation* (ed. by P. L. Forey, C. J. Humphries and R. I. Vane-Wright), pp. 1–13. Oxford University Press.

Reid, W., Barber, C. and Miller, K. (1992) *Global biodiversity strategy. Guidelines for action to save, study and use Earth's biotic wealth sustainably and equitably.* WRI, IUCN and UNEP, Washington D.C.

Scherer, S. (1990) The protein molecular clock. Time for a reevaluation. *Evolutionary Biology*, **24**, 83–106.

Schulze, E.-D. and Mooney, H. A. (eds.) (1993) *Biodiversity and ecosystem function.* Springer-Verlag, Berlin.

Scott, J. M., Davis, F., Csuti, B., Noss, R., Butterfield, B., Groves, C., Anderson, H., Caicco, S., D'Erchia, F., Edwards, T. C., Ulliman, J. and Wright, R. G. (1993) Gap analysis: a geographical approach to protection of biological diversity. *Wildlife Monographs*, **123**, 1–41.

Smith, A. B., Lafay, B. and Christen, R. (1992) Comparative variation of morphological and molecular evolution through geologic time: 28*S* ribosomal RNA versus morphology in echinoids. *Philosophical Transactions of the Royal Society of London*, B, **338**, 365–382.

Solow, A., Polasky, S. and Broadus, J. (1993) On the measurement of biological diversity. *Journal of Environmental Economics and Management*, **24**, 60–68.

Stork, N. E. (1987) Guild structure of arthropods from Bornean rain forest trees. *Ecological Entomology*, **12**, 69–80.

Vane-Wright, R. I., Humphries, C. J. and Williams, P. H. (1991) What to protect?—Systematics and the agony of choice. *Biological Conservation*, **55**, 235–254.

Vrba, E. S. (1980) Evolution, species and fossils: how does life evolve? *South African Journal of Science*, **76**, 61–84.

Walker, B. H. (1992) Biodiversity and ecological redundancy. *Conservation Biology*, **6**, 18–23.

Weitzman, M. L. (1992) On diversity. *Quarterly Journal of Economics*, **107**, 363–405.

Weitzman, M. L. (1993) What to preserve? An application of diversity theory to crane conservation. *Quarterly Journal of Economics*, **108**, 157–183.

Williams, P. H. (1992) *WORLDMAP priority areas for biodiversity. Using version 3.* Privately distributed, London.

Williams, P. H. (1993a) Measuring more of biodiversity for choosing conservation areas, using taxonomic relatedness. In *International Symposium on Biodiversity and Conservation* (ed. by T.-Y. Moon), pp. 194–227. Korean Entomological Institute, Seoul.

Williams, P. H. (1993b) *WORLDMAP priority areas for biodiversity. Using version 3.07.* Privately distributed, London.

Williams, P. H. and Gaston, K. J. (1994) Measuring more of biodiversity: can higher-taxon richness predict wholesale species richness? *Biological Conservation*, 67, 211–217.

Williams, P. H. and Humphries, C. J. (1994) Biodiversity, taxonomic relatedness and endemism in conservation. In *Systematics and conservation evaluation* (ed. by P. L. Forey, C. J. Humphries and R. I. Vane-Wright), pp. 269–287. Oxford University Press.

Williams, P. H., Humphries, C. J. and Vane-Wright, R. I. (1991) Measuring biodiversity: taxonomic relatedness for conservation priorities. *Australian Systematic Botany*, 4, 665–679.

Williams, P. H., Humphries, C. J. and Gaston, K. J. (1994) Centres of seed-plant diversity: the family way. *Proceedings of the Royal Society of London*, 256, 67–70.

Williams, P. H., Gaston, K. J. and Humphries, C. J. (1994) Do conservationists and molecular biologists value differences between organisms in the same way? *Biodiversity Letters*, 2, 67–78.

Wilson, E. O. (1992) *The diversity of life*. Penguin Press, London.

15

Utilising genetic information in plant conservation programmes

Richard A. Ennos

Introduction

Over the last 15 years there has been an explosion of interest in measuring genetic variation within plant species. On the one hand, a series of increasingly sophisticated methods have been applied to detect differences among individuals at the molecular level. Such variation has been dubbed 'genetic marker variation' (Avise 1994). In parallel with these developments, quantitative genetic analysis first developed in animal and plant breeding has been applied to natural populations to elucidate and quantify the genetic basis of phenotypic variation (Lawrence 1984). The objective of this chapter is to consider how studies of both genetic marker variation and quantitative genetic variation, when carefully interpreted, can be used to guide the development of conservation strategies for threatened plant species, and to illustrate this with reference to an ongoing conservation programme involving native pine in Scotland.

Genetic marker variation

Genetic marker variation represents differences in the genetic information of plants within (or between) species. Genetic marker loci may be detected at the DNA level (RFLP, RAPD, microsatellite, minisatellite loci) (Burke *et al.* 1992), or by analysing the primary products (isozymes) (Hamrick 1989) or secondary products (alkaloids, monoterpenes) (Tobolski and Hanover 1971) of genes. Genetic markers may be located in any of the three genomes (nuclear, chloroplast and mitochondrial) possessed by plants. For the purposes of initial interpretation, the effects of these forms of variation are assumed to be selectively neutral, though this may not always be the case.

Using a stratified sampling scheme, surveys of genetic marker variation can be used to build up a picture of the 'genetic structure' of plant populations involved in conservation programmes. Genetic structure is quantified in terms of gene diversity H_s within populations, the parameters F_{is} (inbreeding coefficient) and D (linkage disequilibrium) describing the arrangement of genetic variation within individuals in each population and a measure of the distribution of genetic diversity among populations and regions (F_{st} or G_{st}) (Wright 1951, Nei 1973). What inferences that can be drawn from analyses of genetic structure which are of relevance to conservation?

Gene diversity within populations

Gene diversity is of intrinsic interest in conservation since the ability of a population to adapt to environmental change is governed by the level of available genetic variation for relevant characters. Measuring gene diversity for marker loci could provide a way of quantifying this potential. However, we must be aware of a number of limitations in equating diversity for marker loci directly with the adaptive potential of populations.

The first of these is that gene diversity levels are relative rather than absolute. Valid comparisons can be made within, but not between, classes of genetic marker. This occurs because the different methods used to score genetic markers detect different proportions of the total variation in DNA base sequences. Moreover, the rates of mutation vary widely between different classes of marker. At drift-mutation equilibrium, different levels of gene diversity are expected for different classes of marker within the same population. Levels of genetic variation expected are also affected by the genomic location of the marker (nuclear, chloroplast or mitochondrial genome) since these genomes differ in their mode of inheritance and recombination rates.

The second problem is that it is not clear whether levels of variation for genetic markers are valid measures of levels of variation for loci involved in future adaptation. If future adaptation relies on genetic variation which, in the current environment, is selectively neutral, then a measure of the level of variation for neutral genetic markers in a population will be a relevant measure of the future adaptive potential of that population. On the other hand, if future adaptation relies on genetic variation that is currently under selection, the correlation between genetic marker variation and future adaptive potential may be poor. If the important variation is under balancing selection, genetic markers may underestimate the potential for adaptation and, if the variation is under directional selection, genetic markers will overestimate the adaptive potential of the population.

Finally, even where future adaptation does rely on currently selectively neutral variation, there are situations where a discrepancy may occur between levels of variation for single marker loci and levels of additive genetic variance for quantitative characters, the true determinant of the adaptive potential of a population (Lande and Barrowclough 1987). This is because the rate at which variation for quantitative characters is replaced after a bottleneck is much faster than the rate of replenishment of genetic variation at individual marker loci with low mutation rates. Many different loci may mutate to generate additive variance for a single character, whereas for a single locus a rare mutation at that specific locus is needed. Populations that have recently expanded after a severe bottleneck may possess sufficient additive genetic variation for adaptation, yet show no gene diversity for isozyme loci. Thus, gene diversity as measured by genetic markers may, under a number of conditions, be a poor indicator of the adaptive potential of a population.

Arrangement of variation within individuals

The arrangement of allelic variation within individuals in a population is governed by the mating system of the plant (Allard 1975). Plant mating systems are described genetically in terms of the 'mixed mating model'. Here it is assumed that seeds are produced either by selfing at a rate s, or by random outcrossing at a rate $t = (1-s)$. At equilibrium between outcrossing and selfing, the inbreeding coefficient in the population $F_{is} = s/(2-s)$. By calculating F_{is} from observed genotype frequencies, a rough estimate of the mating system can be obtained.

The mating system also influences the multilocus structure of the population such that inbreeding retards the breakdown of associations among alleles at different loci. Predominantly outcrossing populations show little or no linkage disequilibrium (D) among loci, whereas linkage disequilibrium is pronounced in selfing species. Thus, by analysing single and multilocus genotype frequencies, a qualitative picture of the plant's mating system can be obtained. More detailed measurements of mating systems can be found by comparing genetic marker distributions in parents and progeny, and fitting appropriate mating system models (Brown 1989).

An understanding of the mating system is crucial for the development of effective conservation policy in plants. In the context of in situ conservation, a knowledge of the mating system may facilitate an assessment of the relative genetic risks faced by plant species within a community. This stems from the fact that highly outcrossing species are expected to carry a much greater load of deleterious recessive mutations than inbreeding species. Hence, they are likely to suffer far more severely than predominantly inbreeding species from inbreeding depression if their population size contracts rapidly, or if they are forced to produce a high proportion of selfed offspring as a consequence of reduced population densities, loss of pollinators etc. As such, mating system estimation may highlight those species at highest genetic risk within a threatened community.

Strategies for ex situ conservation of plants are also very much dependent upon the mating system that they possess. In highly homozygous selfing species, little or no genetic variation is found within individuals or families and strong correlations are found across loci, whereas in outcrossing species substantial variation is distributed within both individuals and families, and variation at different loci is essentially uncorrelated. More plants will therefore have to be sampled from an inbreeding than an outcrossing population to capture the same proportion of genetic variation.

On the other hand, the maintenance and production of seed for reintroduction is far simpler in the case of highly selfing species than for outcrossers.

Selfing species can be left to produce offspring without the need for pollination management, and the genetic integrity of accessions will be retained from one generation to the next. In outcrossing species, however, management of pollination may be necessary. This will be required to ensure cross-pollination in the absence of specialist pollinators, to guard against increases in selfing rate in small *ex situ* populations containing limited numbers of partners, and to prevent crosses either between accessions from contrasting parts of the species range, or with related species growing in the same area. Without such pollination management the value of the seed for population reestablishment following *ex situ* conservation will be severely compromised.

Distribution of variation among populations

Where plants have been sampled according to a stratified scheme, variation for genetic markers can be split into its 'within' and 'among' population components. The proportion of total diversity due to differences among populations can be quantified using genetic differentiation statistics such as F_{st} or G_{st} (Wright 1951, Nei 1973). Interpretation of these statistics requires an appreciation of the processes leading to genetic differentiation for selectively neutral markers.

For markers with low mutation rates (e.g. isozymes, RFLP markers), the equilibrium distribution of genetic variation within and among populations is governed by a balance between genetic drift and interpopulation gene migration. Differentiation among populations increases when effective population sizes and/or interpopulation migration rates decrease. One vital point to note is that any spatial structuring of variation is related to the history of the populations and the particular patterns of past interpopulation gene flow. Populations with extensive gene exchange in the past will be less differentiated from one another than those that were genetically isolated over that time. No causal association between patterns of genetic marker variation and environmental variation is expected.

The implication for plant conservation is that although analysis of population differentiation for selectively neutral markers can provide insights into past levels and patterns of interpopulation gene flow, it may tell us nothing about adaptive differentiation among plant populations. High population differentiation for genetic markers indicates species that have historically been split into relatively genetically isolated populations, whereas low levels reflect genetically cohesive species (Hamrick 1989). By analysing the patterns of differentiation among populations using genetic distance measures (Nei 1972), individual populations that have been genetically isolated in the past may be distinguished. Such populations may be candidates for genetic supplementation in a conservation programme, if measures of genetic diversity indicate that they are genetically depauperate. Patterns of increasing differentiation with distance may be interpreted as resulting from migration from source populations (Lagercrantz and Ryman 1990). This type of analysis may be very important in identifying the locations of refugia which may then be targeted for protection in conservation programmes.

One practical limitation in deducing past history from measures of population differentiation is that in many outcrossing plant species, interpopulation gene flow has been so large that levels of differentiation are too small for informative analysis. Measures of differentiation are normally obtained from biparentally inherited nuclear markers, that migrate among populations both by pollen and seed. In the future, analysis of differentiation for maternally inherited organelle markers may prove more fruitful (Petit *et al.* 1993*a*, Ennos 1994) . Such markers can only be dispersed between populations by seed. They are therefore expected to show greater degrees of population differentiation than nuclear markers (Dong and Wagner 1993, Strauss *et al.* 1993), and may prove valuable in conservation for investigating the historical origins of populations, their seed migration routes, and the refugia from which they were derived (Ferris *et al.* 1993, Petit *et al.* 1993*b*).

Distribution of variation among regions

Within the geographic range of a plant species, sets of populations may occur in different regions which are so geographically isolated that gene flow among them can be regarded as negligible. Differentiation among these groups for genetic markers will result from genetic drift since the time when the regional groups were separated. Differences could be substantial if any of the groups were founded by small numbers of individuals, or if regional groups had, at any time, passed through a period of small population size in isolated refugia (Tobolski and Hanover 1971). If, however, neither of these situations had occurred, levels of differentiation between regional groups could be very small. Data on regional differentiation for neutral markers is of relevance to conservation in as much as it allows us to elucidate the historical origins of threatened species. However it gives us no insight into the adaptive differences between regional groups.

Quantitative genetic variation

The phenotype of a plant results from an interaction between its genotype and the environment. If groups of genetically related individuals are grown in a common environment, phenotypic differences between these groups indicate the presence of genetic differences between them. Quantitative genetic analysis provides a way of estimating the size of these genetic differences from the data on phenotypic measurements (Mitchell-Olds and Rutledge 1986). Two applications of this analysis are relevant to plant conservation: estimating levels of genetic variation within populations, and measuring the extent of genetic differentiation among populations.

Genetic variation within populations

The standard approach for estimating quantitative genetic variation within a population involves collecting open pollinated seed from a number of parents, growing the resulting families in a fully randomised design under the same environmental conditions, and measuring the characters of interest. The between-family component of variance is estimated from an analysis of variance. If the mating system of the population is known from studies of genetic markers (see above), the heritability of variation, h^2, can be estimated (Lawrence 1984). Although often used as such, heritability is not a measure of genetic variation. The most appropriate parameter for comparing genetic variation among characters and populations is the coefficient of genetic variation, $C_G = (h^2 \cdot V_p)^{0.5}/\bar{x}$, where V_p is the phenotypic variance for the character and $\bar{x}$ is its mean value in the population (Houle 1992).

This quantitative genetic approach appears to provide a direct measure of the additive genetic variability, and hence the adaptive potential of populations for ecologically relevant traits. However, it has a number of serious limitations. The first of these is that the value and precision of C_G estimates is dependent upon the environment in which measurements are made. When estimated in natural populations, environmental variation may be so high that C_G values possess excessively large standard errors (Schwaegerle and Levin 1991, Schoen et al. 1994) and are of little value for estimating the adaptive potential of the population. If experiments are conducted under more controlled environmental conditions (greenhouse, common garden), phenotypes may be expressed that would not be seen in nature. Estimates of C_G may be inflated as a consequence of novel genotype–environment interactions. However, with a judicious choice of controlled environmental conditions, meaningful measurements of the adaptive potential of populations should be possible. These data can be used in conjunction with gene diversity measures to highlight genetically depauperate populations requiring an imput of genetic variability, or genetically variable populations suitable as sources for reintroduction programmes.

Genetic differentiation among populations

Plant conservation programmes commonly involve the transfer of genetic material between sites, for the purpose of genetic supplementation, population reestablishment etc. In order to guide such transfers, an understanding of the extent of adaptive genetic differentiation among populations is essential. Reciprocal transplant experiments provide the most efficient means for estimating the nature and extent of adaptive differentiation to both abiotic and biotic components of the environment (Schmidt and Levin 1985, Mitchell-Olds 1986). The results of such experiments can be used directly to establish zones within which population transfer may be undertaken without loss of environmental adaptation. This methodology is well established for determining the choice of suitable seed sources in forestry (Campbell, 1986).

While reciprocal transplant studies are highly desirable, it must be acknowledged that the time and expense involved in their establishment and maintenance is likely to be prohibitive in most plant conservation programmes. On what basis should guidelines for population movement be followed in the absence of reciprocal transplant results? As I have emphasised, patterns of genetic differentiation for selectively neutral genetic markers reflect historical patterns of colonisation and gene flow. On the other hand, differentiation for adaptively important genetic variation will largely conform to patterns of environmental variation among populations, even when gene flow is high (Jain and Bradshaw 1966, Davies and Snaydon 1976). The results of ecological assessment of environmental similarity among sites, rather than data on neutral genetic markers, should therefore be used in deciding on population transfer guidelines and seed zones for conservation purposes.

Application of genetic studies

In the foregoing discussion I considered the potential contribution of genetic analysis to plant conservation. This is a necessary first step for the integration of genetics into the discipline of conservation. The value of the approach can only be judged, however, by practical results. With this in mind, I wish to describe an ongoing programme whose aim is to conserve and reestablish native pinewoods in Scotland. I will indicate the role that genetics has played in its development, and the potential contribution of genetic studies in the future.

The native pinewoods of Scotland

The native pinewoods of Scotland represent one of the few remaining natural woodland ecosystems in Britain (Steven and Carlisle 1959). Scots pine, *Pinus sylvestris*, together with birch, *Betula pendula* and *B. pubescens*, constitute the major forest tree components in the ecosystem. As a result of ruthless exploitation and mismanagement, this ecosystem has been reduced to a tiny fraction of its former area. However, a growing awareness of the value of native pinewoods has led to the development of conservation programmes aimed both at the rehabilitation and extension of existing native pinewoods, and their reestablishment on sites from which they have been lost (Forestry Commission 1989).

Though the ultimate objective of the programme is to restore a complete ecosystem, the conservation effort has concentrated on regenerating and planting native *P. sylvestris*, the dominant woodland component. It is anticipated that once the trees are on the ground, reestablishment of the associated flora and fauna will follow. One of the priorities of the programme must be to ensure that the reestablished Scots pine is both of the highest genetic quality and appropriately adapted to the site conditions on which it is planted. To anticipate the difficulties in achieving these objectives, an appreciation of the history and ecological amplitude of native pine populations is essential.

History of pine in Scotland

The story of *P. sylvestris* in Britain can be traced back to 10000 BP when the last glaciers retreated from Scotland (Birks 1989). According to pollen records the species spread rapidly north from continental Europe to reach what is now the Scotland–England border by 8000 BP. At the same time, a disjunct pine population, originating from long distance dispersal or (more controversially) from trees that had survived glaciation, began to expand in northwest Scotland (Kinloch *et al.* 1986). Seed dispersal from Ireland into southwest Scotland may also have taken place some time after this period. Thus, there are many possible origins for pine in Scotland. Pine populations in England and Ireland were subsequently lost, while pine in Scotland spread to occupy freely draining sites of low nutrient status. Following adverse climate change and peat formation, many populations on the west of Scotland were eliminated at around 4000 BP.

The area occcupied by pinewoods remained relatively static until about 350 years ago when clearfelling for timber, charcoal, ships' masts, etc. took place (Anderson 1967). Deer and sheep grazing were encouraged, effectively preventing regeneration. This resulted in a dramatic decrease in the abundance of pinewoods to some 5% of their former area. In the nineteenth century, large-scale planting of Scots pine was practised, utilising native seed sources at first, but turning to large-scale importation of material from France and Germany in later years. Planting has continued throughout this century, though there has been an attempt by the Forestry Commission to ensure that seed was of native sources. Currently in Scotland the native pinewoods constitute some 8% (11000 ha) of the total area of Scots pine, the remainder being plantations of unknown origin.

Ecological amplitude of native pine

Although the geographic region occupied by native pine in Scotland is relatively small (roughly 200 × 200 km), environmental conditions vary widely among sites. The most dramatic differences are found between the west and the east of the region, reflecting differences in precipitation derived chiefly from Atlantic weather systems arriving from the west. Rainfall varies dramatically from over 2 m/year in the mild oceanic climate of the west coast to less than 800 mm/year in the more continental climate of the Cairngorm region. Associated with these differences in climate are differences in soil type, wind exposure, flora, fauna and prevalence of pests and diseases (Bunce 1977).

Contribution of genetic studies

The picture of native pine emerging from a consideration of history and ecology is of a once widespread species of possible multiple origin that has been relatively recently fragmented into a series of small stands. Native stands are now outnumbered by a far larger plantation population of Scots pine, partially derived from non-native sources. Native populations in the west and east of the distribution grow under very different abiotic and biotic environmental conditions. If seed from native stands is to be used for the regeneration of existing populations and the establishment of new stands, a number of important questions immediately suggest themselves:

(1) Has fragmentation of the native population led to the loss of genetic variability?

(2) Has a reduction in stocking density of native stands led to changes in the breeding system, particularly increases in selfing rates?

(3) Are levels of interpopulation pollen migration sufficient to allow gene flow from plantations to native stands?

(4) Is there any genetic evidence of a multiple origin for native Scots pine?

(5) Does adaptive genetic differentiation occur between native and non-native sources of Scots pine?

(6) Is there significant adaptive genetic differentiation among native populations of Scots pine from different sites within Scotland?

Results derived from studies of both genetic marker variation and quantitative genetic variation can be used to address these questions (Ennos 1991).

Analysis of genetic markers   Data on genetic marker variation in native pine are available for both isozymes and resin monoterpenes (Kinloch *et al.* 1986). Of these, isozyme data are most readily analysed by population genetic methods while interpretation of monoterpene variation in terms of allelic variation at specific loci is problematic. The mean gene diversity recorded for isozyme markers within the fragmented native pine populations, H_s^- is 0.309. This value is high, and comparable to values in Scots pine populations in Sweden and China ($H_s = 0.268$ and 0.235 respectively) that have not suffered from massive reductions in population size (Wang *et al.* 1991). This suggests that effective population sizes in the recent past have been high, but that recent fragmentation, occurring only 1 or 2 generations ago, has so far led to no significant reduction in genetic diversity within the remnant populations.

Results of mating system analysis using isozyme markers have shown that outcrossing rates in the extensive continental populations of Scots pine lie between 0.81 and 0.95, indicating predominant outcrossing (Muona 1990). Increases in the selfing rate of conifers have been associated with reductions in population densities like those which have occurred in the native pinewoods (Knowles *et al.* 1987). However, a study in the heavily exploited pinewood at Loch Maree revealed a multilocus outcrossing rate that did not differ significantly from 1.0 (Helgason and Ennos 1991), indicating no evidence for an increase in the proportion of selfed seed in the native populations. The low density of populations may be leading to a reduction in the amount of viable seed, but not to a higher proportion of inbred progeny, an important result to know if the seed is to form the basis of the regeneration and replanting programme.

Analysis of the distribution of neutral genetic variation within and among populations reveals that only 4% of total variability resides among populations (Kinloch *et al.* 1986), implying levels of gene flow among native populations, at least in the recent past, were very high. Long-distance gene flow by pollen is well established in pine, with successful pollination ocurring over at least tens of kilometers (Millar 1983). Low differentiation among populations for neutral markers is therefore not unexpected. Therefore, high levels of gene flow from the numerically superior plantation population of Scots pine into the native populations is not only possible, but likely. Rates of pollen contamination could be high where the two populations are in close proximity (Nagasaka and Szmidt 1985). Whether this has a deleterious effect on seed collected from native trees will depend on the extent of adaptive differences between native and planted stock (discussed in the next section).

Although measures of genetic divergence for neutral markers are generally low among populations, one significant regional difference is apparent within Scotland. Populations from the northwest are significantly different from the remainder in their resin monoterpene profiles, and to a lesser extent in their isozyme allele frequencies (Kinloch *et al.* 1986). This result adds some weight to the suggestion of a multiple origin for Scots pine in Scotland, though the details of this multiple origin remain obscure.

Analysis of quantitative genetic variation Measurement of quantitative genetic variation within native pine populations come from a single experiment containing ten open pollinated families from each of four populations planted at a single site close to Edinburgh, outside the natural range of native pine (Ennos, unpublished). Measurements on tree height, needle length and branch angle were made after 5 years. Significant genetic variation for height was detected within the populations, but only marginally significant genetic variation was found for needle length and branch angle. Coefficients of genetic variation were 0.18 (height), 0.08 (needle length) and 0.10 (branch angle). Quantitative genetic variation is therefore expressed when the trees are grown in a novel environment, but the extent of variation is dependent on the character measured. From the point of view of conservation, lack of additive genetic variation is unlikely to be a hinderance to future adaptation.

Data on adaptive genetic differentiation among populations from a series of long-term provenance experiments (Lines and Mitchell 1964, Worrell 1992) suggest that populations from outside Scotland perform less well under Scottish conditions than do native populations. Provenances from France and Germany show a mean reduction in growth rate of at least 10% compared with Scottish origins. Stem form is also poorer, and survival rates of

French sources are sometimes only 50% of native populations. It is difficult to avoid the conclusion that if continental material is present in the plantation populations of Scots pine, pollen flow into native stands will result in the production of less fit progeny. It is interesting to note that in the trial of four native populations mentioned above, significant differences in tree height were found among populations. The population showing the lowest height growth, some 11% less than the mean, and the lowest survival rate, was derived from a stand adjacent to a plantation of known German origin. This plantation is currently being harvested to minimise risks of future gene flow.

Fewer experiments have been performed on adaptive differentiation within Scotland than on differentiation over larger geographic scales. Anecdotal evidence of plantation failures on the west coast of Scotland, when planted with seed from the east, suggests substantial local adaptation. In exposed sites, needle retention of western seed sources is better than those from the east. Populations transferred from westerly populations to eastern sites tend to show lower productivity and less resistance to the rust pathogen, *Peridermium pini*, than local plants (Lines and Mitchell 1964). Though the information is very incomplete, adaptive differentiation to the very diverse environmental conditions within Scotland is indicated from these results.

For conservation purposes, guidelines are required to restrict transfer of native seed within Scotland so as to ensure that the pine being planted is well adapted to the site. Longitudinal transfers must clearly be prevented. In the absence of results from reciprocal transplant experiments, further subdivision into seed zones within the east and the west should be based on underlying patterns of environmental variation. Seed zones have indeed been designated by the Forest Authority which largely conform to these recommendations (Forestry Commission 1989). However, the boundaries between zones have been drawn so as to reflect patterns of variation for neutral genetic markers, the monoterpenes. As I have argued above, this is unjustified in view of the lack of correlation between genetic markers and patterns of adaptive differentiation. The fact that the resulting seed zones conform to a large extent with patterns of environmental variation is a fortunate coincidence. Changes to the boundaries of the seed zones are recommended so that they reflect the underlying pattern of environmental variation.

Conclusion

Studies of genetic marker variation are capable of making a valuable contribution to the development of a plant conservation programme. From relatively simple and rapid surveys of genetic variation, strong conclusions about the mating system, gene flow, and population history of the species can be drawn. However, genetic markers are rather less informative about the overall levels of genetic variation within populations and the adaptive potential of these populations.

Genetic analysis of quantitative variation, on the other hand, can provide data both on levels of genetic variation within populations and adaptive differences among populations. There are, however, problems with interpreting data when experiments are conducted in common garden or greenhouse conditions, rather than at the native sites. The practical difficulties and extent of resources needed to conduct quantitative genetic analysis on large numbers of populations are likely to limit their application in conservation.

Genetic studies, when properly interpreted, may form a valuable part of a practical conservation programme. They can highlight those areas in which genetic factors may be important, and suggest simple steps that can be used to overcome potential problems. In the case of native pinewood conservation, genetic studies indicate the need to establish appropriate seed zones to take account of adaptive differentiation, and have led to the recognition that pollen flow from non-native stands may be a significant problem. On the other hand, they have reassured us that loss of genetic variability in the remnant populations, and undesirable increases in the frequency of self-fertilised seeds, are not serious threats to the conservation programme.

Future research priorities are to understand more about the patterns of adaptive differentiation within native pine by establishing reciprocal transplant experiments and studying physiological differences among populations from different climatic regions. A second priority is to develop genetic markers that can be used to distinguish native and non-native pine so that the genetic composition of plantation populations can be determined and monitored. Such markers could also be used to confirm the identity of stands suspected to be of native origin, for which conservation measures would then be justified. The value of these genetic studies will ultimately be judged by their contribution to the reestablishment of native pinewoods.

References

Allard, R. W. (1975). The mating system and microevolution. *Genetics*, **79**, 115–126.

Anderson, M. L. (1967). *A history of Scottish forestry*. Nelson, London.

Avise, J. C. (1994). *Molecular markers, natural history and evolution*, pp. 44–91. Chapman and Hall, London.

Birks, H. J. B. (1989). Holocene isochrone maps and patterns of tree-spreading in the British Isles. *Journal of Biogeography*, **16**, 503–540.

Brown, A. H. D. (1989). Genetic characterization of plant mating systems. In *Plant population genetics, breeding, and genetic resources* (ed. A. H. D. Brown, M. T. Clegg, A. L. Kahler and B. S. Weir), pp. 145–162. Sinauer, Sunderland, MA.

Bunce, R. G. H. (1977). The range of variation within the pinewoods. In *Native pinewoods of Scotland*. Proceedings of the Aviemore Symposium, 1975 (ed. R. G. H. Bunce and J. N. R. Jeffers), pp. 10–25. Institute of Terrestrial Ecology, Cambridge.

Burke, T., Rainey, W. E. and White, T. J. (1992). Molecular variation and ecological problems. In *Genes in ecology*. 33rd symposium of the British ecological society. (ed. R. J. Berry, T. J. Crawford and G. M. Hewitt), pp. 229–254. Blackwell, Oxford.

Campbell, R. K. (1986). Mapped genetic variation of Douglas-fir to guide seed transfer in southwest Oregon. *Silvae Genetica*, **35**, 85–96.

Davies, M. S. and Snaydon, R. W. (1976). Rapid population differentiation in a mosaic environment. III. Measures of selection pressures. *Heredity*, **36**, 59–66.

Dong, J. and Wagner, D. B. (1993). Taxonomic and population differentiation of mitochondrial diversity in *Pinus banksiana* and *Pinus contorta*. *Theoretical and Applied Genetics*, **86**, 573–578.

Ennos, R. A. (1991). Genetic variation in Caledonian pine populations: origins, exploitation and conservation. In *Genetic variation of forest tree populations in Europe* (ed. G. Muller-Starck and M. Ziehe), pp. 235–249. Sauerlander, Frankfurt.

Ennos, R. A. (1994). Estimating the relative rates of pollen and seed migration among plant populations. *Heredity*, **72**, 250–259.

Ferris, C., Oliver, R. P., Davy, A. J. and Hewitt, G. M. (1993). Native oak chloroplasts reveal an ancient divide across Europe. *Molecular Ecology*, **2**, 337–344.

Forestry Commission (1989). *Native pinewoods grants and guidelines*. Forestry Commission, Edinburgh.

Hamrick, J. L. (1989). Isozymes and analysis of genetic structure of plant populations. In *Isozymes in plant biology*. (ed. D. Soltis and O. Soltis), pp. 87–105. Dioscorides Press, Washington D. C.

Helgason, T. and Ennos, R. A. (1991). The outcrossing rate and gene frequencies in a native Scots pinewood population determined using isozyme markers. *Scottish Forestry*, **45**, 111–119.

Houle, D. (1992). Comparing evolvability and variability of quantitative traits. *Genetics*, **130**, 195–204.

Jain, S. K. and Bradshaw, A. D. (1966). Evolutionary divergence among adjacent plant populations. I. The evidence and its theoretical analysis. *Heredity*, **21**, 407–441.

Kinloch, B. B., Westfall, R. D. and Forrest, G. I. (1986). Caledonian Scots pine: origins and genetic structure. *New Phytologist*, **104**, 703–729.

Knowles, P., Furnier, G. R., Aleksiuk, M. A. and Perry, D. J. (1987). Significant levels of self fertilisation in natural populations of tamarack. *Canadian Journal of Botany*, **65**, 1087–1091.

Lagercrantz, U. and Ryman, N. (1990). Genetic structure of Norway spruce (*Picea abies*): concordance of morphological and allozymic variation. *Evolution*, **44**, 38–53.

Lande, R. and Barrowclough, G. F. (1987). Effective population size, genetic variation, and their use in population management. In *Viable populations for conservation* (ed. M. E. Soule), pp. 87–123. Cambridge University Press.

Lawrence, M. J. (1984). The genetical analysis of ecological traits. In *Evolutionary ecology. 33rd symposium of the British Ecological Society* (ed. B. Shorrocks), pp. 27–63. Blackwell, Oxford.

Lines, R. and Mitchell, A. F. (1964). Results of some older Scots pine provenance experiments. In *Report on forest research for the year ended March, 1964*, pp. 171–194. H. M. S. O., London.

Millar, C. I. (1983). A steep cline in *Pinus muricata*. *Evolution*, **37**, 311–319.

Mitchell-Olds, T. (1986). Quantitative genetics of survival and growth in *Impatiens capensis*. *Evolution*, **40**, 107–116.

Mitchell-Olds, T. and Rutledge, J. J. (1987). Quantitative genetics in natural plant populations: a review of the theory. *American Naturalist*, **127**, 379–402.

Muona, O. (1990). Population genetics in forest tree improvement. In *Plant population genetics, breeding, and genetic resources* (ed. A. H. D. Brown, M. T. Clegg, A. L. Kahler and B. S. Weir), pp. 282–298. Sinauer, Sunderland, MA.

Nagasaka, K. and Szmidt, A. E. (1985). Multilocus analysis of external pollen contamination of a Scots pine (*Pinus sylvestris* L.) seed orchard. In *Population genetics in forestry. Lecture notes in biomathematics 60* (ed. H.-R. Gregorius), pp. 134–138. Springer-Verlag, Berlin.

Nei, M. (1972). Genetic distance between populations. *American Naturalist*, **106**, 283–292.

Nei, M. (1973). Analysis of gene diversity in subdivided populations. *Proceedings of the National Academy of Sciences U.S.A.*, **70**, 3321–3323.

Petit, R. J., Kremer, A. and Wagner, D. B. (1993*a*). Finite island model for organelle and nuclear genes in plants. *Heredity*, **71**, 630–641.

Petit, R. J., Kremer, A. and Wagner, D. B. (1993*b*). Geographic structure of chloroplast DNA polymorphisms in European oaks. *Theoretical and Applied Genetics*, **87**, 122–128.

Schmidt, K. P. and Levin, D. A. (1985). The comparative demography of reciprocally sown populations of *Phlox drummondii* Hook. I. Survivorship, fecundities and finite rates of increase. *Evolution*, **39**, 396–404.

Schoen, D. J., Bell, G. and Lechowicz, M. J. (1994). The ecology and genetics of fitness in forest plants IV. Quantitative genetics of fitness components in *Impatiens pallida* (Balsaminaceae). *American Journal of Botany*, **81**, 232–239.

Schwaegerle, K. E. and Levin, D. A. (1991). Quantitative genetics of fitness traits in a wild population of *Phlox. Evolution*, **45**, 169–177.

Steven, H. M. and Carlisle, A. (1959). *The native pinewoods of Scotland.* Oliver and Boyd, Edinburgh.

Strauss, S. H., Hong, Y-P. and Hipkins, V. D. (1993). High levels of population differentiation for mitochondrial DNA haplotypes in *Pinus radiata, muricata* and *attenuata. Theoretical and Applied Genetics*, **86**, 605–611.

Tobolski, J. J. and Hanover, J. W. (1971). Genetic variation in the monoterpenes of Scotch pine. *Forest Science*, **17**, 293–299.

Wang, X.-R., Szmidt, A. E. and Lindgren, D. (1991). Allozyme differentiation among populations of *Pinus sylvestris* (L.) from Sweden and China. *Hereditas*, **114**, 219–226.

Worrell, R. (1992). A comparison between European continental and British provenances of some British native trees: growth, survival and stem form. *Forestry*, **65**, 253–280.

Wright, S. (1951). The genetical structure of populations. *Annals of Eugenics*, **15**, 313–354.

16

Essential ingredients of real metapopulations, exemplified by the butterfly *Plebejus argus*

Chris D. Thomas

Introduction

Metapopulations are groups of local populations connected by occasional dispersal, in which persistence and patch occupancy are determined by the balance between local extinctions and colonizations (Gilpin and Hanski 1991). The development of metapopulation theory has been important in both population dynamics and conservation, since it has expanded the concept of population persistence from the local scale to whole regions. Many models have been constructed, of varying realism, and these have provided insight into factors which may be important to the persistence of groups of local populations. As these models are increasingly being applied to conservation, the time has come to assess critically the applicability of metapopulation theory to real populations, particularly those of rare and endangered species (Harrison 1994).

The metapopulation approach is attractive to conservation biologists and managers for several reasons. First, local population extinctions can be thought of as 'normal' in metapopulations. If a species dies out from a whole region, there is no specific blame to attach to the manager of a particular reserve. Second, the metapopulation approach gives hope in the face of failure—despite local extinctions from small habitat fragments, a particular species may persist in the general area and recolonize later. Third, it explains why empty habitats can be found, and suggests that human-mediated (re)establishment might counteract local extinctions. Fourth, and most important, it gives conservationists a basis for arguing that habitat loss should be resisted strongly, since loss of a few habitat patches could lead eventually to total population extinction in an entire region. Unfortunately, the first three of these 'attractions' of metapopulation theory, particularly the first two, may generate ill-founded complacency and are often based on misinterpretations of the theory.

Many of the assumptions and conclusions of most existing metapopulation models are apparently violated by most field systems, and extreme caution is needed before the theoretical conclusions can be used to protect or manage populations of endangered species (Thomas 1994*a*; Harrison 1994). My purpose here is to assess what assumptions might be needed if models are to be used to make specific predictions in real metapopulations. This is done by examining populations of the butterfly *Plebejus argus* L. (Lycaenidae), a species which is rare and declining in most of its remaining British range (Thomas 1993). *Plebejus argus* is unusually well fitted to a metapopulation approach: it is patchily distributed, has low rates of dispersal between local populations, and exhibits extinction/colonization dynamics.

The basic Levins model

The idea that the local distributions of many organisms change through time is fundamental to ecology. Many organisms inhabit patchy environments, and there are local extinctions and colonizations. That extinctions and colonizations could be in balance was formalized by Levins (1969, 1970) in a simple model in which he assumed that

(1) the environment is binary (patch or matrix);

(2) there is an infinite number of habitat patches;

(3) all occupied patches have an equal probability of becoming extinct;

(4) there are no internal population dynamics within patches (a patch is either populated or not);

(5) the patches of habitat persist forever (regardless of whether they are occupied);

(6) local populations become extinct independently of one another; and

(7) all vacant habitat patches have an equal probability of being colonized.

Assumptions (3) and (7) encompass many subsidiary assumptions, such as that patches are of the same size, quality and isolation as one another.

Levins related the fraction of habitat patches occupied by a particular species (p) to two parameters, the extinction rate e and the migration/colonization rate m,

$$\frac{\mathrm{d}p}{\mathrm{d}t} = mp(1-p) - ep. \tag{16.1}$$

Provided that $m > e$, this gives a positive equilibrium p^*

$$p^* = 1 - \frac{e}{m} \tag{16.2}$$

Equation 16.1 is the basis for alternative and more complex models (Hanski and Gilpin 1991) for which the existence of one or more equilibria has been and still is a major focus of theoretical metapopulation dynamics. Gilpin and Hanski (1991) outline the basic development of the theory, following Levins, and show the consequences of varying one or more of these basic assumptions. In the remainder of this chapter, I will examine how the assumptions listed above need to be varied if metapopulation models are to be realistic descriptions of natural systems, using *P. argus* as an example. This is not intended as a criticism of Levins' important model, nor of subsequent models which have relaxed one or more of these assumptions: my aim is to use the assumptions as a framework to organise and identify factors which should be included in models if we are to make specific predictions about the conservation and management of real metapopulations.

The habitat template

Ask a naturalist, conservation manager or most ecologists why a species occurs in a particular place, and the usual answer will be that 'the habitat is right'. Ask why the same species varies in density (where present) from place to place, and the likely answer will be that 'the sites differ in habitat quality' (or in management, often a determinant of habitat quality). Species occur in particular habitats and are likely to be most abundant where the habitat is most suitable for them. These two generalizations form the basis of most conservation measures, and should not be abandoned in the drive to adopt metapopulation theory. An accurate ability to identify suitable habitat regardless of a species' presence is a prerequisite for metapopulation studies.

Limestone   Most metapopulation models assume that the environment is binary (assumption (1)): it is either suitable habitat for a particular species (patches) or uninhabitable (the matrix). Where *P. argus* occurs on limestone grassland in North Wales, it is possible to distinguish between areas where *P. argus* occurs (habitat) and does not occur (presumed to be non-habitat when adjacent to existing populations) on the basis of aspect, slope, vegetation cover, vegetation height, and the density of *Lasius* ants—these ants being mutualists of the immature stages of *P. argus* (Thomas 1985*a,b*; Thomas and Harrison 1992; Jordano and Thomas 1992; Jordano *et al.* 1992). Introductions of *P. argus* to apparently suitable, but unoccupied habitat have produced thriving new local populations, confirming that empty habitat can be identified and then mapped (Thomas and Harrison 1992). Habitat quality does vary from patch to patch, but the coarse distinction of limestone areas into habitat and non-habitat is useful in that it allows us to examine the effects of patch area and isolation on the probability that a habitat patch will support a local population of *P. argus* (Fig. 16.1). The accuracy of distinguishing habitat from non-habitat could potentially be tested by mapping an unknown area of limestone solely on the basis of the habitat (if available) and predicting the distribution of the butterfly.

Heathland   *P. argus* also occurs on heathlands, but in this vegetation type I have been unable to separate habitat and non-habitat sufficiently accurately to map suitable, but unoccupied patches. This stems not so much from a lack of knowledge of the ecological requirements of *P. argus* on heathland, but from the fact that habitat quality appears to be continuously variable with respect

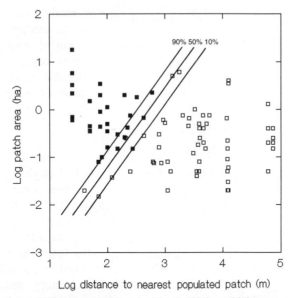

Fig. 16.1 Patches of limestone habitat in North Wales which were occupied by *P. argus* (solid) or vacant (open) in 1990, in relation to $\log_{10}$ patch isolation (axis 10 m to 100 km) and $\log_{10}$ area (axis 0.001 to 100 ha). The lines show the combinations of isolation and area which give 90%, 50% and 10% probabilities that a patch will be occupied. These are fitted from a logistic regression which explains 85% ($\rho^2 = 0.85$) of the variation for patch occupancy in relation to the two axes (Thomas *et al.* 1992). This increases to 91.6% if a non-significant area*isolation interaction is included, but the interaction effect is not plotted because the total model log Likelihood χ^2 is not significantly increased. Redrawn from Thomas and Harrison (1992) and Thomas *et al.* (1992).

to vegetation structure, elevation, and the density of mutualistic ants (Thomas 1985*a,b*; Ravenscroft 1990; Jordano and Thomas 1992; Jordano *et al.* 1992; Fig. 16.2). Some areas of unoccupied, but suitable habitat can be identified without difficulty, and successful introductions have been made to them (Thomas and Harrison 1992; Ravenscroft 1992*a,b*), but other areas are of intermediate quality—absence of *P. argus* might be due either to local extinction from a generally suitable habitat patch that has not yet been recolonized (a metapopulation explanation), or to a habitat quality too low ever to support a local population (a habitat/niche explanation). For heathland, there is no point along the continuum of habitat quality where a researcher can draw a line between habitat and non-habitat. Even if a notional point can be defined at a particular time, it is likely to vary from year to year depending on climate-related breeding success. When this problem arises, some of the conceptual messages of metapopulation theory may be useful, but even spatially explicit metapopulation models (those in which patch sizes and their locations are specified) are unlikely to be directly applicable. A habitat gradient approach is needed instead (Hochberg *et al.* 1994).

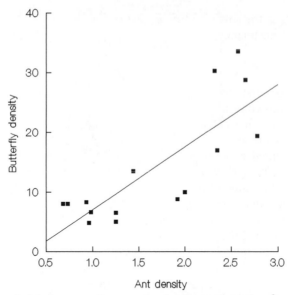

Fig. 16.2   Correlation between *P. argus* density (butterflies/400 m²) and ant density (ant nests/100 m²) within a heathland. Plotted from data in Ravenscroft (1990).

One way of incorporating habitat quality into spatially-explicit meta-population models without resorting to extra parameters has been suggested by Hanski (1994; see also Verboom *et al.* 1991*a*). Local population size is the product of area and population density. If the probability of local extinction is determined by local population size (Fig. 16.3), then physical patch areas could be reduced *in the model* by the extent to which the habitat is sub-optimal, and simulation models would be applied to 'effective' rather than 'real' areas. This should certainly be pursued, although this modification may produce complications if rates of emigration or immigration also depend on area or habitat quality.

Habitat quality forms the template for metapopulation theories. If these theories are to be tested properly, quantification of the distribution and quality of habitat is essential. Coarse measures such as 'heathland' and 'limestone grassland' are inadequate—only a subset of such vegetation is suitable for a particular species. For *P. argus*, local breeding areas within a single heathland fragment may be separated from one another as effectively by mature heather as by a ploughed field. Building a dynamic model which counts an entire fragment of heathland or limestone grassland as a patch, regardless of aspect, slope, vegetation height, host plant density (not a pro-blem for this polyphagous butterfly) or density of ants (in the case of *P. argus*) is of little value, unless it can be shown that the entire fragment can be populated by a particular species. Large areas may be more likely to be populated simply because they are more likely to contain *some* suitable habitat: if this is the sole cause of an association between patch area and the probability of occupancy, a metapopulation approach is unnecessary.

Patch number

The assumption that the number of patches is infinite (assumption (2)) may be adequate and convenient if (1) there is a very low probability of all local populations becoming extinct simultaneously as a result of chance colonization–extinction processes, and (2) regional events (e.g. a 'bad year') do not produce correlated extinctions throughout finite areas (Nisbet and Gurney 1982; Hanski 1991). However, patch numbers are usually small for endangered species. For limestone grassland habitats in North Wales, there were 20, 16 and 16 patches of habitat in three regions that were occupied by *P. argus*, and 11, 10, 4, 4, 4, 2, 2 and 1 patches in eight unoccupied regions. With this range, infinite patch number is likely to be an oversimplification, and more realistic models will be necessary (Hanski 1991).

Extinction probability

Patch size

In the simplest models, each local population has the same probability of extinction (assumption (3)), suggesting homogeneities in patch size and shape. However, patch sizes often vary over orders of magnitude, with local extinctions occurring predominantly in the smallest patches (e.g. Schoener and Spiller 1987; Schoener 1991; Harrison 1991; 1994; Thomas and Jones 1993). *P. argus* is no exception. For documented sites, populated patches vary from < 0.05 ha to at least 60 ha, a 1000-fold difference (Thomas and Harrison 1992; Thomas 1994*a*; Fig. 16.1 for limestone grassland). Clearly, extinction probabilities in this system vary enormously. Models not incorporating variation in the probability of local extinction, although useful in the general development of the subject, are not directly applicable to field situations and may lead to false conclusions. Possible solutions include using explicit variation in patch areas combined with area-dependent probabilities of extinction, or including variation in extinction probabilities among patches directly (cf. Verboom *et al.* 1991*a*; Gyllenberg and Hanski 1992; McKelvey *et al.* 1992; Hanski and Thomas 1994; Hanski 1994).

Causes of extinction

When habitat patches vary in size, local population sizes will vary too. In *P. argus*, local extinctions were confined to relatively small (< 0.9 ha) patches that supported relatively small (≤1200 adults) local populations (Thomas and Harrison 1992; Thomas 1994a: Fig. 16.3). To assume (assumption (3)) that all local populations have the same probability of extinction is therefore invalid.

This problem can be overcome by varying the probability of extinction among patches. The probability of extinction as a result of demographic stochasticity (chance birth and death events which affect each individual independently) declines rapidly with increasing local population size, and the probability of extinction from environmental stochasticity (environmental events which affect birth and death probabilities for all individuals simultaneously) declines gradually with increasing local population size (e.g. Leigh 1981). There are various ways of modelling demographic and environmental stochasticities, but they all generate the same general pattern of a decreasing probability of extinction with increasing population size. I will call these 'stochastic' models. Stochastic models of local dynamics can be incorporated into metapopulation models, but the inclusion of local dynamics within patches comes at the cost of a new set of assumptions.

In stochastic models, there are two common assumptions: a mean population size prior to extinction (there are no long-term trends; assumption (4)), and extinctions are caused by temporary or chance events (e.g. several bad years, a random walk to extinction), not by permanent deterioration of local conditions (i.e. not by loss of the patch; assumption (5)). These assumptions are rarely applicable to scarce and declining species (Thomas 1994*a,b*). Most local extinctions of British butterflies are caused by the deterioration of local habitat conditions, and changes in habitat also cause long-term trends in population sizes (Pollard 1982; J.A. Thomas 1991; Thomas 1994*a*) (Table 16.1). The result of local habitat deterioration is 'deterministic' extinction: the population fails to replace itself each generation and eventually becomes extinct. In terms of conservation management,

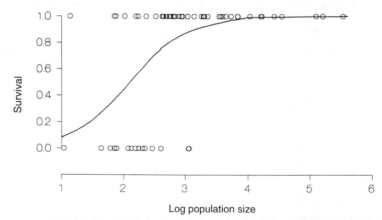

Fig. 16.3 Relationship between $\log_{10}$ *P. argus* local population size in 1983 (peak numbers × 3) and population persistence to 1990. Limestone and heathland populations pooled. Local populations that became extinct have a survival value of 0, those that persisted have the value 1. The line gives the fitted probability of survival for 7 years, from a logistic regression. From Thomas (1994*a*), courtesy of Intercept Ltd.

Table 16.1 Vegetation changes in North Wales associated with *Plebejus argus* local population survival, extinction and colonization between 1983 and 1990. For limestone grassland, habitat suitability can be defined quite clearly (Thomas 1985*a,b*; Jordano *et al.* 1992). For heathland, deterioration refers to localities where the heathland habitat was clearly less suitable for *P. argus* in 1990 than in 1983. Improvement refers to localities where the 1990 habitat was clearly closer to ideal than the 1983 habitat. Data refer only to entire patches. Survived = *P. argus* present in 1983 and 1990. Extinct = present 1983, absent 1990. Colonized = absent 1983, present 1990.

(a) Limestone grassland

	Vegetation suitability in 1983 and 1990		
	Vegetation suitable 1983 and 1990	Suitable 1983, too short* 1990	Too short 1983, suitable 1990
Survived	24	0	0
Extinct	1	2	0
Colonized	0	0	6†

(b) Heathland

	No overall change	Deterioration		Improvement	
		Loss of heath veg.	Becoming too long	Too short ⇒suitable	Too long ⇒suitable
Survived	17	0	8	1	0
Extinct	2	2	6	0	0
Colonized	3	0	0	13	6

*< 3 cm: changes in limestone vegetation height were caused by changes in domestic livestock and rabbit grazing. Following a cessation of grazing, limestone grassland can also become too overgrown for *P. argus*, but it may take > 7 years for this to happen.

†5 limestone patch colonizations are reported by Thomas and Harrison (1992), after the 1990 survey. A visit in 1991 has resulted in one area that had been colonized by 1990 being re-classified from 'patch expansion' to 'colonization.'

the difference between stochastic and deterministic extinction is crucial. If habitat change is the major cause of population trends and extinction, management may be able to prevent extinctions, halt declines or facilitate population increases. Under the stochasticity assumptions, there is little a conservation manager can do other than ensure that the original population size is large, and hope for the best.

The processes of extinction have further repercussions if we are to predict future dynamics and population persistence. In stochastic models with different carrying capacities in different local populations, and no long-term trends in population size, local populations which start off large have low probabilities of extinction in all subsequent years. This is not necessarily the case when the causes of extinction are examined in detail. In many cases, deterministic extinction is caused by successional vegetation dynamics which generate long-term trends in the local carrying capacity. In heathland, suitable habitat gradually deteriorates for *P. argus* as the vegetation becomes tall and overgrown, such that the probability of extinction will depend on the age of the habitat. Local populations which have the lowest probabilities of extinction in one time period (patches passing from early to mid succession) may have the highest probabilities of local extinction in the next time period (the same patches now passing from middle to late succession).

In contrast, limestone habitats can potentially be maintained indefinitely by grazing, and probabilities of survival may be independent of time. Knowing the vegetation processes that lead to extinction therefore suggests different conservation strategies for *P. argus* on limestone (maintain existing large populations) and heathland (provide continuity of seral conditions by initiating new successions) (Thomas 1985*b*).

Spatial scale of extinction

In theory, and in most empirical studies, the boundaries of patches are regarded as fixed (cf. assumptions (5) and (6)). But if extinction is habitat-driven, the shapes and sizes of habitat patches may also vary. This means that extinction rates may have rather little meaning if they are measured simply as the proportion of local populations lost. If a particular system has many small patches, local populations will tend to become extinct and the extinction rate will appear high, whereas a very large patch that sustains the same loss of breeding area will not become completely extinct, but will shrink and fragment. The patterns shown by *P. argus* on heathland (especially) are consistent with this: it disappears from small units of land, but populations occupying large patches tend to contract and fragment (Fig. 16.4). This has important conservation implications. The loss of an organism from half of the area it occupies should give cause for concern regardless of whether the loss is half of each large patch or extinction from 50% of the small patches.

Colonization probability

Colonization distances

In simple metapopulation models, all empty habitat patches share the same probability of being colonized (assumption (7)). This is clearly not true in real butterfly metapopulations: the probability of colonization depends on isolation, and the most isolated habitat patches remain vacant (Thomas and Harrison 1992; Thomas 1994*a*; Thomas *et al.* 1992) (Figs 16.1 and 16.5). Hanski (1994) used spatially explicit models (see also Hanski and Thomas 1994) to estimate the probabilities that different patches will be colonized,

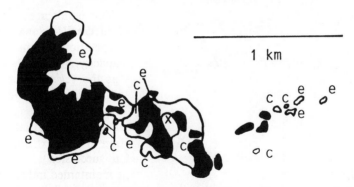

Fig. 16.4 Changing distribution of areas populated by *P. argus* at South Stack Range (Royal Society for the Protection of Birds reserve). Filled areas were populated in both 1983 and 1990. Empty outlines were populated at one time only: e = presumed extinction (1983 only), c = presumed colonization (1990 only). x = not populated in either period. From Thomas and Harrison (1992), courtesy of Blackwell Scientific Publications.

depending on their proximity to existing population sources of different sizes. These have been applied to *P. argus* with some success in that they can reproduce existing spatial patterns of patch occupancy quite well. One difficulty in interpreting the success of this approach is that the (unmeasured) dispersal parameters are calculated principally by minimizing the difference between 'predicted' and observed spatial patterns of patch occupancy. Nonetheless, parameters which give a good fit of patch occupancy in a metapopulation close to equilibrium also predict fairly accurately the successful establishment and spread of *P. argus* away from single localities where they have been introduced.

Per-generation (i.e. per year) probabilities of colonization by *P. argus* will vary from close to 1, for new habitat adjacent to existing local populations, to approximately zero, for those >5 km from a population source (Fig. 16.5). Distance-dependent dispersal/colonization is an essential component of models of real systems (e.g. Verboom *et al.* 1991*a*,*b*; McKelvey *et al.* 1992; Hanski 1994; Hanski and Thomas 1994), and the task of modellers and field ecologists alike will be to find useful dispersal or colonization parameters which can be measured as well as modelled.

Habitat and colonization

Most documented colonizations by British butterflies, including *P. argus* (Table 16.1), are of areas of habitat which have recently become suitable for that species, and which are within dispersal range (Thomas 1994*a*,*b*). When fresh successional habitat is created on heathland, or the grazing intensity changes in areas of limestone grassland, *P. argus* typically either colonizes within a few generations or fails to colonize indefinitely. This may result in

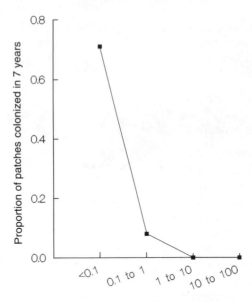

Distance from population source (km)

Fig. 16.5 Colonization of patches of limestone habitat in North Wales, in relation to distance from the nearest population source. Sample sizes were 7 (<0.1 km), 12, 24 and 19 (>10 km) habitat patches for the four distance categories. Five colonizations were observed in the <0.1 km category—the two patches that remained vacant in this distance category were so small (0.02 and 0.015 ha) that they were probably vacant in 1990 because of rapid extinction rather than failed colonization. Most observed colonizations were of habitat patches that became suitable between 1983 and 1990 (Table 16.1), so patches would not have been available for colonization for the full seven years. Most of the more distant patches (> 1 km) were available for colonization throughout the period. Redrawn from Thomas (1994*a*).

the founding of new local populations, but when the new habitat is adjacent to, or between, existing local populations, patches may simply expand or coalesce (Fig. 16.4).

Rapid colonization following habitat creation or improvement provides a challenge for modellers. In existing models, the number of habitat patches is fixed (finite or infinite) and the locations and sizes of patches do not change, even in spatially-explicit models. In terms of general theoretical conclusions, this may not matter provided that the overall density and size distribution of habitat patches remain approximately constant. But patch numbers, sizes and distributions are unlikely to remain constant for most endangered species in most modern landscapes.

In terms of conservation, it is crucial to understand the habitat changes which precede colonization. In conventional models, previously populated patches are available for colonization (assumption (5)). Therefore, colonization rates cannot be increased easily, other than by increasing extinction rates! If real colonizations usually follow an improvement in habitat quality, as they seem to, then rates of extinction and colonization are decoupled in time. This is good news for a manager, who can attempt to increase rates of colonization by creating new habitat patches or enlarging existing ones.

Integrated spatial dynamics—of what value is metapopulation theory?

The spatial dynamics of *P. argus* populations illustrate some of the factors that must be built into models if they are to be used in practical conservation planning. In addition to these factors, extinctions may be correlated in space and time (assumption (3); Harrison and Quinn 1989; Gilpin 1990), immigration may affect the probability of extinction (Hanski 1985), and some metapopulations may lack long-term equilibria (Thomas and Jones 1993). Metapopulation theorists have explored the consequences of varying a number of assumptions (e.g. in Gilpin and Hanski 1991), usually one or two at a time. That all assumptions cannot easily be relaxed simultaneously does not mean that the metapopulation approach should be abandoned, but we should not expect highly accurate predictions from these models in the continuously variable, natural world. The benefit from the approach is that it focuses attention on spatial processes, long-term dynamics, and on colonizations and local extinctions. It has also helped identify influential parameters which we might seek to quantify. For these reasons the metapopulation concept does seem useful, provided the term is used loosely to mean groups of local populations connected by dispersal, without necessarily implying that there is a balance between local extinctions and colonizations (Harrison 1994).

One major focus of theoretical metapopulation studies is p, the proportion of habitat patches which are occupied within a metapopulation. Despite the theoretical interest, this parameter appears to be of little empirical value. To start with, it may be difficult or impossible to decide which patches are sufficiently close together to count as a single metapopulation (good long-term colonization data are required): the greater the number of isolated patches or groups of patches that are (perhaps arbitarily) included, the lower the value of p. Even if this can be done, p will depend on the existing range of patch sizes (variable among metapopulations and species), on the range of patch sizes included (tiny fragments of habitat are excluded for pragmatic reasons—I could include any number of vacant 1 m^2 patches), and on the range of isolation scores for a particular metapopulation. The smallest and most isolated patches will (almost) always be vacant, and the largest and least isolated patches will (almost) always be populated (Fig. 16.1; see Hanski 1992, 1994; Thomas *et al.* 1992; Thomas 1994*a*). What we want is incidence functions, not a single value. We want to know *how* patch quality, area, age and isolation affect occupancy; overall occupancy across all patches has little meaning in natural systems.

The idea of an overall equilibrium should also be questioned, particularly if colonizations and extinctions follow the shifting spatial mosaic of the habitat (Thomas 1994a,b; Thomas and Jones 1993; Harrison 1994). The existence of an equilibrium will depend on whether (1) rates of loss of existing habitat and creation of fresh habitat are themselves in equilibrium, rarely the case for the habitats of endangered species, and (2) a species is successfully able to track changes in the distribution of habitat via dispersal. *P. argus* becomes locally extinct where the habitat becomes unsuitable, and it colonizes fresh habitat near to existing local populations. Continuity of suitable habitat is critical to persistence. *P. argus* is usually lost from limestone outcrops with less than about 30 ha of potential habitat (not all of which is suitable at one time), and from heathlands of < 5–10 ha (Thomas 1985b, Thomas 1994a). The smaller areas for heathlands probably reflect that heaths have been reduced and fragmented over the past 100 years, and that they are not at equilibrium (in the absence of management, heathland metapopulations will probably become extinct in areas < 30–100 ha).

Conclusion

Spatial processes are extremely important to this rather sedentary butterfly, but realistic assumptions will have to be made if metapopulation models are to be precise enough to guide practical conservation. I believe that the following will be needed in models:

(1) Variation in patch size and patch quality. This will generate variation in local population density, population size, extinction probability and numbers of emigrants produced.

(2) Variation in patch isolation, and distance-dependent dispersal. This will generate variation in immigration rates and colonization probabilities. For patches that are very close together, there may be substantial exchanges of individuals each generation.

(3) Temporal variation in patch size, isolation and quality.

(4) Deterministic extinction as well as stochastic extinction.

(5) A spatially dynamic habitat mosaic on which to superimpose traditional metapopulation dynamics. This dynamic mosaic will produce local extinctions, generate habitat patches of variable age, and hence create age-dependent probabilities of patch occupancy. In the special case of species occupying successional habitats, the probability of extinction will also depend on the age of a habitat patch.

The emphasis of modelling should be to:

(1) Concentrate on transient dynamics rather than hypothetical equilibria–this will help managers decide whether or not actions taken within their

working lifetimes are likely to result in an increase or decrease of a particular species over the same period.

(2) Avoid predicting the proportion of occupied patches. Instead, predict the probability that patches of particular areas, isolation, age and quality will be occupied, and use these relationships to predict the numbers and sizes of local populations in a particular network of habitat patches.

Acknowledgements

Ilkka Hanski, Susan Harrison and two other referees made many helpful suggestions to improve the manuscript.

References

Gilpin, M. E. (1990). Extinction of a finite metapopulation in correlated environments. In: *Living in a patchy environment* (eds. B. Shorrocks and I. Swingland), pp 177–86. Oxford Scientific Publications.

Gilpin, M. and Hanski, I. (eds.) (1991). *Metapopulation dynamics: empirical and theoretical investigations*. Academic Press, London.

Gyllenberg, M. and Hanski, I. (1992). Single-species metapopulation dynamics: a structured model. *Theoretical Population Biology*, **42**, 35–61.

Hanski, I. (1985). Single-species spatial dynamics may contribute to long-term rarity and commonness. *Ecology*, **66**, 335–43.

Hanski, I. (1991). Single-species metapopulation dynamics: concepts, models and observations. *Biological Journal of the Linnean Society*, **42**, 17–38.

Hanski, I. (1992). Inferences from ecological incidence functions. *American Naturalist*, **139**, 657–62.

Hanski, I. (1994). A practical model of metapopulation dynamics. *Journal of Animal Ecology*, **63**, 151–62.

Hanski, I. and Gilpin, M. (1991). Metapopulation dynamics: brief history and conceptual domain. *Biological Journal of the Linnean Society*, **42**, 3–16.

Hanski, I. and Thomas, C. D. (1994). Metapopulation dynamics and conservation: a spatially explicit model applied to butterflies. *Biological Conservation*, **68**, 167–180.

Harrison, S. and Quinn, J. F. (1989). Correlated environments and the persistence of metapopulations. *Oikos*, **56**, 293–298.

Harrison, S. (1991). Local extinction in a metapopulation context: an empirical evaluation. *Biological Journal of the Linnean Society*, **42**, 73–88.

Harrison, S. (1994) Metapopulations and conservation. *Symposium of the British Ecological Society*, **35**, 111–128.

Hochberg, M. E., Clarke, R. T., Elmes, G. W. and Thomas, J. A. (1994). Interactions between a large blue butterfly and its red ant host, and their effects on interspecific ant competition. *Journal of Animal Ecology*, **63**, 375–391.

Jordano, D., Rodríguez, J., Thomas, C. D. and Fernández Haeger, J. (1992). The distribution and density of a lycaenid butterfly in relation to *Lasius* ants. *Oecologia*, **91**, 439–46.

Jordano, D. and Thomas, C. D. (1992). Specificity of an ant–lycaenid interaction. *Oecologia*, **91**, 431–38.

Levins, R. (1969). Some demographic and genetic consequences of environmental heterogeneity for biological control. *Bulletin of the Entomological Society of America*, **15**, 237–40.

Levins, R. (1970). Extinction. *American Maths Society*, **2**, 77–107.

Leigh, E. G. (1981). The average lifetime of a population in a varying environment. *Journal of Theoretical Biology*, **90**, 213–39.

McKelvey, K., Noon, B. R. and Lamberson, R. H. (1992). Conservation planning for species occupying fragmented landscapes: the case of the northern spotted owl. In: *Biotic Interactions and Global Change* (eds. P. M. Kareiva, J. G. Kingsolver and R. B. Huey). Ch. 26. Sinauer, Sunderland, MA.

Nisbet, R. M. and Gurney, W. S. C. (1982). *Modelling fluctuating populations*. John Wiley, New York.

Pollard, E. (1982). Monitoring butterfly abundance in relation to the management of a nature reserve. *Biological Conservation*, **24**, 317–28.

Ravenscroft, N. O. M. (1990). The ecology and conservation of the silver-studded blue butterfly *Plebejus argus* L. on the Sandlings of East Anglia. *Biological Conservation*, **53**, 21–36.

Ravenscroft, N. O. M. (1992a). The use of introductions for the conservation of a fragmented population of *Plebejus argus* (Lepidoptera: Lycaenidae) in Suffolk, England. In: *Future of butterflies in Europe: strategies for survival* (eds. T. Pavlieek-van Beek, A. H. Ova and J. E. van der Made), pp 213–21. Wageningen, Netherlands.

Ravenscroft, N. O. M. (1992b). The fortunes of the silver-studded blue, *Plebejus argus* (L.) (Lepidoptera: Lycaenidae) at artificial sites on the Sandlings, Suffolk. *Entomologist's Gazette*, **43**, 157–61.

Schoener, T. W. (1991). Extinction and the nature of the metapopulation: a case system. *Acta Oecologia*, **12**, 53–75.

Schoener, T. W. and Spiller, D. A. (1987). High population persistence in a system with high turnover. *Nature*, **330**, 474–77.

Thomas, C. D. (1985a). Specializations and polyphagy of *Plebejus argus* (Lepidoptera: Lycaenidae) in North Wales. *Ecological Entomology*, **10**, 325–40.

Thomas, C. D. (1985b). The status and conservation of the butterfly *Plebejus argus* (Lepidoptera: Lycaenidae) in North West Britain. *Biological Conservation*, **33**, 29–51.

Thomas, C. D. (1991). Spatial and temporal variability in a butterfly population. *Oecologia*, **87**, 577–80.

Thomas, C. D. (1993) The silver-studded blue, *Plebejus argus* L. In: *Conservation biology of Lycaenidae (Butterflies)* (ed. T. R. New), pp 97–99. IUCN, Gland, Switzerland.

Thomas, C. D. (1994a). Local extinctions, colonizations and distributions: habitat tracking by British butterflies. In: *Individuals, Populations and Patterns in Ecology* (S. R. Leather, A. D. Watt, N. J. Mills and K. F. A. Walters, eds.), pp. 319–336. Intercept Ltd., Andover.

Thomas, C. D. (1994b). Extinction, colonization and metapopulations: environmental tracking by rare species. *Conservation Biology*, **8**, 373–378.

Thomas, C. D. and Harrison, S. (1992). Spatial dynamics of a patchily-distributed butterfly species. *Journal of Animal Ecology*, **61**, 437–46.

Thomas, C. D. and Jones, T. M. (1993). Partial recovery of a skipper butterfly (*Hesperia comma*) from population refuges: lessons for conservation in a fragmented landscape. *Journal of Animal Ecology*, **62**, 472–81.

Thomas, C. D., Thomas, J. A. and Warren, M. S. (1992). Distributions of occupied and vacant butterfly habitats in fragmented landscapes. *Oecologia*, **92**, 563–67.

Thomas, J. A. (1991). Rare species conservation: case studies of European butterflies. *Symposium of the British Ecological Society*, **31**, 149–97.

Verboom, J., Schotman, A., Opdam, P. and Metz, J. A. J. (1991*a*). European nuthatch metapopulations in a fragmented agricultural landscape. *Oikos*, **61**, 149–56.

Verboom, J., Lankester, K. and Metz, J. A. J. (1991*b*). Linking local and regional dynamics in stochastic metapopulation models. *Biological Journal of the Linnean Society*, **42**, 39–55.

INDEX

Note: figures and tables are indicated by *italic* page numbers.